Principles of Statistical Analysis

This compact course is written for the mathematically literate reader who wants to learn to analyze data in a principled fashion. The language of mathematics enables clear exposition that can go quite deep, quite quickly, and naturally supports an axiomatic and inductive approach to data analysis. Starting with a good grounding in probability, the reader moves to statistical inference via topics of great practical importance – simulation and sampling, as well as experimental design and data collection – that are typically displaced from introductory accounts. The core of the book then covers both standard methods and such advanced topics as multiple testing, meta-analysis, and causal inference.

E R Y   A R I A S - C A S T R O   is a professor in the Department of Mathematics and in the Halıcıoğlu Data Science Institute at the University of California, San Diego, where he specializes in theoretical statistics and machine learning. His education includes a bachelor's degree in mathematics and a master's degree in artificial intelligence, both from École Normale Supérieure de Cachan (now École Normale Supérieure Paris-Saclay) in France, as well as a Ph.D. in statistics from Stanford University in the United States.

"With the rapid development of data-driven decision making, statistical methods have become indispensable in countless domains of science, engineering, and management science, to name a few. Ery Arias-Castro's excellent text gives a self-contained and remarkably broad exposition of the current diversity of concepts and methods developed to tackle the challenges of data science. Simply put, everyone serious about understanding the theory behind data science should be exposed to the topics covered in this book."

—Philippe Rigollet, Professor
Department of Mathematics, Massachusetts Institute of Technology

"A course on statistical modeling and inference has been a staple of many first-year graduate engineering programs. While there are many excellent textbooks on this subject, much of the material is inspired by models of physical systems, and as such these books deal extensively with parametric inference. The emerging data revolution, on the other hand, requires an engineering student to develop an understanding of statistical inference rooted in problems inspired by data-driven applications, and this book fills that need. Arias-Castro weaves together diverse concepts such as data collection, sampling, and inference in a unified manner. He lucidly presents the mathematical foundations of statistical data analysis, and covers advanced topics on data analysis. With over 700 problems and computer exercises, this book will serve the needs of beginner and advanced engineering students alike."

—Venkatesh Saligrama, Professor
Data Science Faculty Fellow, Department of Electrical and Computer Engineering,
Department of Computer Science (by courtesy), Boston University

"In this book, aimed at senior undergraduates or beginning graduate students with a reasonable mathematical background, the author proposes a self-contained and yet concise introduction to statistical analysis. By putting a strong emphasis on the randomization principle, he provides a coherent and elegant perspective on modern statistical practice. Some of the later chapters also form a good basis for a reading group. I will be recommending this excellent book to my collaborators."

—Nicolas Verzelen, Associate Professor
Mathematics, Computer Science, Physics, and Systems Department,
University of Montpellier

"This text is highly recommended for undergraduate students wanting to grasp the key ideas of modern data analysis. Arias-Castro achieves something that is rare in the art of teaching statistical science – he uses mathematical language in an intelligible and highly helpful way, without surrendering key intuitions of statistics to formalism and proof. In this way, the reader can get through an impressive amount of material without, however, ever getting into muddy waters."

—Richard Nickl, Professor
Statistical Laboratory, Cambridge University

INSTITUTE OF MATHEMATICAL STATISTICS
TEXTBOOKS

Editorial Board
Nancy Reid (University of Toronto)
John Aston (University of Cambridge)
Arnaud Doucet (University of Oxford)
Ramon van Handel (Princeton University)

ISBA Editorial Representative
Peter Müller (University of Texas at Austin)

IMS Textbooks give introductory accounts of topics of current concern suitable for advanced courses at master's level, for doctoral students and for individual study. They are typically shorter than a fully developed textbook, often arising from material created for a topical course. Lengths of 100–290 pages are envisaged. The books typically contain exercises.

In collaboration with the International Society for Bayesian Analysis (ISBA), selected volumes in the IMS Textbooks series carry the "with ISBA" designation at the recommendation of the ISBA editorial representative.

Other Books in the Series (*with ISBA)

1. *Probability on Graphs*, by Geoffrey Grimmett
2. *Stochastic Networks*, by Frank Kelly and Elena Yudovina
3. *Bayesian Filtering and Smoothing*, by Simo Särkkä
4. *The Surprising Mathematics of Longest Increasing Subsequences*, by Dan Romik
5. *Noise Sensitivity of Boolean Functions and Percolation*, by Christophe Garban and Jeffrey E. Steif
6. *Core Statistics*, by Simon N. Wood
7. *Lectures on the Poisson Process*, by Günter Last and Mathew Penrose
8. *Probability on Graphs (Second Edition)*, by Geoffrey Grimmett
9. *Introduction to Malliavin Calculus*, by David Nualart and Eulàlia Nualart
10. *Applied Stochastic Differential Equations*, by Simo Särkkä and Arno Solin
11. *Computational Bayesian Statistics*, by M. Antónia Amaral Turkman, Carlos Daniel Paulino, and Peter Müller
12. *Statistical Modelling by Exponential Families*, by Rolf Sundberg
13. *Two-Dimensional Random Walk: From Path Counting to Random Interlacements*, by Serguei Popov
14. *Scheduling and Control of Queueing Networks*, by Gideon Weiss

Principles of Statistical Analysis

Learning from Randomized Experiments

ERY ARIAS-CASTRO

University of California, San Diego

CAMBRIDGE
UNIVERSITY PRESS

University Printing House, Cambridge CB2 8BS, United Kingdom

One Liberty Plaza, 20th Floor, New York, NY 10006, USA

477 Williamstown Road, Port Melbourne, VIC 3207, Australia

314–321, 3rd Floor, Plot 3, Splendor Forum, Jasola District Centre,
New Delhi – 110025, India

103 Penang Road, #05–06/07, Visioncrest Commercial, Singapore 238467

Cambridge University Press is part of the University of Cambridge.

It furthers the University's mission by disseminating knowledge in the pursuit of
education, learning, and research at the highest international levels of excellence.

www.cambridge.org
Information on this title: www.cambridge.org/9781108489676
DOI: 10.1017/9781108779197

First published 2022

A catalogue record for this publication is available from the British Library.

ISBN 978-1-108-48967-6 Hardback
ISBN 978-1-108-74744-8 Paperback

I would like to dedicate this book to some professors that have, along the way, inspired, supported, and mentored me in my studies and academic career, and to whom I am eternally grateful:

David L. Donoho
my doctoral thesis advisor

Persi Diaconis
my first co-author on a research article

Yves Meyer
my master's thesis advisor

Robert Azencott
my undergraduate thesis advisor

In controlled experimentation it has been found not difficult to introduce explicit and objective randomization in such a way that the tests of significance are demonstrably correct. In other cases we must still act in the faith that Nature has done the randomization for us. [...] We now recognize randomization as a postulate necessary to the validity of our conclusions, and the modern experimenter is careful to make sure that this postulate is justified.

Ronald A. Fisher
International Statistical Conferences, 1947

Contents

Part II   Practical Considerations

Part III   Elements of Statistical Inference

Preface

This book is intended for the mathematically literate reader who wants to understand how to analyze data in a principled fashion. The language of mathematics allows for a more concise, and arguably clearer exposition that can go quite deep, quite quickly, and naturally accommodates an axiomatic and inductive approach to data analysis, which is the raison d'être of the book. To elaborate, the book starts with a preliminary foundation in probability theory, continues with an intermezzo of sampling and data collection, and finally moves to statistical inference – the core of the book which includes, in addition to standard topics, more advanced ones such as multiple testing, meta-analysis, and causal inference. The book thus provides a self-contained exposition of fundamental principles and methods of statistical analysis, covering topics which are typically displaced from introductory, general accounts. Emphasis is on inference, and more exploratory approaches to data analysis such as clustering and dimensionality reduction are not covered.

The compact treatment is grounded in mathematical theory and concepts, and is fairly rigorous, even though measure theoretic matters are kept in the background, and most proofs are left as problems. In fact, much of the learning is accomplished through embedded problems – around 700 of them! Some problems call for mathematical derivations, and assume a certain comfort with calculus, or even real analysis. Other problems require basic programming on a computer.

Structure

The book is divided into three parts. The introduction to probability, in Part I, stands as the mathematical foundation for statistical inference. Indeed, without a solid foundation in probability and, in particular, a good understanding of how experiments are modeled, there is no clear distinction between descriptive and inferential analyses. The exposition there is quite standard. It starts by introducing Kolmogorov's axioms, which are instantiated in the context of discrete sample spaces. The narrative then transitions to a comprehensive discussion of distributions on the real line, both discrete and continuous, and also multivariate. This is followed by an introduction of the basic concentration inequalities and limit theorems. (A construction of the Lebesgue integral is not included, and measure-theoretic matters are mostly avoided.) Part I ends with a brief discussion of Markov chains and related stochastic processes.

Some utilitarian, but absolutely critical, aspects of probability and statistics are discussed in Part II. These include probability sampling and pseudo-random number generation – the practical side of randomness; as well as survey sampling and experimental design – the practical side of data collection.

Part III is the core of the book. It attempts to build a theory of statistical inference from first principles. The foundation is randomization, either controlled by design or assumed to be natural. In either case, randomization provides the essential randomness needed to justify probabilistic modeling. It naturally leads to conditional inference, and allows for causal inference. In this framework, permutation tests play a special, almost canonical role. Monte Carlo sampling, performed on a computer, is presented as an alternative to complex mathematical derivations, and the bootstrap is then introduced as an accommodation when the sampling distribution is not directly available and has to be estimated.

What is not here

I do not find normal models to be particularly compelling: unless there is a central limit theorem at play, there is no real reason to believe numerical data are normally distributed. Normal models are thus mentioned only in passing. More generally, parametric models are not emphasized – except for those that arise naturally in some experiments.

The usual emphasis on parametric inference is, I find, misplaced and misleading, as it can be (and often is) introduced independently of how the data were gathered, thus creating a chasm that separates the design of experiments and the analysis of the resulting data. Bayesian modeling is, consequently, not covered beyond basic definitions in the context of average risk optimality. Linear models and time series are not discussed in any detail. As is typically the case for an introductory book, especially of this length and at this level, there is only a hint of abstract decision theory, and multivariate analysis is omitted entirely.

How to use this Book

The idea for this book arose from a dissatisfaction with how statistical analysis is typically taught at the undergraduate and master's levels, coupled with an inspiration for weaving a narrative, which I find more compelling.

This narrative was formed over years of teaching statistics at the University of California, San Diego, in particular an undergraduate-level course on *computational statistics* focusing on resampling methods of inference. As it stands, however, the book is perhaps best used for independent study.

The reader is invited to progress through the book in the order in which the material is presented, working on the problems as they come, and saving those that seem harder for later. If an experienced instructor or tutor is available as an occasional guide, it is worthwhile to tackle even the harder problems when they are encountered.

Although the text emphasizes a conceptual understanding of data analysis, it is also grounded in practice. A large number of articles in the applied sciences are cited with the intention of providing the reader with a sense of how statistics is used in real life. In addition, a companion R notebook is provided to facilitate the transition from theory to practice. It is available from the author's webpage [https://math.ucsd.edu/~eariasca].

Intention

The book introduces, what I believe, are essential concepts that I would want a student graduating with a bachelor's or master's degree in statistics to have been exposed to, even if only in passing.

My main hope in writing this book is that it seduces mathematically minded people into learning more about statistical analysis, at least for their personal enrichment, particularly in this age of artificial intelligence, machine learning, and data science more broadly.

Acknowledgements

Ian Abramson, Emmanuel Candès, Persi Diaconis, David Donoho, Larry Goldstein, Gábor Lugosi, Dimitris Politis, Philippe Rigollet, Joseph Romano, Jason Schweinsberg, and Nicolas Verzelen, provided feedback or pointers on particular topics. Clément Berenfeld, Zexin Pan, and Yunhao Sun, as well as two anonymous referees and a copyeditor, Benjamin Johnson, retained by Cambridge, helped proofread the manuscript (all remaining errors are, of course, mine). Diana Gillooly, and later Becca Grainger, were my main contacts at Cambridge, offering expert guidance through the publishing process. I am grateful for all this generous support.

I looked at a number of textbooks in Probability and Statistics, and also lecture notes. The most important ones are cited in the text.

I used a number of freely available resources contributed by countless people all over the world. The manuscript was typeset in LaTeX (using TeXShop, TeXstudio, TeXworks, BibDesk, and JabRef) and the figures were produced using R via RStudio. The companion R notebook was written using Bookdown. I made extensive use of Google Scholar, Wikipedia, and some online discussion lists such as StackExchange.

Part I

Elements of Probability Theory

1

Axioms of Probability Theory

Probability theory is the branch of mathematics that models and studies random phenomena. Although randomness has been the object of much interest over many centuries, the theory only reached maturity with *Kolmogorov's axioms*[1] in the 1930s [195].

As a mathematical theory founded on Kolmogorov's axioms, *Probability Theory* is essentially uncontroversial at this point. However, the notion of probability (i.e., chance) remains somewhat controversial. We will adopt here the frequentist notion of probability [193], which defines the chance that a particular experiment results in a given outcome as the limiting frequency of this event as the experiment is repeated an increasing number of times. The problem of giving probability a proper definition as it concerns real phenomena is discussed in [67] (with a good dose of humor).

1.1 Elements of Set Theory

Kolmogorov's formalization of probability relies on some basic notions of *Set Theory*.

A *set* is simply an abstract collection of 'objects', sometimes called *elements* or *items*. Let Ω denote such a set. A *subset* of Ω is a set made of elements that belong to Ω. In what follows, a set will be a subset of Ω.

We write $\omega \in \mathcal{A}$ when the element ω belongs to the set $\mathcal{A}$. And we write $\mathcal{A} \subset \mathcal{B}$ when set $\mathcal{A}$ is a subset of set $\mathcal{B}$. This means that $\omega \in \mathcal{A} \Rightarrow \omega \in \mathcal{B}$. A set with only one element ω is denoted $\{\omega\}$ and is called a *singleton*. Note that $\omega \in \mathcal{A} \Leftrightarrow \{\omega\} \subset \mathcal{A}$. The *empty set* is defined as a set with no elements and is denoted $\varnothing$. By convention, it is included in any other set.

Problem 1.1 Prove that $\subset$ is transitive, meaning that if $\mathcal{A} \subset \mathcal{B}$ and $\mathcal{B} \subset \mathcal{C}$, then $\mathcal{A} \subset \mathcal{C}$.

---

[1] Named after Andrey Kolmogorov (1903–1987).

3

The following are some basic set operations.

- *Intersection and disjointness*   The intersection of two sets A and B is the set with all the elements belonging to both A and B, and is denoted $A \cap B$. A and B are said to be *disjoint* if $A \cap B = \varnothing$.
- *Union*   The union of two sets A and B is the set with elements belonging to A or B, and is denoted $A \cup B$.
- *Set difference and complement*   The set difference of B minus A is the set with elements those in B that are not in A, and is denoted $B \smallsetminus A$. It is sometimes called the complement of A in B. The complement of A in the whole set Ω is often denoted A^c.
- *Symmetric set difference*   The symmetric set difference of A and B is defined as the set with elements either in A or in B, but not in both, and is denoted $A \bigtriangleup B$.

Sets and set operations can be visualized using a *Venn diagram*. See Figure 1.1 for an example.

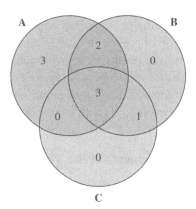

Figure 1.1 A Venn diagram helping visualize the sets $A = \{1,2,4,5,6,7,8,9\}$, $B = \{2,3,4,5,7,9\}$, and $C = \{3,4,5,9\}$. The numbers shown in the figure represent the size of each subset. For example, the intersection of these three sets contains 3 elements, since $A \cap B \cap C = \{4,5,9\}$.

Problem 1.2 Prove that $A \cap \varnothing = \varnothing$, $A \cup \varnothing = A$, and $A \smallsetminus \varnothing = A$. What is $A \bigtriangleup \varnothing$?

Problem 1.3 Prove that the complement is an involution, i.e., $(A^c)^c = A$.

Problem 1.4 Show that the set difference operation is not symmetric in the sense that $B \smallsetminus A \neq A \smallsetminus B$ in general. In fact, prove that $B \smallsetminus A = A \smallsetminus B$ if and only if $A = B = \varnothing$.

Proposition 1.5. *The following are true:*

(i) *The intersection operation is commutative, meaning $A \cap B = B \cap A$, and associative, meaning $(A \cap B) \cap C = A \cap (B \cap C)$. The same is true for the union operation.*

(ii) *The intersection operation is distributive over the union operation, meaning $(A \cup B) \cap C = (A \cap C) \cup (B \cap C)$.*

(iii) *It holds that $(A \cap B)^c = A^c \cup B^c$. More generally, $C \smallsetminus (A \cap B) = (C \smallsetminus A) \cup (C \smallsetminus B)$.*

We thus may write $A \cap B \cap C$ and $A \cup B \cup C$, that is, without parentheses, as there is no ambiguity. More generally, for a collection of sets $\{A_i : i \in I\}$, where I is some index set, we can therefore refer to their intersection and union, denoted

$$\text{(intersection)} \quad \bigcap_{i \in I} A_i, \qquad \text{(union)} \quad \bigcup_{i \in I} A_i .$$

Remark 1.6 For the reader seeing these operations for the first time, it can be useful to think of $\cap$ and $\cup$ in analogy with the product $\times$ and sum $+$ operations on the integers. In that analogy, $\varnothing$ plays the role of 0.

Problem 1.7 Prove Proposition 1.5. In fact, prove the following identities:

$$\Big(\bigcup_{i \in I} A_i \Big) \cap B = \bigcup_{i \in I} (A_i \cap B),$$

and

$$\Big(\bigcup_{i \in I} A_i \Big)^c = \bigcap_{i \in I} A_i^c, \quad \text{as well as} \quad \Big(\bigcap_{i \in I} A_i \Big)^c = \bigcup_{i \in I} A_i^c,$$

for any collection of sets $\{A_i : i \in I\}$ and any set B.

1.2 Outcomes and Events

Having introduced some elements of Set Theory, we use some of these concepts to define a probability experiment and its possible outcomes.

1.2.1 Outcomes and the Sample Space

In the context of an *experiment*, all the possible *outcomes* are gathered in a *sample space*, denoted Ω henceforth. In mathematical terms, the sample space is a set and the outcomes are elements of that set.

Example 1.8 (Flipping a coin) Suppose that we flip a coin three times in sequence. Assuming the coin can only land heads (H) or tails (T), the sample space Ω consists of all possible ordered sequences of length 3, which in lexicographic order can be written as

$$\Omega = \Big\{ \text{HHH, HHT, HTH, HTT, THH, THT, TTH, TTT} \Big\}.$$

Example 1.9 (Drawing from an urn) Suppose that we draw two balls from an urn in sequence. Assume the urn contains red (R), green (G), and (B) blue balls. If the urn contains at least two balls of each color, or if at each trial the ball is returned to the urn, the sample space Ω consists of all possible ordered sequences of length 2, which in the RGB order can be written as

$$\Omega = \Big\{ \text{RR, RG, RB, GR, GG, GB, BR, BG, BB} \Big\}. \tag{1.1}$$

If the urn (only) contains one red ball, one green ball, and two or more blue balls, and a ball drawn from the urn is not returned to the urn, the number of possible outcomes is reduced and the resulting sample space is now

$$\Omega = \Big\{ \text{RG, RB, GR, GB, BR, BG, BB} \Big\}.$$

Problem 1.10 What is the sample space when we flip a coin five times? More generally, can you describe the sample space, in words and/or mathematical language, corresponding to an experiment where the coin is flipped n times? What is the size of that sample space?

Problem 1.11 Consider an experiment that consists in drawing two balls from an urn that contains red, green, blue, and yellow balls. However, yellow balls are ignored, in the sense that if such a ball is drawn then it is discarded. How does that change the sample space compared to Example 1.9?

While in the previous examples the sample space is finite, the following is an example where it is (countably) infinite.

Example 1.12 (Flipping a coin until the first heads) Consider an experiment where we flip a coin repeatedly until it lands heads. The sample space in this case is

$$\Omega = \Big\{ \text{H, TH, TTH, TTTH, } \ldots \Big\}.$$

Problem 1.13 Describe the sample space when the experiment consists in drawing repeatedly without replacement from an urn with red, green, and blue balls, three of each color, until a blue ball is drawn.

Remark 1.14 A sample space is in fact only required to contain all possible outcomes. For instance, in Example 1.9 we may always take the sample space to be (1.1) even though in the second situation that space contains outcomes that will never arise.

1.2.2 Events

Events are subsets of Ω that are of particular interest. We say that an event *happens* when the experiment results in an outcome that belongs to the event.

Example 1.15 In the context of Example 1.8, consider the event that the second toss results in heads. As a subset of the sample space, this event is defined as

$$\mathcal{E} = \Big\{ \text{HHH, HHT, THH, THT} \Big\}.$$

Example 1.16 In the context of Example 1.9, consider the event that the two balls drawn from the urn are of the same color. This event corresponds to the set

$$\mathcal{E} = \Big\{ \text{RR, GG, BB} \Big\}.$$

Example 1.17 In the context of Example 1.12, the event that the number of total tosses is even corresponds to the set

$$\mathcal{E} = \Big\{ \text{TH, TTTH, TTTTTH, \dots} \Big\}.$$

Problem 1.18 In the context of Example 1.8, consider the event that at least two tosses result in heads. Describe this event as a set of outcomes.

1.2.3 Collection of Events

Recall that we are interested in particular subsets of the sample space Ω and that we call these 'events'. Let Σ denote the collection of events. We assume throughout that Σ satisfies the following conditions:

• The entire sample space is an event, meaning

$$\Omega \in \Sigma. \tag{1.2}$$

- The complement of an event is an event, meaning

$$\mathcal{A} \in \Sigma \;\Rightarrow\; \mathcal{A}^c \in \Sigma. \tag{1.3}$$

- A countable union of events is an event, meaning

$$\mathcal{A}_1, \mathcal{A}_2, \cdots \in \Sigma \;\Rightarrow\; \bigcup_{i \geq 1} \mathcal{A}_i \in \Sigma. \tag{1.4}$$

A collection of subsets that satisfies these conditions is called a *σ-algebra*.[2]

Problem 1.19 Suppose that Σ is a σ-algebra. Show that $\varnothing \in \Sigma$ and that a countable intersection of subsets of Σ is also in Σ.

From now on, Σ will denote a σ-algebra over Ω unless otherwise specified. (Note that such a σ-algebra always exists: an example is $\{\varnothing, \Omega\}$.) The pair (Ω, Σ) is then called a *measurable space*.

Remark 1.20 (The power set) The *power set* of Ω, often denoted 2^Ω, is the collection of all its subsets. (Problem 1.49 provides a motivation for this name and notation.) The power set is trivially a σ-algebra. In the context of an experiment with a discrete sample space, it is customary to work with the power set as σ-algebra, because this can always be done without loss of generality (Chapter 2). When the sample space is not discrete, the situation is more complex and the σ-algebra needs to be chosen with more care (Section 3.2).

1.3 Probability Axioms

Before observing the result of an experiment, we speak of the probability that an event will happen. The Kolmogorov axioms formalize this assignment of probabilities to events. This has to be done carefully so that the resulting theory is both coherent and useful for modeling randomness.

A *probability distribution* (aka *probability measure*) on (Ω, Σ) is any real-valued function $\mathbb{P}$ defined on Σ satisfying the following properties or axioms:[3]

- *Non-negativity*

$$\mathbb{P}(\mathcal{A}) \geq 0, \quad \forall \mathcal{A} \in \Sigma.$$

- *Unit measure*

$$\mathbb{P}(\Omega) = 1.$$

---

[2] This refers to the algebra of sets presented in Section 1.1.

[3] Throughout, we will often use 'distribution' or 'measure' as shorthand for 'probability distribution'.

- *Additivity on disjoint events*   For any discrete collection of disjoint events $\{A_i : i \in I\}$,

$$\mathbb{P}\left(\bigcup_{i \in I} A_i\right) = \sum_{i \in I} \mathbb{P}(A_i). \tag{1.5}$$

A triplet $(\Omega, \Sigma, \mathbb{P})$ with Ω a sample space (a set), Σ a σ-algebra over Ω, and $\mathbb{P}$ a distribution on Σ, is called a *probability space*. We consider such a triplet in what follows.

Problem 1.21 Show that $\mathbb{P}(\varnothing) = 0$ and that

$$0 \le \mathbb{P}(A) \le 1, \quad A \in \Sigma.$$

Thus, although nominally a probability distribution takes values in $\mathbb{R}_+$, in fact it takes values in $[0, 1]$.

Proposition 1.22 (Law of Total Probability). *For any two events A and B,*

$$\mathbb{P}(A) = \mathbb{P}(A \cap B) + \mathbb{P}(A \cap B^c). \tag{1.6}$$

Problem 1.23 Prove Proposition 1.22 using the 3rd axiom.

The 3rd axiom applies to events that are disjoint. The following is a corollary that applies more generally. (In turn, this result implies the 3rd axiom.)

Proposition 1.24 (Law of Addition). *For any two events A and B, not necessarily disjoint,*

$$\mathbb{P}(A \cup B) = \mathbb{P}(A) + \mathbb{P}(B) - \mathbb{P}(A \cap B). \tag{1.7}$$

In particular,

$$\mathbb{P}(A^c) = 1 - \mathbb{P}(A), \tag{1.8}$$

and,

$$A \subset B \implies \mathbb{P}(B \smallsetminus A) = \mathbb{P}(B) - \mathbb{P}(A). \tag{1.9}$$

Proof   We first observe that we can get (1.9) from the fact that B is the disjoint union of A and $B \smallsetminus A$ and an application of the 3rd axiom.

We now use this to prove (1.7). We start from the disjoint union

$$A \cup B = (A \smallsetminus B) \cup (B \smallsetminus A) \cup (A \cap B).$$

Applying the 3rd axiom yields

$$\mathbb{P}(A \cup B) = \mathbb{P}(A \smallsetminus B) + \mathbb{P}(B \smallsetminus A) + \mathbb{P}(A \cap B).$$

Then $\mathcal{A} \setminus \mathcal{B} = \mathcal{A} \setminus (\mathcal{A} \cap \mathcal{B})$, and applying (1.9), we get

$$\mathbb{P}(\mathcal{A} \setminus \mathcal{B}) = \mathbb{P}(\mathcal{A}) - \mathbb{P}(\mathcal{A} \cap \mathcal{B}),$$

and exchanging the roles of $\mathcal{A}$ and $\mathcal{B}$,

$$\mathbb{P}(\mathcal{B} \setminus \mathcal{A}) = \mathbb{P}(\mathcal{B}) - \mathbb{P}(\mathcal{A} \cap \mathcal{B}).$$

After some cancellations, we obtain (1.7), which then immediately implies (1.8). □

Problem 1.25 (Uniform distribution) Suppose that Ω is finite. For $\mathcal{A} \subset \Omega$, define $\mathbb{U}(\mathcal{A}) = |\mathcal{A}|/|\Omega|$, where $|\mathcal{A}|$ denotes the number of elements in $\mathcal{A}$. Show that $\mathbb{U}$ is a probability distribution on Ω (equipped with its power set, as usual).

1.4 Inclusion-Exclusion Formula

The inclusion-exclusion formula is an expression for the probability of a discrete union of events. We start with some basic inequalities that are directly related to the inclusion-exclusion formula and useful on their own.

Boole's Inequality

Also know as the *union bound*, this inequality[4] is arguably one of the simplest, yet also one of the most useful, inequalities of Probability Theory.

Problem 1.26 (Boole's inequality) Prove that for any countable collection of events $\{\mathcal{A}_i : i \in I\}$,

$$\mathbb{P}\left(\bigcup_{i \in I} \mathcal{A}_i\right) \leq \sum_{i \in I} \mathbb{P}(\mathcal{A}_i). \tag{1.10}$$

Note that the right-hand side can be larger than 1 or even infinite. [One possibility is to use a recursion on the number of events, together with Proposition 1.24, to prove the result for any finite number of events. Then pass to the limit to obtain the result as stated.]

Bonferroni's Inequalities

These inequalities[5] comprise Boole's inequality. For two events, we saw the Law of Addition (Proposition 1.24), which is an exact expression for the probability of their union. Consider now three events $\mathcal{A}, \mathcal{B}, \mathcal{C}$. Boole's

---

[4] Named after George Boole (1815–1864).
[5] Named after Carlo Emilio Bonferroni (1892–1960).

inequality (1.10) gives

$$\mathbb{P}(\mathcal{A} \cup \mathcal{B} \cup \mathcal{C}) \le \mathbb{P}(\mathcal{A}) + \mathbb{P}(\mathcal{B}) + \mathbb{P}(\mathcal{C}).$$

The following provides an inequality in the other direction.

Problem 1.27 Show that

$$\mathbb{P}(\mathcal{A} \cup \mathcal{B} \cup \mathcal{C}) \ge \mathbb{P}(\mathcal{A}) + \mathbb{P}(\mathcal{B}) + \mathbb{P}(\mathcal{C})$$
$$- \mathbb{P}(\mathcal{A} \cap \mathcal{B}) - \mathbb{P}(\mathcal{B} \cap \mathcal{C}) - \mathbb{P}(\mathcal{C} \cap \mathcal{A}).$$

[Drawing a Venn diagram will prove useful.]

In the proof, one typically proves first the identity

$$\mathbb{P}(\mathcal{A} \cup \mathcal{B} \cup \mathcal{C}) = \mathbb{P}(\mathcal{A}) + \mathbb{P}(\mathcal{B}) + \mathbb{P}(\mathcal{C})$$
$$- \mathbb{P}(\mathcal{A} \cap \mathcal{B}) - \mathbb{P}(\mathcal{B} \cap \mathcal{C}) - \mathbb{P}(\mathcal{C} \cap \mathcal{A})$$
$$+ \mathbb{P}(\mathcal{A} \cap \mathcal{B} \cap \mathcal{C}),$$

which generalizes the Law of Addition to three events.

Proposition 1.28 (Bonferroni's inequalities). *Consider any collection of events* $\mathcal{A}_1, \ldots, \mathcal{A}_n$, *and define*

$$S_k := \sum_{1 \le i_1 < \cdots < i_k \le n} \mathbb{P}(\mathcal{A}_{i_1} \cap \cdots \cap \mathcal{A}_{i_k}).$$

Then

$$\mathbb{P}(\mathcal{A}_1 \cup \cdots \cup \mathcal{A}_n) \le \sum_{j=1}^{k} (-1)^{j-1} S_j, \quad k \ odd;$$

$$\mathbb{P}(\mathcal{A}_1 \cup \cdots \cup \mathcal{A}_n) \ge \sum_{j=1}^{k} (-1)^{j-1} S_j, \quad k \ even.$$

Problem 1.29 Write down all of Bonferroni's inequalities for the case of four events $\mathcal{A}_1, \mathcal{A}_2, \mathcal{A}_3, \mathcal{A}_4$.

Inclusion-Exclusion Formula

The last Bonferroni inequality (at $k = n$) is in fact an equality, the so-called *inclusion-exclusion formula*,

$$\mathbb{P}(\mathcal{A}_1 \cup \cdots \cup \mathcal{A}_n) = \sum_{j=1}^{n} (-1)^{j-1} S_j. \tag{1.11}$$

(In particular, the last inequality in Problem 1.29 is an equality.)

1.5 Conditional Probability and Independence

1.5.1 Conditional Probability

Conditioning on an event B restricts the sample space to B. In other words, although the experiment might yield other outcomes, conditioning on B focuses the attention on the outcomes that made B happen. In what follows we assume that $\mathbb{P}(B) > 0$.

Problem 1.30 Show that $\mathbb{Q}$, defined for $A \in \Sigma$ as $\mathbb{Q}(A) = \mathbb{P}(A \cap B)$, is a probability distribution if and only if $\mathbb{P}(B) = 1$.

To define a bona fide probability distribution we renormalize $\mathbb{Q}$ to have total mass equal to 1 (required by the 2nd axiom) as follows:

$$\mathbb{P}(A \mid B) = \frac{\mathbb{P}(A \cap B)}{\mathbb{P}(B)}, \quad \text{for } A \in \Sigma.$$

We call $\mathbb{P}(A \mid B)$ the *conditional probability* of A given B.

Problem 1.31 Show that $\mathbb{P}(\cdot \mid B)$ is indeed a probability distribution on Ω.

Problem 1.32 In the context of Example 1.8, assume that any outcome is equally likely. Then what is the probability that the last toss lands heads if the previous tosses landed heads? Answer that same question when the coin is tossed $n \geq 2$ times, with n arbitrary and possibly large. [Regardless of n, the answer is $1/2$.]

The conclusions of Problem 1.32 may surprise some readers. And indeed, conditional probabilities can be rather unintuitive. We will come back to Problem 1.32, which is an example of the *Gambler's Fallacy*. Here is another famous example.

Example 1.33 (Monty Hall Problem) This problem is based on a television show in the US called *Let's Make a Deal* and named after its longtime presenter, Monty Hall. The following description is taken from a *New York Times* article [189]:

> Suppose you're on a game show, and you're given the choice of three doors: Behind one door is a car; behind the others, goats. You pick a door, say No. 1, and the host, who knows what's behind the other doors, opens another door, say No. 3, which has a goat. He then says to you, "Do you want to pick door No. 2?" Is it to your advantage to take the switch?

Not many problems in probability are discussed in the *New York Times*, to say the least. This problem is so simple to state and the answer so counter-intuitive that it generated quite a controversy (read the article). The problem can mislead anyone, including professional mathematicians, let alone the layperson appearing on television!

There is an entire book on the Monty Hall Problem [154]. The textbook [84] discusses this problem among other paradoxes arising when dealing with conditional probabilities.

1.5.2 Independence

Two events A and B are said to be *independent* if knowing that B happens does not change the chances (i.e., the probability) that A happens. This is formalized by saying that the probability of A conditional on B is equal to its (unconditional) probability, or in formula,

$$\mathbb{P}(A \mid B) = \mathbb{P}(A). \tag{1.12}$$

The wording in English would imply a symmetric relationship, and it is indeed the case that (1.12) is equivalent to $\mathbb{P}(B \mid A) = \mathbb{P}(B)$. The following equivalent definition of independence makes the symmetry transparent.

Proposition 1.34. *Two events A and B are independent if and only if*

$$\mathbb{P}(A \cap B) = \mathbb{P}(A)\mathbb{P}(B). \tag{1.13}$$

The identity (1.13) is often taken as a definition of independence.

Problem 1.35 Show that any event that never happens (i.e., having zero probability) is independent of any other event. In particular, $\varnothing$ is independent of any event.

Problem 1.36 Show that any event that always happens (i.e., having probability one) is independent of any other event. In particular, Ω is independent of any event.

The identity (1.13) only applies to independent events. However, it can be generalized as follows. (Note the parallel with the Law of Addition (1.7).)

Problem 1.37 (Law of Multiplication) Prove that, for any events A and B,

$$\mathbb{P}(A \cap B) = \mathbb{P}(A \mid B)\mathbb{P}(B). \tag{1.14}$$

Problem 1.38 (Independence and disjointness) The notions of independence and disjointness are often confused by the novice, even though they are very different. For example, show that two disjoint events are independent only when at least one of them either never happens or always happens.

Problem 1.39 Combine the Law of Total Probability (1.6) and the Law of Multiplication (1.14) to get

$$\mathbb{P}(\mathcal{A}) = \mathbb{P}(\mathcal{A}\,|\,\mathcal{B})\,\mathbb{P}(\mathcal{B}) + \mathbb{P}(\mathcal{A}\,|\,\mathcal{B}^c)\,\mathbb{P}(\mathcal{B}^c) \qquad (1.15)$$

Problem 1.40 Suppose we draw without replacement from an urn with r red balls and b blue balls. At each stage, every ball remaining in the urn is equally likely to be picked. Use (1.15) to derive the probability of drawing a blue ball on the 3rd trial.

1.5.3 Mutual Independence

One may be interested in several events at once. Some events, $\mathcal{A}_i, i \in I$, are said to be *mutually independent* (or *jointly independent*) if

$$\mathbb{P}(\mathcal{A}_{i_1} \cap \cdots \cap \mathcal{A}_{i_k}) = \mathbb{P}(\mathcal{A}_{i_1}) \times \cdots \times \mathbb{P}(\mathcal{A}_{i_k}),$$
$$\text{for any } k\text{-tuple } 1 \leq i_1 < \cdots < i_k \leq r.$$

They are said to be *pairwise independent* if

$$\mathbb{P}(\mathcal{A}_i \cap \mathcal{A}_j) = \mathbb{P}(\mathcal{A}_i)\,\mathbb{P}(\mathcal{A}_j), \quad \text{for all } i \neq j.$$

Obviously, mutual independence implies pairwise independence. The reverse implication is false, as the following counter-example shows.

Problem 1.41 Consider the uniform distribution on

$$\big\{(0,0,0), (0,1,1), (1,0,1), (1,1,0)\big\}.$$

Let $\mathcal{A}_i$ be the event that the ith coordinate is 1. Show that these events are pairwise independent but not mutually independent.

The following generalizes the Law of Multiplication (1.14). It is sometimes referred to as the *Chain Rule*.

Proposition 1.42 (General Law of Multiplication). *For any collection of events,* $\mathcal{A}_1, \ldots, \mathcal{A}_r,$

$$\mathbb{P}(\mathcal{A}_1 \cap \cdots \cap \mathcal{A}_r) = \prod_{k=1}^{r} \mathbb{P}\big(\mathcal{A}_k\,|\,\mathcal{A}_1 \cap \cdots \cap \mathcal{A}_{k-1}\big). \qquad (1.16)$$

For example, for any events $\mathcal{A}, \mathcal{B}, \mathcal{C}$,

$$\mathbb{P}(\mathcal{A} \cap \mathcal{B} \cap \mathcal{C}) = \mathbb{P}(\mathcal{C}\,|\,\mathcal{A} \cap \mathcal{B})\,\mathbb{P}(\mathcal{B}\,|\,\mathcal{A})\,\mathbb{P}(\mathcal{A}).$$

Problem 1.43 In the same setting as Problem 1.32, show that the result of the tosses are mutually independent. That is, define A_i as the event that the ith toss results in heads and show that $A_1, \ldots, A_n$ are mutually independent. In fact, show that the distribution is the uniform distribution (Problem 1.25) if and only if the tosses are fair and mutually independent.

1.5.4 Bayes Formula

The *Bayes formula*[6] can be used to "turn around" a conditional probability.

Proposition 1.44 (Bayes formula). *For any two events A and B,*

$$\mathbb{P}(A \mid B) = \frac{\mathbb{P}(B \mid A)\,\mathbb{P}(A)}{\mathbb{P}(B)}. \tag{1.17}$$

Proof By (1.14),

$$\mathbb{P}(A \cap B) = \mathbb{P}(A \mid B)\,\mathbb{P}(B),$$

and also

$$\mathbb{P}(A \cap B) = \mathbb{P}(B \mid A)\,\mathbb{P}(A),$$

which yield the result when combined. □

The denominator in (1.17) is sometimes expanded using (1.15) to get

$$\mathbb{P}(A \mid B) = \frac{\mathbb{P}(B \mid A)\,\mathbb{P}(A)}{\mathbb{P}(B \mid A)\,\mathbb{P}(A) + \mathbb{P}(B \mid A^c)\,\mathbb{P}(A^c)}. \tag{1.18}$$

This form is particularly useful when $\mathbb{P}(B)$ is not directly available.

Problem 1.45 Suppose we draw without replacement from an urn with r red balls and b blue balls. What is the probability of drawing a blue ball on the 1st trial when drawing a blue ball on the 2nd trial?

Base Rate Fallacy

Consider a medical test for the detection of a rare disease. There are two types of mistakes that the test can make:

- *False positive* when the test is positive even though the subject does not have the disease;
- *False negative* when the test is negative even though the subject has the disease.

---

[6] Named after Thomas Bayes (1701–1761).

Let α denote the probability of a false positive; $1 - \alpha$ is sometimes called the *sensitivity*. Let β denote the probability of a false negative; $1 - \beta$ is sometimes called the *specificity*. For example, the study reported in [143] evaluates the sensitivity and specificity of several HIV tests.

Suppose that the incidence of a certain disease is π, meaning that the disease affects a proportion π of the population of interest. A person is chosen at random from the population and given the test, which turns out to be positive. What are the chances that this person actually has the disease? Ignoring the *base rate* (i.e., the disease's prevalence) would lead one to believe these chances to be $1 - \beta$. This is an example of the *Base Rate Fallacy*.

Indeed, define the events

$$\mathcal{A} = \text{'the person has the disease'},$$

$$\mathcal{B} = \text{'the test is positive'}.$$

Thus, our goal is to compute $\mathbb{P}(\mathcal{A}\,|\,\mathcal{B})$. Because the person was chosen at random from the population, we know that $\mathbb{P}(\mathcal{A}) = \pi$. We know the test's sensitivity, $\mathbb{P}(\mathcal{B}^c\,|\,\mathcal{A}^c) = 1 - \alpha$, and its specificity, $\mathbb{P}(\mathcal{B}\,|\,\mathcal{A}) = 1 - \beta$. Plugging this into (1.18), we get

$$\mathbb{P}(\mathcal{A}\,|\,\mathcal{B}) = \frac{(1-\beta)\pi}{(1-\beta)\pi + \alpha(1-\pi)}. \tag{1.19}$$

Mathematically, the Base Rate Fallacy arises from confusing $\mathbb{P}(\mathcal{A}\,|\,\mathcal{B})$ (which is what we want) with $\mathbb{P}(\mathcal{B}\,|\,\mathcal{A})$. We saw that the former depends on the latter *and* on the base rate $\mathbb{P}(\mathcal{A})$.

Problem 1.46 Show that $\mathbb{P}(\mathcal{A}\,|\,\mathcal{B}) = \mathbb{P}(\mathcal{B}\,|\,\mathcal{A})$ if and only if $\mathbb{P}(\mathcal{A}) = \mathbb{P}(\mathcal{B})$.

Example 1.47 (Finding terrorists) In a totally different setting, Sageman [160] makes the point that a system for identifying terrorists, even if 99% accurate, cannot be ethically deployed on an entire population.

Fallacies in the Courtroom

Suppose that in a trial for murder in the US, some blood of type O- was found on the crime scene, matching the defendant's blood type. That blood type has a prevalence of about 1% in the US.[7] This leads the prosecutor to conclude that the suspect is guilty with 99% chance. But this is an example of the *Prosecutor's Fallacy*.

---

[7] https://redcrossblood.org/learn-about-blood/blood-types.html

In terms of mathematics, the error is the same as in the Base Rate Fallacy. In practice, the situation is even worse here because it is not even clear how to define the base rate. (Certainly, the base rate cannot be the unconditional probability that the defendant is guilty.)

In the same hypothetical setting, the defense could argue that, assuming the crime took place in a city with a population of about half a million, the defendant is only one among five thousand people in the region with the same blood type and that therefore the chances that he is guilty are $1/5000 = 0.02\%$. The argument is actually correct if there is no other evidence and it can be argued that the suspect was chosen more or less uniformly at random from the population. Otherwise, in particular if the latter is doubtful, this is is an example of the *Defendant's Fallacy*.

Example 1.48 *People v. Collins* is a robbery case[8] that took place in Los Angeles, California in 1968. A witness had seen a Black male with a beard and mustache together with White female with a blonde ponytail fleeing in a yellow car. The Collins (a married couple) exhibited all these attributes. The prosecutor argued that the chances of another couple matching the description were 1 in 12000000. This lead to a conviction. However, the California Supreme Court overturned the decision. This was based on the questionable computations of the base rate as well as the fact that the chances of another couple in the Los Angeles area (with a population in the millions) matching the description were much higher.

For more on the use of statistics in the courtroom, see [187].

1.6 Additional Problems

Problem 1.49 Show that if $|\Omega| = N$, then the collection of all subsets of Ω (including the empty set) has cardinality 2^N. This motivates the notation 2^Ω for this collection and also its name, as it is often called the power set of Ω.

Problem 1.50 Let $\{\Sigma_i, i \in I\}$ denote a family of σ-algebras over a set Ω. Prove that $\bigcap_{i\in I} \Sigma_i$ is also a σ-algebra over Ω.

Problem 1.51 Let $\{\mathcal{A}_i, i \in I\}$ denote a family of subsets of a set Ω. Show that there is a unique smallest (in terms of inclusion) σ-algebra over Ω that contains each of these subsets. This σ-algebra is said to be *generated* by the family $\{\mathcal{A}_i, i \in I\}$.

---

[8] https://courtlistener.com/opinion/1207456/people-v-collins/

Problem 1.52 (General Base Rate Fallacy) Assume that the same diagnostic test is performed on m individuals to detect the presence of a certain pathogen. Due to variation in characteristics, the test performed on Individual i has sensitivity $1 - \alpha_i$ and specificity $1 - \beta_i$. Assume that a proportion π of these individuals have the pathogen. Show that (1.19) remains valid as the probability that an individual chosen uniformly at random has the pathogen given that the test is positive, with $1 - \alpha$ defined as the average sensitivity and $1 - \beta$ defined as the average specificity, meaning $\alpha = \frac{1}{m} \sum_{i=1}^{m} \alpha_i$ and $\beta = \frac{1}{m} \sum_{i=1}^{m} \beta_i$.

2

Discrete Probability Spaces

We consider in this chapter the case of a probability space $(\Omega, \Sigma, \mathbb{P})$ with discrete sample space Ω. As we noted in Remark 1.20, it is customary to let Σ be the power set of Ω. We do so anytime we are dealing with a discrete probability space, as this can be done without loss of generality.

2.1 Probability Mass Functions

Given a probability distribution $\mathbb{P}$, define its *mass function* as

$$f(\omega) := \mathbb{P}(\{\omega\}), \quad \omega \in \Omega. \tag{2.1}$$

In general, we say that f is a mass function on Ω if it is a real-valued function on Ω satisfying the following conditions:

- *Non-negativity*

$$f(\omega) \geq 0, \quad \text{for any } \omega \in \Omega.$$

- *Unit measure*

$$\sum_{\omega \in \Omega} f(\omega) = 1.$$

Necessarily, such an f takes values in $[0, 1]$.

Problem 2.1 Show that (2.1) defines a probability mass function on Ω. Conversely, show that a probability mass function f on Ω defines a probability distribution on Ω as follows:

$$\mathbb{P}(\mathcal{A}) := \sum_{\omega \in \mathcal{A}} f(\omega), \quad \text{for } \mathcal{A} \subset \Omega. \tag{2.2}$$

Problem 2.2 Show that, if $\mathbb{P}$ has mass function f, and $\mathcal{B}$ is an event with $\mathbb{P}(\mathcal{B}) > 0$, then the conditional distribution given $\mathcal{B}$, meaning $\mathbb{P}(\cdot \,|\, \mathcal{B})$, has mass function

$$f(\omega \mid \mathcal{B}) := \frac{f(\omega)}{\mathbb{P}(\mathcal{B})} \{\omega \in \mathcal{B}\}, \quad \text{for } \omega \in \Omega,$$

where $\{\omega \in \mathcal{B}\} = 1$ if $\omega \in \mathcal{B}$ and $= 0$ otherwise.

2.2 Uniform Distributions

Assume the sample space Ω is finite. For a set $\mathcal{A}$, denote its cardinality (meaning the number of elements it contains) by $|\mathcal{A}|$. The *uniform distribution* on Ω is defined as

$$\mathbb{P}(\mathcal{A}) = \frac{|\mathcal{A}|}{|\Omega|}, \quad \text{for } \mathcal{A} \subset \Omega. \tag{2.3}$$

In Problem 1.25, we saw that this is indeed a probability distribution on Ω (equipped with its power set).

Equivalently, the uniform distribution on Ω is the distribution with constant mass function. Because of the requirements a mass function satisfies by definition, this necessarily means that the mass function is equal to $1/|\Omega|$ everywhere, meaning

$$f(\omega) := \frac{1}{|\Omega|}, \quad \text{for } \omega \in \Omega.$$

The resulting probability space models an experiment where all outcomes are equally likely.

Example 2.3 (Rolling a die) Consider an experiment where a die, with six faces numbered 1 through 6, is rolled once and the result is recorded. The sample space is $\Omega = \{1, 2, 3, 4, 5, 6\}$. The usual assumption that the die is fair is modeled by taking the distribution to be the uniform distribution, which puts mass $1/6$ on each outcome.

Remark 2.4 (Combinatorics) The uniform distribution is intimately related to *Combinatorics*, which is the branch of Mathematics dedicated to counting. This is because of its definition in (2.3), which implies that computing the probability of an event $\mathcal{A}$ reduces to computing its cardinality $|\mathcal{A}|$, meaning counting the number of outcomes in $\mathcal{A}$.

2.3 Bernoulli Trials

Consider an experiment where a coin is tossed repeatedly. We speak of *Bernoulli trials*[9] when the probability of heads remains the same at each toss regardless of the previous tosses. We consider a biased coin with probability of landing heads equal to $p \in [0, 1]$. We will call this a *p-coin*.

Problem 2.5 (The roulette) An American roulette has 38 slots: 18 are colored black (B), 18 are colored red (R), and 2 slots colored green (G). A ball is rolled, and eventually lands in one of the slots. One way a player can gamble is to bet on a color, red or black. If the ball lands in a slot with that color, the player doubles his bet. Otherwise, the player loses his bet. Assuming the wheel is fair, show that the probability of winning in a given trial is $p = 18/38$ regardless of the color the player bets on. (Note that $p < 1/2$, and $1/2 - p = 1/38$ is the casino's margin.)

Remark 2.6 (Beating the roulette) In the game of roulette, the odds are, of course, against the player. We will see later (in Section 9.2.1) that this guarantees any gambler will lose his fortune if he keeps on playing. This is so if the mathematics underlying this statement are an accurate description of how the game is played in real life. But this is not necessarily the case. For one thing, the equipment can be less than perfect. This was famously exploited in the late 1940s by Hibbs and Walford, and led to casinos using higher-quality roulettes. Still, Thorp and Shannon took on the challenge of beating the roulette in the late 1950s and early 1960s. For that purpose, they built one of the first wearable computers to help them predict where the ball would end based on an appraisal of the ball's position and speed at a certain time. Their system afforded them advantageous odds against the casino [86].

2.3.1 Probability of a Sequence of Given Length

Assume we toss a *p*-coin *n* times (or simply focus on the first *n* tosses if the coin is tossed an infinite number of times). In this case, in contrast with the situation in Problem 1.43, the distribution over the space of *n*-tuples of heads and tails is not the uniform distribution, that is, unless $p = 1/2$. We derive the distribution in closed form for an arbitrary *p*. We do this for the sequence HTHHT (so that $n = 5$) to illustrate the main arguments. Let $\mathcal{A}_i$ be the event that the *i*th trial results in heads. Then, applying the Chain Rule

[9] Named after Jacob Bernoulli (1655–1705).

(1.16), we have

$$\mathbb{P}(\text{HTHHT}) = \mathbb{P}(\mathcal{A}_5^c \mid \mathcal{A}_1 \cap \mathcal{A}_2^c \cap \mathcal{A}_3 \cap \mathcal{A}_4)$$
$$\times \mathbb{P}(\mathcal{A}_4 \mid \mathcal{A}_1 \cap \mathcal{A}_2^c \cap \mathcal{A}_3)$$
$$\times \mathbb{P}(\mathcal{A}_3 \mid \mathcal{A}_1 \cap \mathcal{A}_2^c) \times \mathbb{P}(\mathcal{A}_2^c \mid \mathcal{A}_1) \times \mathbb{P}(\mathcal{A}_1). \quad (2.4)$$

By assumption, a toss results in heads with probability p regardless of the previous tosses, so that

$$\mathbb{P}(\mathcal{A}_5^c \mid \mathcal{A}_1 \cap \mathcal{A}_2^c \cap \mathcal{A}_3 \cap \mathcal{A}_4) = \mathbb{P}(\mathcal{A}_5^c) = 1 - p,$$
$$\mathbb{P}(\mathcal{A}_4 \mid \mathcal{A}_1 \cap \mathcal{A}_2^c \cap \mathcal{A}_3) = \mathbb{P}(\mathcal{A}_4) = p,$$
$$\mathbb{P}(\mathcal{A}_3 \mid \mathcal{A}_1 \cap \mathcal{A}_2^c) = \mathbb{P}(\mathcal{A}_3) = p,$$
$$\mathbb{P}(\mathcal{A}_2^c \mid \mathcal{A}_1) = \mathbb{P}(\mathcal{A}_2^c) = 1 - p,$$
$$\mathbb{P}(\mathcal{A}_1) = p.$$

Plugging this into (2.4), we obtain

$$\mathbb{P}(\text{HTHHT}) = p(1 - p)pp(1 - p) = p^3(1 - p)^2,$$

after rearranging factors.

Problem 2.7 In the same example, show that the assumption that a toss results in heads with the same probability regardless of the previous tosses implies that the events $\mathcal{A}_i$ are mutually independent.

Beyond this special case, the following holds.

Proposition 2.8. *Consider n independent tosses of a p-coin. Regardless of the order, if k denotes the number of heads in a given sequence of n trials, the probability of that sequence is*

$$p^k(1 - p)^{n-k}. \quad (2.5)$$

Problem 2.9 Prove Proposition 2.8 by induction on n.

Remark 2.10 Although n Bernoulli trials do not necessarily result in a uniform distribution, Proposition 2.8 implies that, conditional on the number of heads being k, the distribution is uniform over the subset of sequences with exactly k heads.

Remark 2.11 (Gambler's Fallacy) Consider a casino roulette (Problem 2.5). Assume that you have just observed five spins that all resulted in red (i.e., RRRRR). What color would you bet on? Many a gambler would bet on B in this situation, believing that "it is time for the ball to land black". In fact, unless you have reasons to suspect otherwise, the natural assumption is that each spin of the wheel is fair and independent of the previous ones, and with this assumption the probability of в remains the same as the probability of к (that is, $18/38$).

Example 2.12 The paper [27] studies how the sequence in which unrelated cases are handled affects the decisions that are made in "refugee asylum court decisions, loan application reviews, and Major League Baseball umpire pitch calls", and finds evidence of the Gambler's Fallacy at play.

2.3.2 Number of Heads in a Sequence of Given Length

Suppose again that we toss a p-coin n times independently. In Proposition 2.8, we saw that the number of heads in the sequence dictates the probability of observing that sequence. Thus, it is of interest to study that quantity. In particular, we want to compute the probability of observing exactly k heads, where $k \in \{0, \dots, n\}$ is given.

Factorials

For a positive integer n, define its *factorial* to be

$$n! := \prod_{i=1}^{n} i = n \times (n-1) \times \cdots \times 1.$$

For example, $5! = 5 \times 4 \times 3 \times 2 \times 1 = 120$. By convention, $0! = 1$.

Proposition 2.13. $n!$ *is the number of orderings of n distinct items.*

This can be generalized as follows. For two non-negative integers $k \le n$, define the *falling factorial*

$$(n)_k := n(n-1)\cdots(n-k+1). \tag{2.6}$$

For example, $(5)_3 = 5 \times 4 \times 3 = 60$. By convention, $(n)_0 = 1$. In particular, $(n)_n = n!$.

Proposition 2.14. *Given positive integers $k \le n$, $(n)_k$ is the number of ordered subsets of size k of a set with n distinct elements.*

Proof There are n choices for the 1st position, then $n-1$ remaining choices for the 2nd position, etc., and $n - (k-1) = n - k + 1$ remaining choices for the kth position. These numbers of choices multiply to give the answer.

(Why they multiply and not add is crucial, and is explained at length, for example, in [84].)   □

Binomial Coefficients

For two non-negative integers $k \leq n$, define the *binomial coefficient*

$$\binom{n}{k} := \frac{n!}{k!(n-k)!} = \frac{(n)_k}{k!}. \tag{2.7}$$

The binomial coefficient (2.7) is often read "n choose k", as it corresponds to the number of ways of choosing k distinct items out of a total of n, disregarding the order in which they are chosen.

Proposition 2.15. *Given positive integers $k \leq n$, $\binom{n}{k}$ is the number of (unordered) subsets of size k of a set with n distinct elements.*

Proof Fix k and n, and let N denote the number of (unordered) subsets of size k of a set with n distinct elements. Each such subset can be ordered in $k!$ ways by Proposition 2.13, so that there are $N \times k!$ ordered subsets of size k. Hence, $Nk! = (n)_k$ by Proposition 2.14, resulting in $N = (n)_k/k! = \binom{n}{k}$.   □

Problem 2.16 (Pascal's triangle) Adopt the convention that $\binom{n}{k} = 0$ when $k < 0$ or when $k > n$. Then prove that, for any integers k and n,

$$\binom{n}{k} = \binom{n-1}{k} + \binom{n-1}{k-1}.$$

Do you see how this formula can be used to compute binomial coefficients recursively?

Exactly k Heads out of n Trials

Fix an integer $k \in \{0, \ldots, n\}$. By (2.2) and Proposition 2.8,

$$\mathbb{P}(\text{'exactly } k \text{ heads'}) = Np^k(1-p)^{n-k},$$

where N is the number of sequences of length n with exactly k heads. The trials are numbered 1 through n, and such a sequence is identified by the k trials among these where the coin landed heads. Thus, N is equal to the number of (unordered) subsets of size k of $\{1, \ldots, n\}$, which by Proposition 2.15 corresponds to the binomial coefficient $\binom{n}{k}$. Thus,

$$\mathbb{P}(\text{'exactly } k \text{ heads'}) = \binom{n}{k} p^k(1-p)^{n-k}. \tag{2.8}$$

2.4 Urn Models

We have already discussed an experiment involving an urn in Example 1.9. We consider the more general case of an urn that contains balls of only two colors, say, red and blue. Let r denote the number of red balls and b the number of blue balls. We will call this an (r, b)-urn. The experiment consists in drawing n balls from such an urn. In Example 1.9, we saw that this is enough information to define the sample space. However, the sampling process is not specific enough to define the sampling distribution. We discuss here the two most basic variants: sampling with replacement and sampling without replacement. We make the crucial assumption that at every stage each ball in the urn is equally likely to be drawn.

2.4.1 Sampling with Replacement

As the name indicates, this sampling scheme consists in repeatedly drawing a ball from the urn, every time returning the ball to the urn. Thus, the urn is the same before each draw. Because of our assumptions, this means that the probability of drawing a red ball remains constant, equal to $r/(r + b)$. Thus, we conclude that sampling with replacement from an urn with r red balls and b blue balls is analogous to flipping a p-coin with $p = r/(r + b)$. In particular, based on (2.5), the probability of any sequence of n draws containing exactly k red balls is

$$\left(\frac{r}{r + b}\right)^k \left(\frac{b}{r + b}\right)^{n-k}.$$

Remark 2.17 (From urn to coin) Unlike a general p-coin as described in Section 2.3, where p can be any number in $[0, 1]$, the parameter p that results from sampling with replacement from a finite urn is necessarily a rational number. However, because the rationals are dense in $[0, 1]$, it is possible to use an urn to approximate a p-coin. For that, it simply suffices to choose (integers) r and b be such that $r/(r + b)$ approaches p.

2.4.2 Sampling without Replacement

This sampling scheme consists in repeatedly drawing a ball from the urn, without ever returning the ball to the urn. Thus, the urn changes with each draw. For example, consider an urn with $r = 2$ red balls and $b = 3$ blue balls, and assume we draw a total of $n = 2$ balls from the urn without replacement. On the first draw, the probability of a red ball is $2/(2 + 3) = 2/5$, while the probability of drawing a blue ball is $3/5$.

- After first drawing a red ball, the urn contains 1 red ball and 3 blue balls, so the probability of a red ball on the 2nd draw is $1/4$.
- After first drawing a blue ball, the urn contains 2 red balls and 2 blue balls, so the probability of a red ball on the 2nd draw is $2/4$.

Although the resulting distribution is different than when the draws are executed with replacement, it is still true that all sequences with the same number of red balls have the same probability.

Proposition 2.18. *Assume that $n \le r + b$, for otherwise the sampling scheme is not feasible. For any $0 \le k \le r$, the probability of any sequence of n draws containing exactly k red balls is*

$$\frac{r(r-1)\cdots(r-k+1)b(b-1)\cdots(b-n+k+1)}{(r+b)(r+b-1)\cdots(r+b-n+1)}. \tag{2.9}$$

Remark 2.19 The convention when writing a product like $r(r-1)\cdots(r-k+1)$ is that it is equal to 1 when $k = 0$, equal to r when $k = 1$, equal to $r(r-1)$ when $k = 2$, etc. A more formal way to write such products is using factorials (Section 2.3.2).

We do not provide a formal proof of this result, but examine an example with $n = 5$, $k = 3$, and r and b arbitrary. We consider the outcome $\omega = $ RBRRB. Let A_i be the event that the ith draw is red and note that ω corresponds to $A_1 \cap A_2^c \cap A_3 \cap A_4 \cap A_5^c$. Then, applying the Chain Rule (Proposition 1.42), we have

$$\begin{aligned}
\mathbb{P}(\text{RBRRB}) = {} & \mathbb{P}(A_5^c \mid A_1 \cap A_2^c \cap A_3 \cap A_4) \\
& \times \mathbb{P}(A_4 \mid A_1 \cap A_2^c \cap A_3) \\
& \times \mathbb{P}(A_3 \mid A_1 \cap A_2^c) \times \mathbb{P}(A_2^c \mid A_1) \times \mathbb{P}(A_1).
\end{aligned}$$

Then, for example, $\mathbb{P}(A_3 \mid A_1 \cap A_2^c)$ is the probability of drawing a red after having drawn a red and then a blue. At that point there are $r-1$ reds and $b-1$ blues in the urn, so that probability is $(r-1)/(r-1+b-1) = (r-1)/(r+b-2)$. Reasoning in the same way with the other factors, we obtain

$$\mathbb{P}(\text{RBRRB}) \;=\; \frac{b-1}{r+b-4} \times \frac{r-2}{r+b-3} \times \frac{r-1}{r+b-2} \times \frac{b}{r+b-1} \times \frac{r}{r+b}.$$

Rearranging factors, we recover (2.9) specialized to the present case.

Problem 2.20 Repeat the above with any other sequence of 5 draws with exactly 3 reds. Verify that all these sequences have the same probability of occurring.

Problem 2.21 (Sampling without replacement from a large urn) We already noted that sampling with replacement from an (r, b)-urn amounts to tossing a p-coin with $p = r/(r + b)$. Assume now that we are sampling without replacement, but that the urn is very large. This can be considered in an asymptotic setting where r and b diverge to infinity in such a way that $r/(r + b) \to p$. Show that, in the limit, sampling without replacement from the urn also amounts to tossing a p-coin. Do so by proving that, for any n and k fixed, (2.9) converges to (2.5).

2.4.3 Number of Heads in a Sequence of Given Length

As in Section 2.3.2, we derive the probability of observing exactly k red balls in n draws without replacement. We show that

$$\mathbb{P}(\text{'exactly } k \text{ heads'}) = \binom{r}{k}\binom{b}{n-k} \Big/ \binom{r+b}{n}. \tag{2.10}$$

Indeed, since we assume that each ball is equally likely to be drawn at each stage, it follows that any subset of balls of size n is equally likely. We are thus in the uniform case (Section 2.2), and therefore the probability is given by the number of outcomes in the event, divided by the total number of outcomes.

The denominator in (2.10) is the number of possible outcomes, namely, subsets of balls of size n taken from the urn.[10]

The numerator in (2.10) is the number of outcomes with exactly k red balls. Indeed, any such outcome can be uniquely obtained by first choosing k red balls out of r in total – there are $\binom{r}{k}$ ways to do that – and then choosing $n - k$ blue balls out of b in total – there are $\binom{b}{n-k}$ ways to do that.

2.4.4 Other Urn Models

There are many urn models as, despite their apparent simplicity, their theoretical study is surprisingly rich. We already presented the two most fundamental sampling schemes above. We present a few more below. In each case, we consider an urn with a finite number of balls of different colors.

---

[10] Although the balls could be indistinguishable except for their colors, we use a standard trick in Combinatorics which consists in making the balls identifiable. This is only needed as a thought experiment. One could imagine, for example, numbering the balls 1 through $r + b$.

Pólya Urn Model

In this sampling scheme,[11] after each draw not only is the ball returned to the urn but together with an additional ball of the same color.

Problem 2.22 Consider an urn with r red balls and b blue balls. Show by example, as in Section 2.4.2, that the probability of any outcome sequence of length n with exactly k red balls is

$$\frac{r(r+1)\cdots(r+k-1)b(b+1)\cdots(b+n-k-1)}{(r+b)(r+b+1)\cdots(r+b+n-1)}.$$

(Recall Remark 2.19.) Thus, once again, the central quantity is the number of red balls.

Moran Urn Model

In this sampling scheme,[12] at each stage two balls are drawn: the first ball is returned to the urn together with an additional ball of the same color, while the second ball is not returned to the urn.

Note that if at some stage all the balls in the urn are of the same color, then the urn remains constant forever after. This can be shown to happen eventually and a question of interest is to compute the probability that the urn becomes all red.

Problem 2.23 Argue that if $r = b$, then that probability is $1/2$. In fact, argue that, if $\tau(r,b)$ denotes the probability that the process starting with r red and b blue balls ends up with only red balls, then $\tau(r,b) = 1 - \tau(b,r)$.

Problem 2.24 Derive the probabilities $\tau(1,2)$, $\tau(2,3)$, and $\tau(3,4)$.

Wright–Fisher Urn Model

Assume that the urn contains N balls in total. In this sampling scheme,[13] at each step the entire urn is reconstituted by sampling N balls uniformly at random with replacement from the urn.

Problem 2.25 Start with an urn with r red balls and b blue balls. Give the distribution of the number of red balls that the urn contains after one step.

Remark 2.26 The Wright–Fisher and the Moran urn models were proposed as models of genetic drift, which is the change in the frequency of gene

---

[11] Named after George Pólya (1887–1985).

[12] Named after Patrick Moran (1917–1988).

[13] Named after Sewall Wright (1889–1988) and Ronald Aylmer Fisher (1890–1962).

variants (i.e., alleles) in a given population. In both models, the size of the population remains constant.

2.5 Further Topics

2.5.1 Stirling's Formula

The factorial, as a function on the integers, increases very rapidly.

Problem 2.27 Prove that the factorial sequence $(n!)$ increases faster to infinity than any power sequence, meaning that $a^n/n! \to 0$ for any real number $a > 0$. Moreover, show that $n! \leq n^n$ for $n \geq 1$.

In fact, the size of $n!$ is known very precisely. The following describes the first-order asymptotics. More refined results exist.

Theorem 2.28 (Stirling's formula[14]). *Letting e denote the Euler number,*

$$n! \sim \sqrt{2\pi n}\,(n/e)^n, \quad as \ n \to \infty. \tag{2.11}$$

In fact,

$$1 \leq \frac{n!}{\sqrt{2\pi n}\,(n/e)^n} \leq e^{1/(12n)}, \quad for \ all \ n \geq 1. \tag{2.12}$$

2.5.2 More on Binomial Coefficients

Binomial coefficients appear in many important combinatorial identities. Here are a few examples.

Problem 2.29 Show that there are $\binom{n+1}{k}$ binary sequences with exactly k ones and n zeros such that no two 1's are adjacent.

Problem 2.30 (The binomial identity) Prove that

$$(a+b)^n = \sum_{k=0}^{n} \binom{n}{k} a^k b^{n-k}, \quad \text{for } a, b \in \mathbb{R}.$$

[One way to do so uses the fact that the binomial distribution needs to satisfy the 2nd axiom.]

Problem 2.31 Show that

$$2^n = \sum_{k=0}^{n} \binom{n}{k}.$$

---

[14] Named after James Stirling (1692–1770).

This can be done by interpreting this identity in terms of the number of subsets of $\{1, \ldots, n\}$.

Partitions of an Integer

Consider the number of ways of decomposing a non-negative integer m into a sum of $s \geq 1$ non-negative integers. Importantly, we count different permutations of the same numbers as distinct possibilities. For example, here are the possible decompositions of $m = 4$ into $s = 3$ non-negative integers

$$
\begin{array}{ccccc}
4+0+0 & 3+1+0 & 3+0+1 & 2+2+0 & 2+1+1 \\
2+0+2 & 1+3+0 & 1+2+1 & 1+1+2 & 1+0+3 \\
0+4+0 & 0+3+1 & 0+2+2 & 0+1+3 & 0+0+4
\end{array}
$$

Problem 2.32 Show that this number is equal to $\binom{m+s-1}{s-1}$. How does this change when the integers in the partition are required to be positive?

Catalan Numbers

Closely related to the binomial coefficients are the *Catalan numbers*[15]. The nth Catalan number is defined as

$$
C_n := \frac{1}{n+1} \binom{2n}{n}.
$$

These numbers have many different interpretations of their own. One of them is that C_n is the number of balanced bracket expressions of length $2n$. Here are all such expressions of length 6 ($n = 3$):

$$
()()() \quad ((())) \quad ()(()) \quad (())() \quad (()())
$$

Problem 2.33 Prove that

$$
C_n = \binom{2n}{n} - \binom{2n}{n+1}.
$$

Problem 2.34 Prove the recursive formula

$$
C_0 = 1, \quad C_{n+1} = \sum_{i=0}^{n} C_i C_{n-i}, \quad n \geq 0.
$$

2.5.3 Two Envelopes Problem

Two envelopes containing money are placed in front of you. You are told that one envelope contains double the amount of the other. You are allowed

---

[15] Named after Eugène Catalan (1814–1894).

to choose an envelope and look inside, and based on what you see you have to decide whether to keep the envelope that you just opened or switch for the other envelope. See [136] for an article-length discussion including different perspectives.

A *flawed* reasoning goes as follows:

> If you see x in the envelope, then the amount in the other envelope is either $x/2$ or $2x$, each with probability $1/2$. The average gain if you switch is therefore $(1/2)(x/2) + (1/2)(2x) = (5/4)x$, so you should switch.

The issue is that there are no grounds for the "with probability $1/2$" claim since the distribution that generated x was not specified.

This illustrates the maxim (echoed in [84, Exa 4.28]).

Computing probabilities requires a well-defined probability model.

See Problem 7.105 and Problem 7.106 for two different probability models for this situation that lead to different conclusions.

2.6 Additional Problems

Problem 2.35 A rule of thumb in Epidemiology is that, in the context of examining the safety of a given drug, if one hopes to identify a (severe) side effect affecting 1 out of every 1000 people taking the drug, then a trial needs to include at least 3000 individuals. In that case, what is the probability that in such a trial at least one person will experience the side effect?

Problem 2.36 (With or without replacement) Suppose that we are sampling n balls with replacement from an urn containing N balls numbered $1, \ldots, N$. Compute the probability that all balls drawn are distinct. Now consider an asymptotic setting where $n = n_N$ and $N \to \infty$, and let q_N denote that probability. Show that

$$\lim_{N \to \infty} q_N = \begin{cases} 0 & \text{if } n/\sqrt{N} \to \infty, \\ 1 & \text{if } n/\sqrt{N} \to 0. \end{cases}$$

Problem 2.37 (A stylized Birthday Problem) Compute the minimum number of people, taken at random from those born in a given 365-day year, needed so that at least two share their birthday with probability at least $1/2$. Model the situation using Bernoulli trials and assume that the a person is equally likely to be born any given day.

Problem 2.38 Continuing with Problem 2.37, perform simulations in R to confirm your answer.

Problem 2.39 (More on the Birthday Problem) The use of Bernoulli trials to model the situation in Problem 2.37 amounts to assuming that (1) each person is equally likely to be born any day of the year and (2) that the population is very large. Both are approximations. Keeping (2) in place, show that (1) only makes the number of required people larger.

Problem 2.40 Consider two independent draws with replacement from an urn containing N distinct items. Let p_i denote the probability of drawing item i. What is the probability that the same item is drawn twice? Show that this is minimized when the p_i are all equal, meaning, when drawing uniformly at random. [Use the method of Lagrange multipliers.]

Problem 2.41 Suppose that you have access to a computer routine that takes as input (n, N) and generates n independent draws with replacement from an urn with balls numbered $\{1, \dots, N\}$.

(i)   Explain how you would use that routine to generate n draws from an urn with r red balls and b blue balls with replacement.
(ii)   Explain how you would use that routine to generate n draws from an urn with r red balls and b blue balls without replacement.

First answer these questions in writing. Then answer them by writing a program in R for each situation, using the R function runid as the routine. (This is only meant for pedagogical purposes since the function sample can be directly used to fulfill the purpose in both cases.)

Problem 2.42 (Simpson's reversal) Provide a simple example of a finite probability space $(\Omega, \mathbb{P})$ and events $\mathcal{A}, \mathcal{B}, \mathcal{C}$ such that

$$\mathbb{P}(\mathcal{A} \mid \mathcal{B}) < \mathbb{P}(\mathcal{A} \mid \mathcal{B}^c),$$

while

$$\mathbb{P}(\mathcal{A} \mid \mathcal{B} \cap \mathcal{C}) \geq \mathbb{P}(\mathcal{A} \mid \mathcal{B}^c \cap \mathcal{C})$$

and

$$\mathbb{P}(\mathcal{A} \mid \mathcal{B} \cap \mathcal{C}^c) \geq \mathbb{P}(\mathcal{A} \mid \mathcal{B}^c \cap \mathcal{C}^c).$$

Show that this is *not* possible when $\mathcal{B}$ and $\mathcal{C}$ are independent of each other.

Problem 2.43 (The Two Children) This is a classic problem that appeared in [75]. You are on the airplane and start a conversation with the person next to you. In the course of the conversation, you learn that (1) the person has two children; (2) one of them is a daughter; (3) and she is the oldest.

After (1), what is the probability that the person has two daughters? How does that change after (2)? How does that change after (3)? [Make some necessary simplifying assumptions.]

Problem 2.44 Write an R function polya that takes in a sequence length n, and the composition of the initial urn in terms of numbers of red and blue balls, r and b, and generates a sequence of that length from the Pólya process starting from that urn. Call the function on $(n, r, b) = (200, 5, 3)$ a large number of times, say $M = 10^3$, each time compute the number of red balls in the resulting sequence, and tabulate the fraction of times that this number is equal to k, for all $k \in \{0, \ldots, 200\}$. Plot the corresponding bar chart.

Problem 2.45 Write an R function moran that takes in the composition of the urn in terms of numbers of red and blue balls, r and b, and runs the Moran urn process until the urn is of one color and returns that color and the number of stages that it took to get there. (You may want to bound the number of stages and then stop the process after that, returning a symbol indicating non-convergence.) Use that function to confirm your answers to Problem 2.24 following the guidelines of Problem 2.44.

3

Distributions on the Real Line

Measurements are often numerical in nature, which naturally leads to distributions on the real line. We start our discussion of such distributions in this chapter, and in the process introduce the concept of random variable, which is really a device to facilitate the definition of events (i.e., the writing of probability statements) and the computation of their probabilities.

Our foundation is a probability space, $(\Omega, \Sigma, \mathbb{P})$, modeling a certain experiment.

3.1 Random Variables

Consider a measurement on the outcome of the experiment, which we assume to be numerical and thus represented by a real-valued function on the measurable space (Ω, Σ). At the very minimum, we want to compute the probability that the measurement does not exceed a certain amount. Thus, we say that a real-valued function $X: \Omega \to \mathbb{R}$ – representing a measurement out of the experiment – is a *random variable* on (Ω, Σ) if

$$\{X \leq x\} \in \Sigma, \quad \text{for all } x \in \mathbb{R}. \tag{3.1}$$

This makes it possible to evaluate $\mathbb{P}(X \leq x)$.

Remark 3.1 $\{X \leq x\}$ in (3.1) is shorthand for $\{\omega \in \Omega : X(\omega) \leq x\}$.

3.2 Borel σ-Algebra

A random variable, representing a measurement on the outcome of the underlying experiment, in a sense transfers the randomness from the underlying probability space, $(\Omega, \Sigma, \mathbb{P})$, to the real line, $\mathbb{R}$. But, to speak of a distribution on the real line, we need to equip it with a σ-algebra. At the very minimum, because of (3.1), we require the σ-algebra to include all sets of the form

$$(-\infty, x], \quad \text{for } x \in \mathbb{R}.$$

(This is because $X \le x \;\Leftrightarrow\; X \in (-\infty, x]$.) The *Borel σ-algebra*[16] on $\mathbb{R}$, denoted $\mathcal{B}$ henceforth, is the σ-algebra generated by these intervals, meaning, the smallest σ-algebra over $\mathbb{R}$ that contains all such intervals (Problem 1.51).

Remark 3.2 Although the issue is rather technical, we mention that equipping the real line with its power set would effectively exclude most continuous distributions (see Chapter 5) commonly used in practice to model real-valued observations.

Proposition 3.3. *The Borel σ-algebra $\mathcal{B}$ contains all intervals, as well as all open sets and all closed sets.*

Proof   We only show that $\mathcal{B}$ contains all intervals. For example, take $a < b$. Since $\mathcal{B}$ contains $(-\infty, a]$ and $(-\infty, b]$, it must contain $(-\infty, a]^c$ by (1.2) and also

$$(-\infty, a]^c \cap (-\infty, b],$$

by (1.3). But this is $(a, b]$. Therefore, $\mathcal{B}$ contains all intervals of the form $(a, b]$, where $a = -\infty$ and $b = \infty$ are allowed.

Take an interval of the form $(-\infty, x)$. Note that it is open on the right. Define $\mathcal{U}_n = (-\infty, x - 1/n]$. By assumption, $\mathcal{U}_n \in \mathcal{B}$ for all n. Because of (1.4), $\mathcal{B}$ must also contain their union, and we conclude with the fact that

$$\bigcup_{n \ge 1} \mathcal{U}_n = (-\infty, x).$$

Now that we know that $\mathcal{B}$ contains all intervals of the form $(-\infty, x)$, we can reason as before and show that it must contain all intervals of the form $[a, b)$, where $a = -\infty$ and $b = \infty$ are allowed.

Finally, for any $-\infty < a < b \le \infty$,

$$[a, b] = [a, d) \cup (c, b], \quad \text{and} \quad (a, b) = (a, c] \cup [c, b),$$

for any c, d such that $a < c < d < b$, so that $\mathcal{B}$ must also include intervals of the form $[a, b]$ and (a, b).   □

Problem 3.4 Using Proposition 3.3, show that the Borel σ-algebra is the one generated by the open (resp. closed) sets of $\mathbb{R}$. (This makes it possible

---

[16] Named after Émile Borel (1871–1956).

to define the Borel σ-algebra on any topological space, in particular, on any Euclidean space. See Section 6.1.)

We will say that a function $g: \mathbb{R} \to \mathbb{R}$ is *measurable* if

$$g^{-1}(\mathcal{V}) \in \mathscr{B}, \quad \text{for all } \mathcal{V} \in \mathscr{B}.$$

3.3 Distributions on the Real Line

When considering a probability distribution on the real line, we will always assume that it is defined on the Borel σ-algebra.

The *support* of a distribution P on $(\mathbb{R}, \mathscr{B})$ is the smallest closed set $\mathcal{A}$ such that $P(\mathcal{A}) = 1$.

A random variable X on a probability space $(\Omega, \Sigma, \mathbb{P})$ defines a distribution on $(\mathbb{R}, \mathscr{B})$,

$$P_X(\mathcal{U}) := \mathbb{P}(X \in \mathcal{U}), \quad \text{for } \mathcal{U} \in \mathscr{B}.$$

($\{X \in \mathcal{U}\}$ is sometimes denoted by $X^{-1}(\mathcal{U})$.)

For a random variable X and a distribution P, we write $X \sim P$ when X has distribution P, meaning that $P_X = P$.

Problem 3.5 The *range* of a random variable X on (Ω, Σ) is defined as

$$X(\Omega) := \{X(\omega) : \omega \in \Omega\}.$$

Show that the support of P_X is included in the range of X. When is the inclusion strict?

3.4 Distribution Function

The *distribution function* (aka *cumulative distribution function*) of a distribution P on $(\mathbb{R}, \mathscr{B})$ is defined as

$$F(x) := P((-\infty, x]). \tag{3.2}$$

See Figure 3.1 for an example.

Proposition 3.6. *A distribution is characterized by its distribution function in the sense that two distributions with identical distribution functions must coincide.*

This result relies on rather technical arguments, but for intuition, it comes from the fact that two distributions with identical distribution functions coincide on all intervals of the form $(-\infty, x], x \in \mathbb{R}$, and these intervals

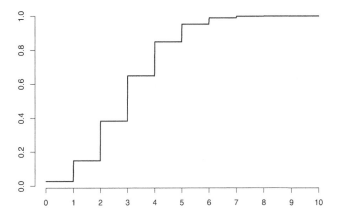

Figure 3.1 A plot of the distribution function of the binomial distribution with parameters $n = 10$ and $p = 0.3$.

generate the Borel σ-algebra – the algebra on which the distributions are defined. The situation is not unlike two continuous functions on the real line that coincide on the rationals: they must be the same.

Problem 3.7 Let F be the distribution function of a distribution P on $(\mathbb{R}, \mathscr{B})$.

(i)   Prove that

$$\mathsf{F} \text{ is non-decreasing,} \quad \lim_{x \to -\infty} \mathsf{F}(x) = 0, \quad \lim_{x \to +\infty} \mathsf{F}(x) = 1. \quad (3.3)$$

(ii)   Prove that F is continuous from the right, meaning

$$\lim_{t \searrow x} \mathsf{F}(t) = \mathsf{F}(x), \quad \text{for all } x \in \mathbb{R}. \quad (3.4)$$

[Use the 3rd probability axiom (1.5).]

(iii)   Prove that F is upper semi-continuous, meaning

$$\limsup_{t \to x} \mathsf{F}(t) \le \mathsf{F}(x), \quad \text{for all } x \in \mathbb{R}. \quad (3.5)$$

It so happens that these properties above define a distribution function, in the sense that any function satisfying these properties is the distribution function of some distribution on the real line.

Theorem 3.8. *Let* $\mathsf{F}\colon \mathbb{R} \to [0,1]$ *satisfy* (3.3)–(3.4). *Then* F *defines a distribution*[17] P *on* $\mathscr{B}$ *via* (3.2). *In particular,*

$$\mathsf{P}((a,b]) = \mathsf{F}(b) - \mathsf{F}(a), \quad \textit{for } -\infty \le a < b \le \infty.$$

Problem 3.9 In the context of the last theorem, for $x \in \mathbb{R}$, define the *left limit* of F at x as

$$\mathsf{F}(x^-) := \lim_{t \nearrow x} \mathsf{F}(t). \tag{3.6}$$

Show that this limit is well-defined. Then prove that

$$\mathsf{F}(x) - \mathsf{F}(x^-) = \mathsf{P}(\{x\}), \quad \text{for all } x \in \mathbb{R}. \tag{3.7}$$

Problem 3.10 Show that a distribution on the real line is discrete if and only if its distribution function is piecewise constant (i.e., staircase) with the set of discontinuities (i.e., jumps) corresponding to the support of the distribution.

Problem 3.11 Show that the set of points where a monotone function $\mathsf{F}\colon \mathbb{R} \to \mathbb{R}$ is discontinuous is countable.

The distribution function of a random variable X is simply the distribution function of its distribution P_X. It can be expressed as

$$\mathsf{F}_X(x) := \mathbb{P}(X \le x).$$

3.5 Survival Function

Consider a distribution P on $(\mathbb{R}, \mathscr{B})$ with distribution function F. The *survival function* of P is defined as

$$\bar{\mathsf{F}}(x) := \mathsf{P}((x,\infty)). \tag{3.8}$$

Problem 3.12 Show that $\bar{\mathsf{F}} = 1 - \mathsf{F}$.

Problem 3.13 Show that a survival function $\bar{\mathsf{F}}$ is non-increasing, continuous from the right, and lower semi-continuous, meaning that

$$\liminf_{t \to x} \bar{\mathsf{F}}(t) \ge \bar{\mathsf{F}}(x), \quad \text{for all } x \in \mathbb{R}.$$

The survival function of a random variable X is simply the survival function of its distribution P_X. It can be expressed as

$$\bar{\mathsf{F}}_X(x) := \mathbb{P}(X > x).$$

---

[17] The distribution P is known as the *Lebesgue–Stieltjes distribution* generated by F.

3.6 Quantile Function

Consider a distribution P on $(\mathbb{R}, \mathscr{B})$ with distribution function F. In Problem 3.10, we saw that F is not necessarily strictly increasing or continuous, in which case it does not admit an inverse in the usual sense. However, as a non-decreasing function, F admits the following form of pseudo-inverse

$$F^-(u) := \min\{x : F(x) \geq u\}, \tag{3.9}$$

sometimes called the *quantile function* of P. (That it is a minimum instead of an infimum is because of (3.5).) See Figure 3.2 for an example.

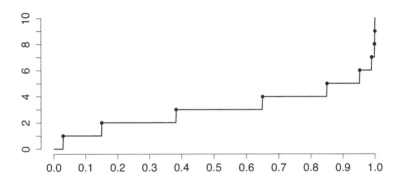

Figure 3.2 A plot of the quantile function of the binomial distribution with parameters $n = 10$ and $p = 0.3$.

Note that F^- is defined on $(0, 1)$, and if we allow it to return $-\infty$ and ∞ values, it can always be defined on $[0, 1]$.

Problem 3.14 Show that F^- is non-decreasing, continuous from the right, and

$$F(x) \geq u \iff x \geq F^-(u).$$

In addition, show that $F(F^-(u)) \geq u$. When is the inequality an equality?

Problem 3.15 Define the following variant of the survival function

$$\widetilde{F}(x) = P([x, \infty)). \tag{3.10}$$

Compare with (3.8), and note that the two definitions coincide when F is continuous. Show that

$$F^-(u) = \inf\{x : \widetilde{F}(x) \leq 1 - u\},$$

and deduce that $\widetilde{F}(F^-(u)) \geq 1 - u$, which in turn implies that

$$\widetilde{F}(x) < 1 - u \implies x > F^-(u).$$

(Unlike (3.9), the infimum above is not necessarily a minimum.)

Quantiles

We say that x is a *u-quantile* of P if

$$F(x) \geq u \quad \text{and} \quad \widetilde{F}(x) \geq 1 - u, \tag{3.11}$$

or equivalently, if X denotes a random variable with distribution P,

$$\mathbb{P}(X \leq x) \geq u \quad \text{and} \quad \mathbb{P}(X \geq x) \geq 1 - u.$$

With this definition, $x \in \mathbb{R}$ is a *u*-quantile for any

$$1 - \widetilde{F}(x) \leq u \leq F(x).$$

Remark 3.16 (Median and other quartiles) A $1/4$-quantile is called a *1st quartile*, a $1/2$-quantile is called *2nd quartile* or more commonly a *median*, and a $3/4$-quantile is called a *3rd quartile*. The quartiles, together with other features, can be visualized using a *boxplot*.

Problem 3.17 Show that for any $u \in (0, 1)$, the set of *u*-quantiles is either a singleton or an interval of the form $[a, b)$ for some $a < b$ that admit a simple characterization, where by convention $[a, a) = \{a\}$.

Problem 3.18 The previous problem implies that there always exists a *u*-quantile when $u \in (0, 1)$. What happens when $u = 0$ or $u = 1$?

Problem 3.19 Show that $F^-(u)$ is a *u*-quantile of F. Thus, (reassuringly) the quantile function returns bona fide quantiles.

Remark 3.20 Other definitions of pseudo-inverse are possible, each leading to a possibly different notion of quantile. For example,

$$F^{\boxminus}(x) := \sup\{x : F(x) \leq u\},$$

and

$$F^{\ominus}(u) := \frac{1}{2}\left(F^-(u) + F^{\boxminus}(u)\right). \tag{3.12}$$

Problem 3.21 Compare F^-, $F^{\boxminus}$, and $F^{\ominus}$. In particular, find examples of distributions where they are different, and also derive conditions under which they coincide.

4

Discrete Distributions

In this chapter, we consider distributions on the real line that have a discrete support. We will call these *discrete distributions*. It is indeed common to count certain occurrences in the outcome of an experiment, and the corresponding counts are invariably integer-valued. In fact, all the major distributions of this type are supported on the (non-negative) integers. We introduce the main ones in this chapter. And, unless otherwise noted, a discrete distribution will be assumed to be supported on the integers.

We consider, as usual, a probability space, $(\Omega, \Sigma, \mathbb{P})$, modeling a certain experiment. This is again our foundation, and all the probability statements that follow are about the outcome of this experiment.

In Chapter 3, we saw that a random variable on that space defines a distribution on the real line equipped with its Borel σ-algebra. A random variable with a discrete distribution is said to be *discrete*.

Problem 4.1 Show that any random variable on a discrete probability space is automatically discrete.

If X is such a random variable, then its distribution P_X is characterized, for example, by its distribution function, $\mathsf{F}_X(x) = \mathbb{P}(X \le x)$, but being a discrete distribution, P_X is also characterized by its mass function, given by

$$f_X(x) := \mathbb{P}(X = x), \quad \text{for } x \in \mathbb{R}.$$

Problem 4.2 Argue that, when dealing with discrete distributions on the real line, equipping $\mathbb{R}$ with its power set is equivalent to equipping it with its Borel σ-algebra.

4.1 Binomial Distributions

Consider the setting of Bernoulli trials, as in Section 2.3, where a p-coin is tossed repeatedly n times. Unless otherwise stated, we assume that the tosses are independent. Letting $\omega = (\omega_1, \ldots, \omega_n)$ denote an element of $\Omega := \{H, T\}^n$, for each $i \in \{1, \ldots, n\}$, define

$$X_i(\omega) = \begin{cases} 1 & \text{if } \omega_i = H, \\ 0 & \text{if } \omega_i = T. \end{cases} \tag{4.1}$$

(Note that X_i is the indicator of the event $\mathcal{A}_i$ defined in Section 2.3.) In particular, the distribution of X_i is given by

$$\mathbb{P}(X_i = 1) = p, \quad \mathbb{P}(X_i = 0) = 1 - p.$$

X_i has the so-called *Bernoulli distribution* with parameter p. We will denote this distribution by $\text{Ber}(p)$.

The X_i are *independent* (discrete) random variables in the sense that

$$\mathbb{P}(X_1 = x_1, \ldots, X_r = x_r) = \mathbb{P}(X_1 = x_1) \times \cdots \times \mathbb{P}(X_r = x_r),$$

$$\forall x_1, \ldots, x_r \in \{0, 1\}, \forall r \geq 2.$$

We will discuss independent random variables in more detail in Section 6.4.

Let Y denote the number of heads in the sequence of n tosses, so that

$$Y = \sum_{i=1}^{n} X_i. \tag{4.2}$$

We note that Y is a random variable on the same sample space Ω. Y has the so-called *binomial distribution* with parameters (n, p). We will denote this distribution by $\text{Bin}(n, p)$.

We already saw in Proposition 2.8 that Y plays a central role in this experiment. And in (2.8), we derived its distribution.

Proposition 4.3 (Binomial distribution). *The binomial distribution with parameters (n, p) has mass function*

$$f(k) = \binom{n}{k} p^k (1 - p)^{n-k}, \quad k \in \{0, \ldots, n\}. \tag{4.3}$$

Discrete mass functions are often drawn as bar plots. See Figure 4.1 for an illustration, where we plot not only the mass function but also the distribution function of a binomial distribution.

4.2 Hypergeometric Distributions

Consider an urn model as in Section 2.4. Suppose, as before, that the urn has r red balls and b blue balls. We sample from the urn n times and, as

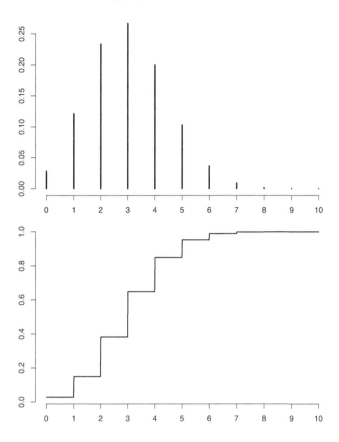

Figure 4.1 A plot of the mass function (top) and distribution function (bottom) of the binomial distribution with parameters $n = 10$ and $p = 0.3$.

before, let $X_i = 1$ if the ith draw is red, and $X_i = 0$ otherwise. (Note that X_i is the indicator of the event $\mathcal{A}_i$ defined in Section 2.4.) If we sample with replacement, we know that the experiment corresponds to Bernoulli trials with parameter $p := r/(r + b)$. We assume therefore that we are sampling without replacement. To be able to sample n times without replacement, we need to assume that $n \leq r + b$.

Let Y denote the number of heads in a sequence of n draws, exactly as in (4.2). The difference is that here the draws (the X_i) are not independent. The distribution of Y is called the *hypergeometric distribution*[18] with parameters (n, r, b). We will denote this distribution by Hyper(n, r, b).

We already computed its mass function in (2.10).

Proposition 4.4 (Hypergeometric distribution)**.** *The hypergeometric distribution with parameters* (n, r, b) *has mass function*

$$f(k) = \binom{r}{k}\binom{b}{n-k} \Big/ \binom{r+b}{n}, \quad k \in \{0, \dots, \min(n, r)\}.$$

4.3 Geometric Distributions

Consider Bernoulli trials as in Section 4.1 but now assume that we toss the p-coin until it lands heads. This experiment was described in Example 1.12. Define the X_i as before, and let Y denote the number of tails until the first heads. For example, $Y(\omega) = 3$ when $\omega = $ TTTH. Note that Y is a random variable on Ω.

It is particularly straightforward to derive the distribution of Y. Indeed, for any integer $k \geq 0$,

$$\begin{aligned}
\mathbb{P}(Y = k) &= \mathbb{P}(X_1 = 0, \dots, X_k = 0, X_{k+1} = 1) \\
&= \mathbb{P}(X_1 = 0) \times \cdots \times \mathbb{P}(X_k = 0) \times \mathbb{P}(X_{k+1} = 1) \\
&= \underbrace{(1 - p) \times \cdots \times (1 - p)}_{k \text{ times}} \times p = (1 - p)^k p,
\end{aligned}$$

using the independence of the X_i in the second line.

The distribution of Y is called the *geometric distribution*[19] with parameter p. We will denote this distribution by Geom(p). It is supported on $\{0, 1, 2, \dots\}$. See Figure 4.2 for an illustration.

Problem 4.5 Because of the Law of Total Probability,

$$\sum_{k=0}^{\infty} (1 - p)^k p = 1, \quad \text{for all } p \in (0, 1). \tag{4.4}$$

Prove this directly.

---

[18] There does not seem to be a broad agreement on how to parameterize this family of distributions. In the notation used here, the outcome is the number of red balls drawn, and for the parameters, the first parameter is the total number of balls drawn, the second is the number of red balls in the urn, while the third is the number of blue balls in the urn.

[19] This distribution is sometimes defined a bit differently, as the number of trials, including the last one, until the first heads. This is the case, for example, in [84].

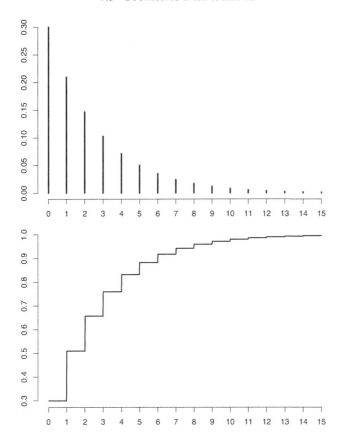

Figure 4.2 A plot of the mass function (top) and distribution function (bottom) of the geometric distribution with parameter $p = 0.3$.

Problem 4.6 Show that $\mathbb{P}(Y > k) = (1 - p)^{k+1}$ for $k \geq 0$ integer. This is the survival function of $\text{Geom}(p)$.

Remark 4.7 (Law of truly large numbers) In [46], Diaconis and Mosteller introduced this principle as one possible source for coincidences. In their own words, the 'law' says that:

> When enormous numbers of events and people and their interactions cumulate over time, almost any outrageous event is bound to occur.

Related concepts include *Murphy's Law*, *Littlewood's Law*, and the *Infinite Monkey Theorem*. Mathematically, the principle can be formalized as the following theorem: *If a p-coin, with $p > 0$, is tossed repeatedly independently, it will land heads eventually.* This theorem is an immediate consequence of (4.4).

Problem 4.8 (Memoryless property) Prove that a geometric random variable Y satisfies

$$\mathbb{P}(Y \geq k + t \mid Y \geq k) = \mathbb{P}(Y \geq t),$$

for all $t \geq 0$ and all $k \geq 0$.

4.4 Other Discrete Distributions

We already saw the families of Bernoulli, binomial, hypergeometric, and geometric distributions. We introduce a few more.

4.4.1 Discrete Uniform Distributions

A *discrete uniform distribution* (on the real line) is a uniform distribution on a finite set of points in $\mathbb{R}$. Thus, the family is parameterized by finite sets of points: such a set, say $\mathcal{X} \subset \mathbb{R}$, defines the distribution with mass function

$$f(x) = \frac{\{x \in \mathcal{X}\}}{|\mathcal{X}|}, \quad x \in \mathbb{R}.$$

The subfamily corresponding to sets of the form $\mathcal{X} = \{1, \ldots, N\}$ plays a special role. This subfamily is obviously much smaller and can be parameterized by the positive integers.

4.4.2 Negative Binomial Distributions

Consider an experiment where we toss a p-coin repeatedly until the mth heads, where $m \geq 1$ is given. Let Y denote the number of tails until we stop. For example, if $m = 3$, then $Y(\omega) = 4$ when $\omega = $ HTHTTTH. Y is clearly a random variable on the same sample space, and has the so-called *negative binomial distribution*[20] with parameters (m, p). We will denote this distribution by NegBin(m, p). It is supported on $k \in \{0, 1, 2, \ldots\}$. Clearly, NegBin$(1, p) = $ Geom(p), so the negative binomial family includes the geometric family.

---

[20] This distribution is sometimes defined a bit differently, as the number of trials, including the last one, until the mth heads. This is the case, for example, in [84].

Proposition 4.9. *The negative binomial distribution with parameters* (m, p) *has mass function*

$$f(k) = \binom{m+k-1}{m-1}(1-p)^k p^m, \quad k = 0, 1, 2, \ldots.$$

Problem 4.10 Prove Proposition 4.9. The arguments are very similar to those leading to Proposition 4.3.

Proposition 4.11. *The sum of m independent random variables, each having the geometric distribution with parameter p, has the negative binomial distribution with parameters (m, p).*

Problem 4.12 Prove Proposition 4.11.

4.4.3 Negative Hypergeometric Distributions

As the name indicates, this distribution arises when, instead of flipping a coin, we draw without replacement from an urn. Assume the urn contains r red balls and b blue balls. Let Y denote the number of blue balls drawn before drawing the mth red ball, where $m \leq r$. Y is a random variable on the same sample space, and has the so-called *negative hypergeometric distribution* with parameters (m, r, b).

Problem 4.13 Derive the mass function of the negative hypergeometric distribution with parameters $(m, r, b) = (3, 4, 5)$.

4.4.4 Poisson Distributions

The *Poisson distribution*[21] with parameter $\lambda \geq 0$ is given by the following mass function

$$f(k) := e^{-\lambda}\frac{\lambda^k}{k!}, \quad k = 0, 1, 2, \ldots.$$

By convention, $0^0 = 1$, so that when $\lambda = 0$, the right-hand side is 1 at $k = 0$ and 0 otherwise.

Problem 4.14 Show that this is indeed a mass function on the non-negative integers. [Recall that $e^x = \sum_{j \geq 0} x^j/j!$.]

Proposition 4.15 (Stability of the Poisson family). *The sum of a finite number of independent Poisson random variables is Poisson.*

---

[21] Named after Siméon Poisson (1781–1840).

The Poisson distribution arises when counting rare events. This is partly justified by Theorem 4.16.

4.5 Law of Small Numbers

Suppose that we are counting the occurrence of a rare phenomenon. For a historical example, Gosset[22] was counting the number of yeast cells using an hemocytometer [181]. This is a microscope slide subdivided into a grid of identical units that can hold a solution. In his experiments, Gosset prepared solutions containing the cells. Each solution was well mixed and spread on the hemocytometer. He then counted the number of cells in each unit. He wanted to understand the distribution of these counts. He performed a number of experiments. One of them is shown in Table 4.1.

Table 4.1 *The following is Table 2 in [181]. There were 103 units with 0 cells, 143 units with 1 cell, etc. See Problem 12.28 for a comparison with what is expected under a Poisson model.*

Number of cells	0	1	2	3	4	5	6
Number of units	103	143	98	42	8	4	2

Mathematically, Gosset reasoned as follows. Let N denote the number of units and n the number of cells. When the solution is well-mixed and evenly spread out over the units, each cell can be assumed to fall in any unit with equal probability $1/N$. Under this model, the number of cells found in a given unit has the binomial distribution with parameters $(n, 1/N)$. Gosset considered the limit where n and N are both large and proved that this distribution is 'close' to the Poisson distribution with parameter n/N when that number is not too large.

This approximation of a binomial distribution with a Poisson distribution is sometimes referred to as the *Law of Small Numbers*, and is formalized in Theorem 4.16. See Figure 4.3 for an illustration indicating that the approximation is already very good for moderate values of n.

Theorem 4.16 (Poisson approximation to the binomial distribution). *Consider a sequence (p_n) with $p_n \in [0, 1]$ and $np_n \to \lambda$ as $n \to \infty$. Then if*

---

[22] William Sealy Gosset (1876–1937) was working at the Guinness brewery, which required that he publish his work anonymously so as not to disclose the fact that he was working for Guinness and that his work could be used in the beer brewing business. Famously, he chose the pseudonym 'Student'.

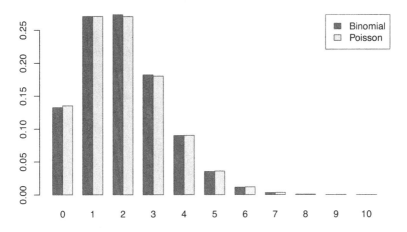

Figure 4.3 A plot of the mass functions of the binomial distribution with parameters $n = 20$ and $p = 1/20$ (dark grey) and of the Poisson distribution with mean $\lambda = 1$ (light grey).

Y_n has distribution $Bin(n, p_n)$ and $k \geq 0$ is an integer,

$$\mathbb{P}(Y_n = k) \to e^{-\lambda}\frac{\lambda^k}{k!}, \quad as \ n \to \infty.$$

Problem 4.17 Prove Theorem 4.16 using Stirling's formula (2.11).

Bateman arrived at the same conclusion in the context of experiments conducted by Rutherford and Geiger in the early 1900s to better understand the decay of radioactive particles. See Table 4.2 for an example of such an experiment.

4.6 Coupon Collector Problem

This problem arises from considering an individual collecting coupons of a certain type, say of players in a certain sports league in a certain sports season. The collector progressively completes his collection by buying envelopes, each containing an undisclosed coupon. With every purchase, the collector hopes the enclosed coupon will be new to his collection. (We assume here that the collector does not trade with others.) If there are N players in the league that season (and therefore that many coupons to collect), how many envelopes would the collector need to purchase in order to complete his collection?

Table 4.2 *The following is part of the table on page 701 of [159].*
The counting of particles was done over 2608 time intervals of
7.5 seconds each. See Problem 12.29 for a comparison with
what is expected under a Poisson model.

Number of particles	Number of intervals
0	57
1	203
2	383
3	525
4	532
5	408
6	273
7	139
8	45
9	27
10	10
11	4
12	0
13	1
14	1
15+	0

In the simplest setting, an envelope is equally likely to contain any one of the N distinct coupons. In that case, the situation can be modeled as a probability experiment where balls are drawn repeatedly with replacement from an urn containing N balls, all distinct, until all the balls in the urn have been drawn at least once. For example, if $N = 10$, the sequence of draws might look like this:

$$3\ 6\ 9\ 9\ 9\ 5\ 7\ 9\ 5\ 8\ 3\ 2\ 1\ 2\ 5\ 2\ 7\ 10\ 3\ 3\ 10\ 1\ 8\ 7\ 9\ 1\ 6\ 4$$

Let T denote the length of the resulting sequence ($T = 28$ in this particular realization of the experiment).

Problem 4.18 Write a function in R taking in N and returning a realization of the experiment. [Use the function **sample** and a **repeat** statement.] Run your function on $N = 10$ a few times to get a sense of how typical outcomes look like.

Problem 4.19 Let $T_0 = 0$, and for $i \in \{1, \ldots, N\}$, let T_i denote the number of balls needed to secure i distinct balls, so that $T = T_N$. Define $W_i = T_i - T_{i-1}$. Show that $W_1, \ldots, W_N$ are independent. Then derive the distribution of W_i.

Problem 4.20 Write a function in R taking in N and returning a realization of the experiment, but this time based on Problem 4.19. Compare this function and that of Problem 4.18 in terms of computational speed. [The function proc.time will prove useful.]

Problem 4.21 For $i \in \{1, \ldots, N\}$, let X_i denote the number of trials it takes to draw ball i. Note that $T = \max\{X_i : 1 \le i \le N\}$.

(i)   Show that

$$T \le t \iff X_i \le t, \ \forall i = 1, \ldots, N.$$

(ii)   What is the distribution of X_i?

(iii)   For $n \ge 1$, and for $k \in \{1, \ldots, N\}$ and any $1 \le i_1 < \cdots < i_k \le N$, compute

$$\mathbb{P}(X_{i_1} \ge n, \ldots, X_{i_k} \ge n).$$

(iv)   Use this and the inclusion-exclusion formula (1.11) to derive the mass function of T in closed form.

4.7 Additional Problems

Problem 4.22 Show that for any $n \ge 1$ integer and any $p \in [0, 1]$,

$$\mathrm{Bin}(n, 1 - p) \text{ coincides with } n - \mathrm{Bin}(n, p). \tag{4.5}$$

Problem 4.23 Let Y be binomial with parameters $(n, 1/2)$. Using the symmetry (4.5), show that

$$\mathbb{P}(Y > n/2) = \mathbb{P}(Y < n/2). \tag{4.6}$$

This means that, when the coin is fair, the probability of getting strictly more heads than tails is the same as the probability of getting strictly more tails than heads. When n is odd, show that (4.6) implies that

$$\mathbb{P}(Y > n/2) = \mathbb{P}(Y < n/2) = \frac{1}{2}.$$

When n is even, show that (4.6) implies that

$$\mathbb{P}(Y > n/2) = \frac{1}{2} + \frac{1}{2}\mathbb{P}(Y = n/2).$$

Then using Stirling's formula (2.11), show that

$$\mathbb{P}(Y = n/2) \sim \sqrt{\frac{2}{\pi n}}, \quad \text{as } n \to \infty.$$

This approximation is in fact very good. Verify this numerically in R.

Problem 4.24 Let $F_{n,\theta}$ denote the distribution function of the binomial distribution with parameters (n, θ). Fix an integer $0 \le y < n$ and show that $\theta \mapsto F_{n,\theta}(y)$ is strictly decreasing, continuous, and one-to-one as a map of $[0, 1]$ to itself. What happens when $y = n$?

Problem 4.25 Continuing with the setting of Problem 4.24, show that for any $y \ge 0$ integer and any $\theta \in [0, 1]$, $n \mapsto F_{n,\theta}(y)$ is non-increasing. In fact, if $0 < \theta < 1$, this function is decreasing over $n \in [y, \infty)$.

Problem 4.26 Let $F^-_{n,\theta}$ denote the quantile function of the binomial distribution with parameters (n, θ). Fix $0 < u < 1$ and show that $\theta \mapsto F^-_{n,\theta}(u)$ is piecewise constant, non-decreasing, and continuous from the left.

Problem 4.27 Suppose that you have access to a computer routine that takes as input a vector of any length k of numbers in $[0, 1]$, say $(q_1, \ldots, q_k)$, and generates $(B_1, \ldots, B_k)$ independent Bernoulli with these parameters (i.e., $B_i \sim \mathrm{Ber}(q_i)$). The question is how to use this routine to generate a random variable from a given mass function f (with finite support). Assume that f is supported on $x_1, \ldots, x_N$ and that $f(x_j) = p_j$.

(i)   Quickly argue that the case $N = 2$ is trivial.

(ii)   Consider the case $N = 3$. Show that the following works. Assume without loss of generality that $p_1 \le p_2 \le p_3$. Apply the routine to $q_1 = p_1$ and $q_2 = p_2/(1 - p_1)$ obtaining (B_1, B_2). If $B_1 = 1$, return x_1; if $B_1 = 0$ and $B_2 = 1$, return x_2; otherwise, return x_3.

(iii)   Extend this procedure to the general case.

Problem 4.28 Show that the sum of independent binomial random variables with same probability parameter p is also binomial with probability parameter p.

Problem 4.29 Prove Proposition 4.15.

Problem 4.30 Suppose that X and Y are two independent Poisson random variables. Show that the distribution of X conditional on $X + Y = t$ is binomial and specify the parameters.

Problem 4.31 For any $p \in (0, 1)$, show that there is $c_p > 0$ such that the following is a mass function on the positive integers

$$f(k) = c_p \frac{p^k}{k}, \quad k \ge 1 \text{ integer.}$$

Derive c_p in closed form. [Recall that $\log(1 - x) = -\sum_{k \geq 1} x^k/k$ for all $x \in (0, 1)$.]

Problem 4.32 For what values of α can one normalize $g(k) = k^{-\alpha}$ into a mass function on the positive integers? Similarly, for what values of α and β can one normalize $g(k) = k^{-\alpha}(\log(k + 1))^{-\beta}$ into a mass function on the positive integers?

Problem 4.33 (The binomial approximation to the hypergeometric distribution) Problem 2.21 asks you to prove that sampling without replacement from an (r, b)-urn amounts, in the limit where $r/(r + b) \to p$, to tossing a p-coin. Argue that, therefore, the hypergeometric distribution with parameters (n, r, b) must approach the binomial distribution with parameters (n, p) when n is fixed and $r/(r + b) \to p$. [Argue in terms of mass functions.]

Problem 4.34 (Game of Googol) Martin Gardner posed the following puzzle in his column in a 1960 edition of *Scientific American*:

> Ask someone to take as many slips of paper as he pleases, and on each slip write a different positive number. The numbers may range from small fractions of 1 to a number the size of a googol or even larger. [A *googol* is defined as 10^{100}.] These slips are turned face down and shuffled over the top of a table. One at a time you turn the slips face up. The aim is to stop turning when you come to the number that you guess to be the largest of the series. You cannot go back and pick a previously turned slip. If you turn over all the slips, then of course you must pick the last one turned.

Let n be the total number of slips. A possible strategy is, for a given $r \in \{1, \ldots, n\}$, to turn r slips, and then keep turning slips until either reaching the last one or stop when the slip shows a number that is at least as large as the largest number among the first r slips.

(i)   Compute the probability that this strategy is correct in terms of (n, r).
(ii) Let r_n denote the optimal choice of r as a function of n. (If there are several optimal choices, it is the smallest.) Compute r_n using R.
(iii) Formally derive r_n to first order when $n \to \infty$.

(This problem has a long history [62].)

5

Continuous Distributions

In some areas of mathematics, physics, and elsewhere, continuous objects and structures are often motivated, or even defined, as limits of discrete objects. For example, in mathematics, the real numbers are defined as the limit of sequences of rationals, and in physics, the laws of thermodynamics arise as the number of particles in a system tends to infinity – the so-called *thermodynamic* or *macroscopic limit*.

In Chapter 4, we introduced and discussed discrete distributions on the real line. Taking these discrete distributions to their continuous limits – which is done by letting their support size increase to infinity in a controlled manner – gives rise to continuous distributions on the real line. In fact, all continuous distributions can be obtained as such limits of discrete distributions.

In what follows, when we make probability statements, we assume that we have in the background a probability space, which by default will be denoted by $(\Omega, \Sigma, \mathbb{P})$. As in Chapter 3, we always equip $\mathbb{R}$ with its Borel σ-algebra, which we denoted by $\mathscr{B}$ in Section 3.2. (Unlike in Chapter 4, where we could as well have equipped $\mathbb{R}$ with its power set, here we need to equip $\mathbb{R}$ with its Borel σ-algebra.)

5.1 From the Discrete to the Continuous

Some of the discrete distributions introduced in Chapter 4 have 'natural' continuous limits when we let the size of their support sets increase. We formalize this passage to the continuum by working with distribution functions. (Recall that a distribution on the real line is characterized by its distribution function.)

5.1.1 From Uniform to Uniform

For a positive integer N, let P_N denote the (discrete) uniform distribution on $\{1, \ldots, N\}$, and let F_N denote its distribution function.

Problem 5.1 Show that, for any $x \in \mathbb{R}$,

$$\lim_{N \to \infty} \mathsf{F}_N(Nx) = F(x) := \begin{cases} 0 & \text{if } x \leq 0, \\ x & \text{if } 0 < x < 1, \\ 1 & \text{if } x \geq 1. \end{cases}$$

The limit function F above is continuous and satisfies the conditions of Theorem 3.8 and so defines a distribution, referred to as the *uniform distribution* on $[0, 1]$; see Section 5.6 for more details.

Remark 5.2 Note that $\mathsf{F}_N(x) \to 0$ for all $x \in \mathbb{R}$, so that scaling x by N is crucial to obtain the limit above.

Remark 5.3 The family of discrete uniform distributions on the real line is much larger. It turns out that it is so large that it is in some sense dense among the class of all distributions on $(\mathbb{R}, \mathscr{B})$. You are asked to prove this in Problem 5.43.

5.1.2 From Binomial to Normal

The following limiting behavior of binomial distributions is one of the pillars of Probability Theory.

Theorem 5.4 (De Moivre–Laplace Theorem[23]). *Fix $p \in (0, 1)$ and let F_n denote the distribution function of the binomial distribution with parameters (n, p). Then, for any $x \in \mathbb{R}$,*

$$\lim_{n \to \infty} \mathsf{F}_n\left(np + x\sqrt{np(1 - p)}\right) = \Phi(x),$$

where

$$\Phi(x) := \int_{-\infty}^{x} \frac{e^{-t^2/2}}{\sqrt{2\pi}} \, dt. \tag{5.1}$$

---

[23] Named after Abraham de Moivre (1667–1754) and Pierre-Simon, Marquis de Laplace (1749–1827).

Proposition 5.5. *The function* Φ *in* (5.1) *satisfies the conditions of Theorem 3.8, in particular because*

$$\int_{-\infty}^{\infty} e^{-t^2/2}\mathrm{d}t = \sqrt{2\pi}.$$

Thus, the function Φ defined in (5.1) defines a distribution, referred to as the *standard normal distribution*; see Section 5.7 for more details.

The theorem above is sometimes referred to as the *normal approximation to the binomial distribution*. See Figure 5.1 for an illustration.

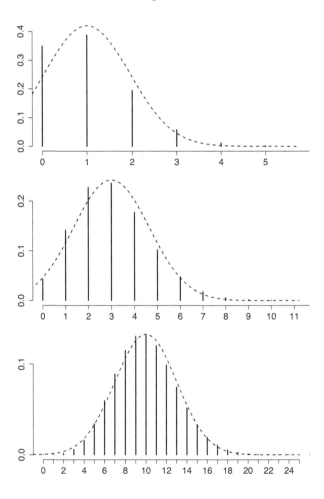

Figure 5.1 An illustration of the normal approximation to the binomial distribution with parameters $n \in \{10, 30, 100\}$ (from left to right) and $p = 0.1$.

As the proof of Theorem 5.4 can be relatively long, we only provide some guidance. Let $\sigma = \sqrt{p(1-p)}$.

Problem 5.6 Let $G_n(x) = F_n(np + x\sigma\sqrt{n})$. Show that it suffices to prove that $G_n(b) - G_n(a) \to \Phi(b) - \Phi(a)$ for all $-\infty < a < b < \infty$.

Problem 5.7 Using (4.3), show that

$$G_n(b) - G_n(a) = \sigma\sqrt{n} \int_a^b \binom{n}{\kappa_n(t)} p^{\kappa_n(t)} (1-p)^{n-\kappa_n(t)} dt + O(1/\sqrt{n}),$$

where $\kappa_n(t) := \lfloor np + t\sigma\sqrt{n} \rfloor$.

Problem 5.8 Show that

$$\sigma\sqrt{n} \binom{n}{\kappa_n(t)} p^{\kappa_n(t)} (1-p)^{n-\kappa_n(t)} \to \frac{e^{-t^2/2}}{\sqrt{2\pi}}, \quad \text{as } n \to \infty,$$

uniformly in $t \in [a, b]$. The rather long, but elementary calculations are based on Stirling's formula in the form of (2.12).

5.1.3 From Geometric to Exponential

Let F_N denote the distribution function of the geometric distribution with parameter $(\lambda/N) \wedge 1$, where $\lambda > 0$ is fixed.

Problem 5.9 Show that, for any $x \in \mathbb{R}$,

$$\lim_{N \to \infty} F_N(Nx) = F(x) := (1 - e^{-\lambda x})\{x > 0\}.$$

The limit function F above satisfies the conditions of Theorem 3.8 and so defines a distribution, referred to as the *exponential distribution* with rate λ; see Section 5.8 for more details.

5.2 Continuous Distributions

A distribution P on $(\mathbb{R}, \mathcal{B})$, with distribution function F, is a *continuous distribution* if F is a continuous function.

Problem 5.10 Show that P is continuous if and only if $P(\{x\}) = 0$ for all $x \in \mathbb{R}$.

Problem 5.11 Show that $F : \mathbb{R} \to \mathbb{R}$ is the distribution function of a continuous distribution if and only if it is continuous, non-decreasing, and satisfies

$$\lim_{x \to -\infty} F(x) = 0, \quad \lim_{x \to \infty} F(x) = 1.$$

We say that a distribution P is a mixture of distributions P_0 and P_1 if there is $b \in [0, 1]$ such that

$$P = (1 - b)P_0 + bP_1. \tag{5.2}$$

Theorem 5.12. *Every distribution on the real line is the mixture of a discrete distribution and a continuous distribution.*

Proof Let P be a distribution on the real line, with distribution function denoted F. Assume that P is neither discrete nor continuous, for otherwise there is nothing to prove.

Let $\mathcal{D}$ denote the set of points where F is discontinuous. By Problem 3.11 and the fact that F is non-decreasing (see (3.3)), $\mathcal{D}$ is countable, and since we have assumed that P is not continuous, $b := P(\mathcal{D}) > 0$. Define

$$F_1(x) = \frac{1}{b} \sum_{t \le x, t \in \mathcal{D}} P(\{t\}).$$

It is easy to see that F_1 defines a discrete distribution supported on $\mathcal{D}$, which is denoted P_1 henceforth. Note that $P_1(\{x\}) = 0$ if $x \notin \mathcal{D}$ and $P_1(\{x\}) = \frac{1}{b}P(\{x\})$ if $x \in \mathcal{D}$.

Define

$$F_0 = \frac{1}{1 - b}(F - bF_1).$$

It is easy to see that F_0 is a distribution function, which therefore defines a distribution, denoted P_0.

By construction, (5.2) holds, and so it remains to prove that F_0 is a continuous. Since F_0 is continuous from the right (see (3.4)), it suffices to show that it is continuous from the left as well, or equivalently, that $F_0(x) - F_0(x^-) = 0$ for all $x \in \mathbb{R}$. (Recall the definition (3.6).) For $x \in \mathbb{R}$, by (3.7), it suffices to establish that $P_0(\{x\}) = 0$. We have

$$P_0(\{x\}) = \frac{1}{1 - b}(P(\{x\}) - bP_1(\{x\})).$$

If $x \notin \mathcal{D}$, then $P(\{x\}) = 0$ and $P_1(\{x\}) = 0$, while if $x \in \mathcal{D}$, $bP_1(\{x\}) = P(\{x\})$, so in any case $P_0(\{x\}) = 0$.   □

5.3 Absolutely Continuous Distributions

A distribution P on $(\mathbb{R}, \mathcal{B})$, with distribution function F, is *absolutely continuous* if F is an absolutely continuous function, meaning that there is an integrable function f such that

$$F(x) = \int_{-\infty}^{x} f(t)dt, \quad \text{for all } x \in \mathbb{R}. \tag{5.3}$$

In that case, we say that f is a *density* of P.

Remark 5.13 (Integrable functions) There are a number of notions of integral. The most natural one in the context of Probability Theory is the *Lebesgue integral*. However, the *Riemann integral* has a somewhat more elementary definition. We will only consider functions for which the two notions coincide and will call these functions *integrable*. This includes *piecewise continuous functions*.[24]

Remark 5.14 (Non-uniqueness of a density) The function f in (5.3) is not uniquely determined by F. For example, if g coincides with f except on a finite number of points, then g also satisfies (5.3). Even then, it is customary to speak of 'the' density of a distribution, and we will do the same on occasion. This is particularly warranted when there is a continuous function f satisfying (5.3). In that case, it is the only one with that property and the most natural choice. More generally, f is chosen as 'simple' possible.

Problem 5.15 Suppose that f and g are such that $\int_{-\infty}^{x} f(t)dt = \int_{-\infty}^{x} g(t)dt$, for all $x \in \mathbb{R}$. Show that if they are both continuous they must coincide.

Problem 5.16 Show that a function f satisfying (5.3), where F is a distribution function, must be non-negative at all of its continuity points.

Problem 5.17 Show that if f satisfies (5.3), where F is a distribution function, then so do $\max(f, 0)$ and $|f|$. Therefore, any absolutely continuous distribution always admits a density function that is non-negative everywhere, and henceforth, we always choose to work with such a density.

Proposition 5.18. *An integrable function $f: \mathbb{R} \to [0, \infty)$ is a density of a distribution if and only if*

$$\int_{-\infty}^{\infty} f(x)dx = 1.$$

In that case, it defines an absolutely continuous distribution via (5.3).

Remark 5.19 Density functions are to absolutely continuous distributions what mass functions are to discrete distributions.

---

[24] A function f is piecewise continuous if its discontinuity points are nowhere dense, or equivalently, if there is a strictly increasing sequence $(a_k : k \in \mathbb{Z})$ with $\lim_{k \to -\infty} a_k = -\infty$ and $\lim_{k \to -\infty} a_k = \infty$ such that f is continuous on (a_k, a_{k+1}).

5.4 Continuous Random Variables

We say that X is a (resp. absolutely) *continuous random variable* on a sample space if it is a random variable on that space, as defined in Chapter 3, and its distribution P_X is (resp. absolutely) continuous, meaning that F_X is (resp. absolutely) continuous as a function. We let f_X denote a density of P_X when one exists.

Problem 5.20 For a continuous random variable X, verify that, for all $a < b$,

$$\mathbb{P}(X \in (a,b]) = \mathbb{P}(X \in [a,b))$$
$$= \mathbb{P}(X \in [a,b]) = \mathbb{P}(X \in (a,b)),$$

and, assuming X is absolutely continuous,

$$\mathbb{P}(X \in (a,b]) = P_X((a,b])$$
$$= F_X(b) - F_X(a)$$
$$= \int_a^b f_X(x)\mathrm{d}x.$$

Problem 5.21 Show that X is a continuous random variable if and only if

$$\mathbb{P}(X = x) = 0, \quad \text{for all } x \in \mathbb{R}.$$

(This is a bit perplexing at first.) In particular, the mass function of X is utterly useless when X is continuous.

Problem 5.22 Assume that X has a density f_X. Show that, for any x where f_X is continuous,

$$\mathbb{P}(X \in [x - h, x + h]) \sim 2h f_X(x), \quad \text{as } h \to 0.$$

5.5 Location/Scale Families of Distributions

Let X be a random variable. Then the family of distributions defined by the random variables $\{X + b : b \in \mathbb{R}\}$, is the *location family of distributions* generated by X. Similarly, the family of distributions defined by the random variables $\{aX : a > 0\}$, is the *scale family of distributions* generated by X, and the family of distributions defined by the random variables $\{aX + b : a > 0, b \in \mathbb{R}\}$, is the *location-scale family of distributions* generated by X.

Problem 5.23 Show that $aX + b$ has distribution function $F_X((\cdot - b)/a)$ and density $\frac{1}{a} f_X((\cdot - b)/a)$.

5.6 Uniform Distributions

The *uniform distribution* on an interval $[a, b]$ is given by the density

$$f(x) = \frac{1}{b-a}\{x \in [a, b]\}.$$

We will denote this distribution by $\mathrm{Unif}(a, b)$. See Figure 5.2 for an illustration.

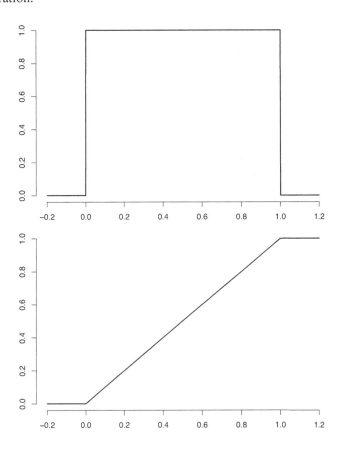

Figure 5.2 A plot of the density function (top) and distribution function (bottom) of the uniform distribution on $[0, 1]$.

In Section 5.1.1, we saw how this sort of distribution arises as a limit of discrete uniform distributions; see also Problem 5.43.

Problem 5.24 Compute the distribution function of $\mathrm{Unif}(a, b)$.

Problem 5.25 (Location-scale family) Show that the family of uniform of distributions, meaning

$$\{\text{Unif}(a,b) : a < b\},$$

is a location-scale family by verifying that

$$\text{Unif}(a,b) \equiv (b-a)\text{Unif}(0,1) + a.$$

Proposition 5.26. *Let U be uniform in* $[0,1]$ *and let* F *be any distribution function with quantile function* F^-. *Then* $\mathsf{F}^-(U)$ *has distribution* F.

Problem 5.27 Prove Proposition 5.26, at least in the case where F is continuous and strictly increasing (in which case F^- is a true inverse).

5.7 Normal Distributions

The *normal distribution* (aka *Gaussian distribution*[25]) with parameters μ and σ^2 is given by the density

$$f(x) = \frac{1}{\sqrt{2\pi\sigma^2}} \exp\left(-\frac{(x-\mu)^2}{2\sigma^2}\right).$$

(That this is a density is due to Proposition 5.5.) We will denote this distribution by $\mathcal{N}(\mu,\sigma^2)$. See Figure 5.3 for an illustration.

In Theorem 5.4, we saw that a normal distribution arises as the limit of binomial distributions, but in fact this limiting behavior is much more general, in particular because of Theorem 8.23, which partly explains why this family is so important.

Problem 5.28 (Location-scale family) Show that the family of normal distributions, meaning

$$\{\mathcal{N}(\mu,\sigma^2) : \mu \in \mathbb{R}, \sigma^2 > 0\},$$

is a location-scale family by verifying that

$$\mathcal{N}(\mu,\sigma^2) \equiv \sigma\mathcal{N}(0,1) + \mu.$$

The distribution $\mathcal{N}(0,1)$ is often called the *standard normal distribution*.

Proposition 5.29 (Stability of the normal family). *The sum of a finite number of independent normal random variables is normal.*

---

[25] Named after Carl Gauss (1777–1855).

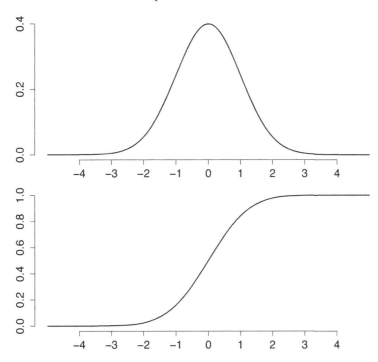

Figure 5.3 A plot of the standard normal density function (top) and distribution function (bottom).

5.8 Exponential Distributions

The *exponential distribution* with rate λ is given by the density

$$f(x) = \lambda \exp(-\lambda x)\, \{x \geq 0\}.$$

We will denote this distribution by $\mathrm{Exp}(\lambda)$. See Figure 5.4 for an illustration.

In Section 5.1.3, we saw how this distribution arises as a continuous limit of geometric distributions; see also Problem 8.48.

Problem 5.30 (Scale family) Show that the family of exponential distributions, meaning

$$\{\mathrm{Exp}(\lambda) : \lambda > 0\},$$

is a scale family by verifying that

$$\mathrm{Exp}(\lambda) \equiv \frac{1}{\lambda}\mathrm{Exp}(1).$$

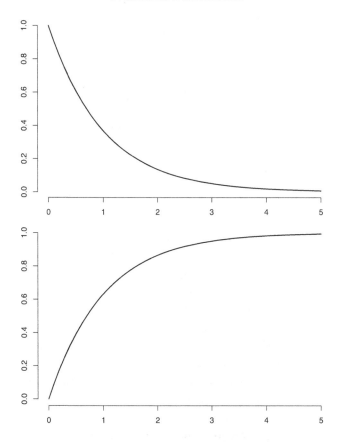

Figure 5.4 A plot of the density function (top) and distribution function (bottom) of the exponential distribution with rate $\lambda = 1$.

Problem 5.31 Compute the distribution function of $\text{Exp}(\lambda)$.

Problem 5.32 (Memoryless property) Show that any exponential distribution has the memoryless property of Problem 4.8.

5.9 Other Continuous Distributions

There are many other continuous distributions and families of such distributions. We introduce a few more below.

5.9.1 Gamma Distributions

The *gamma distribution* with rate λ and shape parameter κ is given by the density

$$f(x) = \frac{\lambda^\kappa}{\Gamma(\kappa)} x^{\kappa-1} \exp(-\lambda x) \{x \geq 0\},$$

where Γ is the so-called *gamma function*. We will denote this distribution by $\text{Gamma}(\lambda, \kappa)$.

Problem 5.33 Show that f above has finite integral if and only if $\lambda > 0$ and $\kappa > 0$.

Problem 5.34 Express the gamma function as an integral. [Use the fact that f above is a density.]

Problem 5.35 (Scale family) Show that the family of gamma distributions with same shape parameter κ, meaning

$$\{\text{Gamma}(\lambda, \kappa) : \lambda > 0\},$$

is a scale family.

It can be shown that a gamma distribution can arise as the continuous limit of negative binomial distributions; see Problem 5.45. The following is the analogue of Proposition 4.11.

Proposition 5.36. *Consider m independent random variables having the exponential distribution with rate λ. Then their sum has the gamma distribution with parameters (λ, m).*

5.9.2 Beta Distributions

The *beta distribution* with parameters (a, b) is given by the density

$$f(x) = \frac{1}{B(\alpha,\beta)} x^{\alpha-1}(1-x)^{\beta-1}\{x \in [0,1]\},$$

where B is the beta function. It can be shown that

$$B(\alpha,\beta) = \frac{\Gamma(\alpha)\Gamma(\beta)}{\Gamma(\alpha+\beta)}.$$

Note that the family includes the uniform distribution on $[0, 1]$.

Problem 5.37 Show that f above has finite integral if and only if $\alpha > 0$ and $\beta > 0$.

Problem 5.38 Express the beta function as an integral.

Problem 5.39 Prove that this is *not* a location and/or scale family of distributions.

5.9.3 Families Related to the Normal Family

A number of families are closely related to the normal family. The following are the main ones.

Chi-Squared Distributions

The *chi-squared distribution* with parameter $m \in \mathbb{N}$ is the distribution of $Z_1^2 + \cdots + Z_m^2$ when $Z_1, \ldots, Z_m$ are independent standard normal random variables. This happens to be a subfamily of the gamma family.

Proposition 5.40. *The chi-squared distribution with parameter m coincides with the gamma distribution with shape $\kappa = m/2$ and rate $\lambda = 1/2$.*

Student Distributions

The *Student distribution*[26] (aka *t-distribution*) with parameter $m \in \mathbb{N}$ is the distribution of $Z/\sqrt{W/m}$ when Z and W are independent, with Z being standard normal and W being chi-squared with parameter m.

Remark 5.41 The Student distribution with parameter $m = 1$ coincides with the *Cauchy distribution*,[27] defined by its density function

$$f(x) := \frac{1}{\pi(1 + x^2)}. \tag{5.4}$$

Fisher Distributions

The *Fisher distribution*[13] (aka *F-distribution*) with parameters (m_1, m_2) is the distribution of $(W_1/m_1)/(W_2/m_2)$ when W_1 and W_2 are independent, with W_j being chi-squared with parameter m_j.

Problem 5.42 Relate the Fisher distribution with parameters $(1, m)$ and the Student distribution with parameter m.

5.10 Additional Problems

Problem 5.43 Let F denote a continuous distribution function, with quantile function denoted F^-. For $1 \leq k \leq N$, define $x_{k:N} = \mathsf{F}^-(k/(N+1))$, and let P_N

---

[26] Named after 'Student', the pen name of Gosset[22].
[27] Named after Augustin-Louis Cauchy (1789–1857).

denote the (discrete) uniform distribution on $\{x_{1:N}, \ldots, x_{N:N}\}$. If F_N denotes the corresponding distribution function, show that $\mathsf{F}_N(x) \to \mathsf{F}(x)$ as $N \to \infty$ for all $x \in \mathbb{R}$.

Problem 5.44 (Bernoulli trials and the uniform distribution) Let $(X_i : i \geq 1)$ be independent with same distribution $\mathrm{Ber}(1/2)$. Show that $Y := \sum_{i \geq 1} 2^{-i} X_i$ is uniform in $[0, 1]$. Conversely, let Y be uniform in $[0, 1]$, and let $\sum_{i \geq 1} 2^{-i} X_i$ be its binary expansion. Show that $(X_i : i \geq 1)$ are independent with same distribution $\mathrm{Ber}(1/2)$.

Problem 5.45 We saw how a sequence of geometric distributions can have as limit an exponential distribution. Show by extension how a sequence of negative binomial distributions can have as limit a gamma distribution. [There is a simple argument based on the fact that a negative binomial (resp. gamma) random variable can be expressed as a sum of independent geometric (resp. exponential) random variables. An analytic proof will resemble that of Theorem 5.4.]

Problem 5.46 Verify Theorem 5.4 by simulation in R. For each $n \in \{10, 10^2, 10^3\}$ and each $p \in \{0.05, 0.2, 0.5\}$, generate $M = 500$ realizations from $\mathrm{Bin}(n, p)$ using the function rbinom and plot the corresponding histogram (with 50 bins) using the function hist. Overlay the graph of the standard normal density.

6

Multivariate Distributions

Some experiments lead to considering not one, but several measurements. As before, each measurement is represented by a random variable, and these are 'stacked' into a *random vector*. For example, in the context of an experiment that consists in flipping a coin n times, we defined n random variables, $X_1, \ldots, X_n$, according to (4.1). These are then concatenated to form the following random vector, $(X_1, \ldots, X_n)$.

In what follows, all the random variables that we consider are defined on the same probability space, denoted $(\Omega, \Sigma, \mathbb{P})$.

6.1 Random Vectors

Let $X_1, \ldots, X_r$ be r random variables on (Ω, Σ), meaning that each X_i satisfies (3.1). Then $X := (X_1, \ldots, X_r)$ is a *random vector* on (Ω, Σ), which is thus a function on Ω with values in $\mathbb{R}^r$,

$$\omega \in \Omega \longmapsto X(\omega) = (X_1(\omega), \ldots, X_r(\omega)) \in \mathbb{R}^r.$$

Problem 6.1 Show that,

$$\{X \in \mathcal{V}\} \in \Sigma,$$

for any set $\mathcal{V}$ of the form

$$(-\infty, x_1] \times \cdots \times (-\infty, x_r], \tag{6.1}$$

where $x_1, \ldots, x_r \in \mathbb{R}$.

We define the Borel σ-algebra of $\mathbb{R}^r$, denoted $\mathcal{B}_r$, as the σ-algebra generated by all hyper-rectangles of the form (6.1). We will always equip $\mathbb{R}^r$ with its Borel σ-algebra. The following generalizes Proposition 3.3.

Proposition 6.2. *The Borel σ-algebra of $\mathbb{R}^r$ contains all hyper-rectangles, as well as all open sets and all closed sets.*

The *support* of a distribution P on $(\mathbb{R}^r, \mathscr{B}_r)$ is the smallest closed set $\mathcal{A}$ such that $P(\mathcal{A}) = 1$. The *distribution function* of a distribution P on $(\mathbb{R}^r, \mathscr{B}_r)$ is defined as

$$F(x_1, \ldots, x_r) := P\big((-\infty, x_1] \times \cdots \times (-\infty, x_r]\big).$$

The distribution of X, also referred to as the *joint distribution* of $X_1, \ldots, X_r$, is defined on the Borel sets

$$P_X(\mathcal{V}) := \mathbb{P}(X \in \mathcal{V}), \quad \text{for } \mathcal{V} \in \mathscr{B}_r.$$

Note that for product sets, meaning when $\mathcal{V} = \mathcal{V}_1 \times \cdots \times \mathcal{V}_r$,

$$\mathbb{P}(X \in \mathcal{V}) = \mathbb{P}(X_1 \in \mathcal{V}_1) \times \cdots \times \mathbb{P}(X_r \in \mathcal{V}_r).$$

The distribution function of X is (of course) the distribution function of P_X, and can be expressed as

$$F_X(x_1, \ldots, x_r) := \mathbb{P}(X_1 \le x_1, \ldots, X_r \le x_r). \tag{6.2}$$

The following generalizes Proposition 3.6.

Proposition 6.3. *A distribution is characterized by its distribution function.*

The distribution of X_i, seen as the ith component of a random vector $X = (X_1, \ldots, X_r)$, is often called the *marginal distribution* of X_i, which is nothing else but its distribution disregarding the other variables.

6.2 Discrete Distributions

We say that a distribution P on $\mathbb{R}^r$ is discrete if it has a countable support set. For such a distribution, it is useful to consider its *mass function*, defined as

$$f(x) := P(\{x\}),$$

or, equivalently,

$$f(x_1, \ldots, x_r) = P(\{(x_1, \ldots, x_r)\}).$$

Problem 6.4 Show that a discrete distribution is characterized by it mass function.

Problem 6.5 Show that when all r variables are discrete, so is the random vector they form.

The mass function of a random vector X is the mass function of its distribution. It can be expressed as

$$f_X(x) := \mathbb{P}(X = x),$$

or, equivalently,

$$f_X(x_1, \ldots, x_r) = \mathbb{P}(X_1 = x_1, \ldots, X_r = x_r).$$

Proposition 6.6. *Let $X = (X_1, \ldots, X_r)$ be a discrete random vector with support on $\mathbb{Z}^r$. Then the (marginal) mass function of X_i can be computed as follows:*

$$f_{X_i}(x_i) = \sum_{j \neq i} \sum_{x_j \in \mathbb{Z}} f_X(x_1, \ldots, x_r), \qquad \text{for } x_i \in \mathbb{Z}.$$

For example, with two random variables, denoted X and Y, both supported on $\mathbb{Z}$,

$$\mathbb{P}(X = x) = \sum_{y \in \mathbb{Z}} \mathbb{P}(X = x, Y = y), \quad \text{for } x \in \mathbb{Z}. \tag{6.3}$$

Problem 6.7 Prove (6.3).

6.2.1 Binary Random Vectors

An r-dimensional binary random vector is a random vector with values in $\{0, 1\}^r$ (sometimes $\{-1, 1\}^r$). Such random vectors are particularly important as they are often used to represent outcomes that are *categorical* in nature (as opposed to numerical). For example, consider an experiment where we roll a die with six faces. Assume without loss of generality that the faces are numbered $1, \ldots, 6$. The fact that the face labels are numbers is typically not relevant, and representing the result of rolling the die with a random variable (with support $\{1, \ldots, 6\}$) could be misleading. We may instead use a binary vector representation for that purpose, as follows:

$$
\begin{aligned}
1 &\rightarrow (1,0,0,0,0,0) \\
2 &\rightarrow (0,1,0,0,0,0) \\
3 &\rightarrow (0,0,1,0,0,0) \\
4 &\rightarrow (0,0,0,1,0,0) \\
5 &\rightarrow (0,0,0,0,1,0) \\
6 &\rightarrow (0,0,0,0,0,1)
\end{aligned}
$$

This allows for the use of vector algebra, and other numerical manipulations that may otherwise make little sense if done directly on the face labels.

6.2.2 Multinomial Distributions

In general, consider a die with $m \geq 2$ faces, thus generalizing a coin. Just like a binomial distribution arises when a coin is tossed a predetermined number of times, the multinomial distribution arises when a die is rolled a predetermined number of times, say n. We assume that the die has faces with distinct labels, say, $1, \ldots, m$, and for $s \in \{1, \ldots, m\}$, we let p_s denote the probability that in a given trial the die shows s. The outcome of the experiment is of the form $\omega = (\omega_1, \ldots, \omega_n)$, where $\omega_i = s$ if the ith roll resulted in face s. We assume that the rolls are independent.

Let $Y_1, \ldots, Y_m$ denote the *counts*

$$Y_s(\omega) := \#\{i : \omega_i = s\}. \tag{6.4}$$

Note that $Y_s \sim \text{Bin}(n, p_s)$. Under the stated circumstances, the random vector of counts $(Y_1, \ldots, Y_m)$ is said to have the *multinomial distribution* with parameters $(n, p_1, \ldots, p_m)$.

Remark 6.8 There is some redundancy in the vector of counts, since $Y_1 + \cdots + Y_m = n$, and also in the parameterization, since $p_1 + \cdots + p_m = 1$. Except for that redundancy, the multinomial distribution with parameters (n, p_1, p_2) (where necessarily $p_2 = 1 - p_1$) corresponds to the binomial distribution with parameters (n, p_1).

Proposition 6.9. *The multinomial distribution with parameters $(n, \boldsymbol{p})$, where $\boldsymbol{p} := (p_1, \ldots, p_m)$, has probability mass function*

$$f_{\boldsymbol{p}}(y_1, \ldots, y_m) := \frac{n!}{y_1! \cdots y_m!} p_1^{y_1} \cdots p_m^{y_m}, \tag{6.5}$$

supported on the m-tuples of integers $y_1, \ldots, y_m \geq 0$ satisfying $y_1 + \cdots + y_m = n$.

Problem 6.10 Suppose that there are n balls, with y_s balls of color s, and that except for their color the balls are indistinguishable. The balls are to be placed in bins numbered $1, \ldots, n$. Show that there are

$$\frac{(y_1 + \cdots + y_m)!}{y_1! \cdots y_m!}$$

different ways of doing so. Then use this combinatorial result to prove Proposition 6.9.

6.2.3 Multivariate Hypergeometric Distributions

Instead of rolling a die, suppose instead that we sample without replacement from an urn containing v_s balls of color $s \in \{1, \ldots, m\}$. This is done n times and, for this to be possible, we assume that $n \leq v := v_1 + \cdots + v_m$. Let $\omega_i = s$ if the ith draw resulted in a ball with color s, and let $(Y_1, \ldots, Y_m)$ be the counts, as before. We say that this vector of counts has the *multivariate hypergeometric distribution* with parameters $(n, v_1, \ldots, v_m)$.

Problem 6.11 Argue as simply as you can that Y_s has (marginally) the hypergeometric distribution with parameters $(n, v_s, v - v_s)$.

Proposition 6.12. *The multivariate hypergeometric distribution with parameters $(n, v_1, \ldots, v_m)$ has probability mass function*

$$f(y_1, \ldots, y_m) := \binom{v_1}{y_1} \times \cdots \times \binom{v_m}{y_m} \Big/ \binom{v}{n},$$

supported on m-tuples of integers $y_1, \ldots, y_m \geq 0$ such that $y_1 + \cdots + y_m = n$.

Problem 6.13 Prove Proposition 6.12.

6.3 Continuous Distributions

We say that a distribution P on $\mathbb{R}^r$ is *continuous* if its distribution function F is continuous as a function on $\mathbb{R}^r$. We say that P is *absolutely continuous* if there is $f : \mathbb{R}^r \to \mathbb{R}$ integrable such that

$$\mathsf{P}(\mathcal{V}) = \int_{\mathcal{V}} f(x) \mathrm{d}x, \quad \text{for all } \mathcal{V} \in \mathscr{B}_r.$$

In that case, we say that f is a *density* of P. Clearly, a distribution is characterized by any one of its density functions.

Remark 6.14 Just as for distributions on the real line, a density is not unique and can always be taken to be non-negative.

We say that the random vector X is (resp. *absolutely*) *continuous* if its distribution is (resp. absolutely) continuous, and will denote a density (when applicable) by f_X, which in particular satisfies

$$\mathbb{P}(X \in \mathcal{V}) = \int_{\mathcal{V}} f_X(x) \mathrm{d}x, \quad \text{for all } \mathcal{V} \in \mathscr{B}_r.$$

See Figure 6.1 for an illustration.

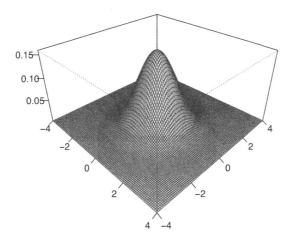

Figure 6.1 A plot of the density function the standard normal distribution in dimension 2, defined as the joint distribution of two independent standard normal random variables.

Proposition 6.15. *Let $X = (X_1, \ldots, X_r)$ be a random vector with density f. Then X_i has a density f_i given by*

$$f_i(x_i) := \int_{-\infty}^{\infty} \cdots \int_{-\infty}^{\infty} f(x_1, \ldots, x_r) \prod_{j \neq i} \mathrm{d}x_j,$$

for $x_i \in \mathbb{R}$.

For example, with two random variables, denoted X and Y for convenience, with joint density $f_{X,Y}$,

$$f_X(x) = \int_{-\infty}^{\infty} f_{X,Y}(x, y) \mathrm{d}y, \quad \text{for } x \in \mathbb{R}.$$

Remark 6.16 Even when all the random variables are continuous with a density, the random vector they define may not have a density in the anticipated sense. This happens, for example, when one variable is a function of the others or, more generally, when the variables are tied by an equation.

Example 6.17 Let T be uniform in $[0, 2\pi]$ and define $X = \cos(T)$ and $Y = \sin(T)$. Then $X^2 + Y^2 = 1$ by construction. In fact, (X, Y) is uniformly distributed on the unit circle.

6.4 Independence

Two random variables, X and Y, are said to be *independent* if

$$\mathbb{P}(X \in \mathcal{U}, Y \in \mathcal{V}) = \mathbb{P}(X \in \mathcal{U})\,\mathbb{P}(Y \in \mathcal{V}),$$
$$\text{for all } \mathcal{U}, \mathcal{V} \in \mathcal{B}.$$

This is sometimes denoted by $X \perp\!\!\!\perp Y$.

Proposition 6.18. *X and Y are independent if and only if, for all $x, y \in \mathbb{R}$,*

$$\mathbb{P}(X \le x, Y \le y) = \mathbb{P}(X \le x)\,\mathbb{P}(Y \le y),$$

or, equivalently, for all $x, y \in \mathbb{R}$,

$$F_{X,Y}(x,y) = F_X(x)F_Y(y).$$

Problem 6.19 Show that, if they are both discrete, X and Y are independent if and only if their joint mass function factorizes as the product of their marginal mass functions, or in formula,

$$f_{X,Y}(x,y) = f_X(x)f_Y(y), \quad \text{for all } x, y \in \mathbb{R}.$$

Problem 6.20 Show that, if (X, Y) has a density, then X and Y are independent if and only if the product of a density for X and a density for Y is a density for (X, Y).

The random variables $X_1, \ldots, X_r$ are said to be *mutually independent* if, for any $\mathcal{V}_1, \ldots, \mathcal{V}_r \in \mathcal{B}$,

$$\mathbb{P}(X_1 \in \mathcal{V}_1, \ldots, X_r \in \mathcal{V}_r) = \mathbb{P}(X_1 \in \mathcal{V}_1) \times \cdots \times \mathbb{P}(X_r \in \mathcal{V}_r).$$

Proposition 6.21. *$X_1, \ldots, X_r$ are mutually independent if and only if, for all $x_1, \ldots, x_r \in \mathbb{R}$,*

$$\mathbb{P}(X_1 \le x_1, \ldots, X_r \le x_r) = \mathbb{P}(X_1 \le x_1) \times \cdots \times \mathbb{P}(X_r \le x_r).$$

Problem 6.22 State and solve the analogue to Problem 6.19.

Problem 6.23 State and solve the analogue to Problem 6.20.

When some variables are said to be 'independent', what is meant by default is that they are mutually independent.

Problem 6.24 Let $X_1, \ldots, X_r$ be independent random variables. Show that $g_1(X_1), \ldots, g_r(X_r)$ are independent random variables for any measurable functions $g_1, \ldots, g_r$.

6.5 Conditional Distribution

Conditional distributions arise when holding some variables fixed and observing how the remaining ones vary. Note that this is different from how the marginal distributions are defined.

6.5.1 Discrete Case

Given two discrete random variables, X and Y, the *conditional distribution* of X given $Y = y$ is defined as the distribution of X conditional on the event $\{Y = y\}$ following the definition given in Section 1.5.1, namely,

$$\mathbb{P}(X \in \mathcal{U} \mid Y = y) = \frac{\mathbb{P}(X \in \mathcal{U}, Y = y)}{\mathbb{P}(Y = y)}. \tag{6.6}$$

For any $y \in \mathbb{R}$ such that $\mathbb{P}(Y = y) > 0$, this defines a discrete distribution, with corresponding mass function

$$f_{X \mid Y}(x \mid y) := \mathbb{P}(X = x \mid Y = y) = \frac{f_{X,Y}(x, y)}{f_Y(y)}.$$

Problem 6.25 (Law of Total Probability revisited) Show the following form of the Law of Total Probability. For two discrete random variables, X and Y, both supported on $\mathbb{Z}$,

$$f_X(x) = \sum_{y \in \mathbb{Z}} f_{X \mid Y}(x \mid y) f_Y(y), \quad \text{for all } x \in \mathbb{Z}.$$

Problem 6.26 (Conditional distribution and independence) Show that two discrete random variables, X and Y, are independent if and only if the conditional distribution of X given $Y = y$ is the same for all y in the support of Y. (And vice versa, as the roles of X and Y can be interchanged.)

6.5.2 Continuous Case

When Y is discrete, the distribution of X given $Y = y$ as defined in (6.6) still makes sense, even when X is continuous. This is no longer the case when Y is continuous, for in that case $\mathbb{P}(Y = y) = 0$ for all $y \in \mathbb{R}$. It is nevertheless possible to make sense of the distribution of X given $Y = y$. We consider the special case where (X, Y) has a density. In that case, the distribution of X given $Y = y$ is defined as the distribution with density

$$f_{X \mid Y}(x \mid y) := \frac{f_{X,Y}(x, y)}{f_Y(y)},$$

with the convention that $0/0 = 0$.

Problem 6.27 Show that $f_{X|Y}(\cdot|y)$ is indeed a density function for any continuity point y of f_Y such that $f_Y(y) > 0$.

Problem 6.28 Show that, for any $x \in \mathbb{R}$,

$$\mathbb{P}(X \le x \mid Y \in [y - h, y + h]) \xrightarrow[h \to 0]{} \int_{-\infty}^{x} f_{X|Y}(t|y)\mathrm{d}t.$$

[For simplicity, assume that $f_{X,Y}, f_X, f_Y$ are all continuous and that $f_Y > 0$ everywhere.]

6.6 Additional Problems

Problem 6.29 Consider an experiment where two fair dice are rolled. Let X_i denote the result of the ith die, $i \in \{1, 2\}$. Assume the variables are independent. Let $X = X_1 + X_2$. Show that X is a bona fide random variable with support $\{0, 1, \ldots, 12\}$, and compute its mass function and its distribution function. (You can use R for the computations. Present the solution in a table.)

Problem 6.30 For $Y = (Y_1, \ldots, Y_m)$, a multinomial random vector with parameters $(n, p_1, \ldots, p_m)$, compute $\mathrm{Cov}(Y_s, Y_t)$ for $s \ne t$.

Problem 6.31 (Generating a multinomial) Say we want to generate an observation from the multinomial distribution with parameters $(n, p_1, \ldots, p_m)$. Complete the following process and show that it does the job: first, generate an observation from $\mathrm{Bin}(n, p_1)$, say y_1; then, independently, generate from $\mathrm{Bin}(n - y_1, p_2/(1 - p_1))$, obtaining y_2; etc.

Problem 6.32 (Uniform distributions) For $\mathcal{V} \in \mathcal{B}_r$, its volume is defined as

$$|\mathcal{V}| := \int \{x \in \mathcal{V}\}\mathrm{d}x = \int_{\mathcal{V}} \mathrm{d}x.$$

When $0 < |\mathcal{V}| < \infty$, we can define the uniform distribution on $\mathcal{V}$ as the distribution with density

$$f_{\mathcal{V}}(x) := \frac{1}{|\mathcal{V}|}\{x \in \mathcal{V}\}.$$

Let $X = (X_1, \ldots, X_r)$ be uniform in $\mathcal{V}$.

(i)   Show that $X_1, \ldots, X_r$ are independent if and only if $\mathcal{V}$ if of the form $\mathcal{V}_1 \times \cdots \times \mathcal{V}_r$ for some $\mathcal{V}_j \in \mathcal{B}$.

(ii)   Suppose that $r = 2$ and that $\mathcal{V}$ is the unit disc. Compute the marginal distribution of X_1, and then compute the distribution of X_1 given $X_2 = x_2$ for some given x_2.

Problem 6.33 (Convolution) Suppose that X and Y have densities and are independent. Show that $Z := X + Y$ has density

$$f_Z(z) := \int_{-\infty}^{\infty} f_X(z - y) f_Y(y) dy. \tag{6.7}$$

This is called the *convolution* of f_X and f_Y, and often denoted by $f_X * f_Y$. State and prove a similar result when X and Y are both supported on $\mathbb{Z}$.

Problem 6.34 Assume that X and Y are independent random variables, with X having a continuous distribution. Show that $\mathbb{P}(X \neq Y) = 1$.

Problem 6.35 Consider an m-by-m matrix with elements being independent continuous random variables. Show that this *random matrix* is invertible with probability 1.

7

Expectation and Concentration

An expectation is simply a (weighted) mean, and means are at the core of Probability Theory and Statistics. While it may not have been clear why random variables were introduced in previous chapters, we will see that they are quite useful when computing expectations. Otherwise, everything we do here can be done directly with distributions instead of random variables.

As usual, our foundation is a probability space $(\Omega, \Sigma, \mathbb{P})$ modeling an experiment of interest. All the events that follow are elements of Σ, and all random variables are measurements on the outcome of this experiment.

7.1 Expectation

The definition and computation of an expectation are based on sums when the underlying distribution is discrete, and on integrals when the underlying distribution is absolutely continuous. We will assume the reader is familiar with the concepts of *absolute summability* and *absolute integrability*, and with the *Fubini–Tonelli Theorem*. (If not, the reader can assume – without much loss of generality in practice – that the distributions or random variables under consideration have bounded support.)

7.1.1 Discrete Expectation

Let X be a discrete random variable with mass function f_X. Its *expectation* or *mean* is defined as

$$\mathbb{E}(X) := \sum_{x \in \mathbb{Z}} x f_X(x), \tag{7.1}$$

so long as the sum converges absolutely.

Example 7.1 When X has a uniform distribution, its expectation is simply the average of the elements belonging to its support. To be more specific, assume that X has the uniform distribution on $\{x_1, \ldots, x_N\}$. Then

$$\mathbb{E}(X) = \frac{x_1 + \cdots + x_N}{N}.$$

Example 7.2 We say that X is *constant* (as a random variable) if there is $x_0 \in \mathbb{R}$ such that $\mathbb{P}(X = x_0) = 1$. In that case, X has an expectation, given by $\mathbb{E}(X) = x_0$.

Proposition 7.3 (Change of variables). *Let X be a discrete random variable and $g: \mathbb{Z} \to \mathbb{R}$. Then, as long as the sum converges absolutely,*

$$\mathbb{E}(g(X)) = \sum_{x \in \mathbb{Z}} g(x) \mathbb{P}(X = x).$$

Problem 7.4 Prove Proposition 7.3.

Problem 7.5 (Summation by parts) Let X be a random variable supported on the integers and with an expectation. Show that

$$\mathbb{E}(X) = \sum_{x \geq 0} \mathbb{P}(X > x) - \sum_{x < 0} \mathbb{P}(X \leq x).$$

7.1.2 Continuous Expectation

Let X be an absolutely continuous random variable with density f_X. Its *expectation* or *mean* is defined as

$$\mathbb{E}(X) := \int_{-\infty}^{\infty} x f_X(x) \mathrm{d}x,$$

as long as the integrand is absolutely integrable.

Problem 7.6 Show that a random variable with the uniform distribution on $[a, b]$ has mean $(a + b)/2$, the midpoint of the support interval.

Proposition 7.7 (Change of variables). *Let X be a random variable with density f_X and g a measurable function on $\mathbb{R}$. Then, as long as the integrand is absolutely integrable,*

$$\mathbb{E}(g(X)) = \int_{-\infty}^{\infty} g(x) f_X(x) \mathrm{d}x.$$

Problem 7.8 Prove Proposition 7.7.

Problem 7.9 (Integration by parts) Let X be a random variable with a well-defined expectation. Show that

$$\mathbb{E}(X) = \int_0^\infty \mathbb{P}(X > x)\mathrm{d}x - \int_{-\infty}^0 \mathbb{P}(X \le x)\mathrm{d}x. \tag{7.2}$$

7.1.3 Properties

Problem 7.10 Show that if X is non-negative and such that $\mathbb{E}(X) = 0$, then X is equal to 0 with probability 1, meaning $\mathbb{P}(X = 0) = 1$.

Problem 7.11 (Linearity of the expectation) Prove that, for a random variable X with well-defined expectation, and $a \in \mathbb{R}$,

$$\mathbb{E}(aX) = a\,\mathbb{E}(X), \text{ and } \mathbb{E}(X + a) = \mathbb{E}(X) + a. \tag{7.3}$$

Problem 7.12 (Monotonicity of the expectation) Show that the expectation is monotone in the sense that, if X and Y are random variables such that $\mathbb{P}(X \le Y) = 1$, then $\mathbb{E}(X) \le \mathbb{E}(Y)$.

Recall that a convex function g on an interval I is a function that satisfies

$$g(ax + (1 - a)y) \le ag(x) + (1 - a)g(y), \tag{7.4}$$
$$\text{for all } x, y \in I \text{ and all } a \in [0, 1].$$

The function g is strictly convex if the inequality is strict whenever $x \ne y$ and $0 < a < 1$.

Theorem 7.13 (Jensen's inequality[28]). *Let X be a continuous random variable and g a convex function on an interval containing the support of X such that both X and $g(X)$ have expectations. Then*

$$g(\mathbb{E}(X)) \le \mathbb{E}(g(X)).$$

If g is strictly convex, the inequality is strict unless X is constant.

For example, for a random variable X with an expectation,

$$|\mathbb{E}(X)| \le \mathbb{E}(|X|). \tag{7.5}$$

Proof sketch We sketch a proof for the case where the variable has finite support. Let the support be $\{x_1, \ldots, x_N\}$ and let $p_j = f_X(x_j)$. Then

$$g(\mathbb{E}(X)) = g\Big(\sum_{j=1}^N p_j x_j \Big),$$

---

[28] Named after Johan Jensen (1859–1925).

and

$$\mathbb{E}(g(X)) = \sum_{j=1}^{N} p_j g(x_j).$$

Thus, we need to prove that

$$g\left(\sum_{j=1}^{N} p_j x_j\right) \le \sum_{j=1}^{N} p_j g(x_j).$$

When $N = 2$, this is a direct consequence of (7.4): simply take $x = x_1$, $y = x_2$, and $a = p_1$, so that $1 - a = p_2$. The general case is proved by induction on N (and is a well-known property on convex functions). □

Proposition 7.14. *For two random variables, X and Y, both having expectations, $X + Y$ has an expectation, which given by*

$$\mathbb{E}(X + Y) = \mathbb{E}(X) + \mathbb{E}(Y).$$

Proof sketch We prove the result when the variables are discrete. Let $g(x, y) = x+y$. Then $X+Y = g(X, Y)$. Using a the analogue of Proposition 7.3 for random vectors, we have

$$\mathbb{E}(g(X, Y)) = \sum_{x \in \mathbb{Z}} \sum_{y \in \mathbb{Z}} g(x, y) \, \mathbb{P}(X = x, Y = y). \tag{7.6}$$

Thus, using that as a starting point, and then interchanging sums as needed (which is possible because of absolute summability), we derive

$$\begin{aligned}
\mathbb{E}(X &+ Y) \\
&= \sum_{x \in \mathbb{Z}} \sum_{y \in \mathbb{Z}} (x + y) \, \mathbb{P}(X = x, Y = y) \\
&= \sum_{x \in \mathbb{Z}} x \sum_{y \in \mathbb{Z}} \mathbb{P}(X = x, Y = y) + \sum_{y \in \mathbb{Z}} y \sum_{x \in \mathbb{Z}} \mathbb{P}(X = x, Y = y) \\
&= \sum_{x \in \mathbb{Z}} x \mathbb{P}(X = x) + \sum_{y \in \mathbb{Z}} y \mathbb{P}(Y = y) \\
&= \mathbb{E}(X) + \mathbb{E}(Y).
\end{aligned}$$

Equation (6.3) justifies the 3rd equality. □

Problem 7.15 Prove by recursion that if $X_1, \ldots, X_r$ are random variables with expectations, then $X_1 + \cdots + X_r$ has an expectation, which is given by

$$\mathbb{E}(X_1 + \cdots + X_r) = \mathbb{E}(X_1) + \cdots + \mathbb{E}(X_r). \tag{7.7}$$

Problem 7.16 (Binomial mean) Show that the binomial distribution with parameters (n, p) has mean np. An easy way to do so uses the definition of

Bin(n, p) as the sum of n independent random variables with distribution Ber(p), as in (4.2), and then using (7.7). A harder way uses the expression for its mass function (4.3) and the definition of expectation (7.1), which leads one to prove the following identity

$$\sum_{k=0}^{n} k \binom{n}{k} p^k (1 - p)^{n-k} = np.$$

Problem 7.17 (Hypergeometric mean) Show that the hypergeometric distribution with parameter (n, r, b) has mean np where $p := r/(r + b)$. The easier way, analogous to that described in Problem 7.16, is recommended.

Proposition 7.18. *For two independent random variables X and Y with expectations, XY has an expectation, which is given by*

$$\mathbb{E}(XY) = \mathbb{E}(X)\,\mathbb{E}(Y).$$

Compare with Proposition 7.14, which does not require independence.

Proof sketch   We prove the result when the variables are discrete. Let $g(x, y) = xy$. Then $XY = g(X, Y)$. Using the analogue of Proposition 7.3 for random vectors, as before, we have (7.6). Using that as a starting point, and then the fact that multiplication distributes over summation,

$$\mathbb{E}(XY) = \sum_{x \in \mathbb{Z}} \sum_{y \in \mathbb{Z}} xy\, \mathbb{P}(X = x, Y = y)$$

$$= \sum_{x \in \mathbb{Z}} \sum_{y \in \mathbb{Z}} xy\, \mathbb{P}(X = x)\, \mathbb{P}(Y = y)$$

$$= \sum_{x \in \mathbb{Z}} x\mathbb{P}(X = x) \sum_{y \in \mathbb{Z}} y\mathbb{P}(Y = y)$$

$$= \mathbb{E}(X)\,\mathbb{E}(Y).$$

We used the independence of X and Y in the 2nd line and absolute convergence in the 3rd line.   □

7.2 Moments

For a random variable X and a non-negative integer k, define the kth *moment* of X as the expectation of X^k, if X^k has an expectation. The 1st moment of a random variable is simply its mean.

Problem 7.19 (Binomial moments) Compute the first four moments of the binomial distribution with parameters (n, p).

Problem 7.20 (Geometric moments) Compute the first four moments of the geometric distribution with parameter p. [Start by proving that, for any $x \in (0, 1)$, $\sum_{j \geq 0} x^j = (1 - x)^{-1}$. Then differentiate up to four times to derive useful identities.]

Problem 7.21 (Uniform moments) Compute the kth moment of the uniform distribution on $[0, 1]$. [Use a recursion on k and integration by parts.]

Problem 7.22 (Normal moments) Compute the first four moments of the standard normal distribution. Verify that they are respectively equal to $0, 1, 0, 3$. Deduce the first four moments of the normal distribution with parameters (μ, σ^2), which in particular has mean μ.

Problem 7.23 Show that if X has a kth moment for some $k \geq 1$, then it has a lth moment for any $l \leq k$. [Use Jensen's inequality.]

Problem 7.24 (Symmetric distributions) Suppose that X and $-X$ have the same distribution. Show that if that distribution has a kth moment and k is odd, then that moment is 0.

The following is one of the most celebrated inequalities in Probability Theory.

Theorem 7.25 (Cauchy–Schwarz inequality[29]). *For two independent random variables X and Y with 2nd moments,*

$$|\mathbb{E}(XY)| \leq \sqrt{\mathbb{E}(X^2)} \sqrt{\mathbb{E}(Y^2)}.$$

Moreover the inequality is strict unless there is $a \in \mathbb{R}$ such that $\mathbb{P}(X = aY) = 1$ or $\mathbb{P}(Y = aX) = 1$.

Proof By Jensen's inequality, in particular (7.5),

$$|\mathbb{E}(XY)| \leq \mathbb{E}(|XY|) = \mathbb{E}(|X||Y|),$$

so that it suffices to prove the result when X and Y are non-negative. The assumptions imply that for any real t, $X + tY$ has a 2nd moment, and we have

$$g(t) := \mathbb{E}\left((X + tY)^2\right)$$
$$= \mathbb{E}(X^2) + 2t\,\mathbb{E}(XY) + t^2\,\mathbb{E}(Y^2),$$

so that g is a polynomial of degree at most 2. Since it is non-negative everywhere it must be non-negative at its minimum. The minimum is at

---

[29] Named after Cauchy [27] and Hermann Schwarz (1843–1921).

$t = -\mathbb{E}(XY)/\mathbb{E}(Y^2)$, so we must have

$$\mathbb{E}(X^2) - 2(\mathbb{E}(XY)/\mathbb{E}(Y^2))\mathbb{E}(XY) + (\mathbb{E}(XY)/\mathbb{E}(Y^2))^2\mathbb{E}(Y^2) \geq 0,$$

which leads to the stated inequality after simplification.

That the inequality is strict unless X and Y are proportional is left as an exercise.   □

Problem 7.26 In the proof, we implicitly assumed that $\mathbb{E}(Y^2) > 0$. Show that the result holds (trivially) when $\mathbb{E}(Y^2) = 0$.

7.3 Variance and Standard Deviation

Assume that X has a 2nd moment. We can then define its *variance* as

$$\mathrm{Var}(X) := \mathbb{E}\left[(X - \mathbb{E}(X))^2\right].$$

The *standard deviation* of a random variable X is the square-root of its variance.

Remark 7.27 (Central moments) Transforming X into $X - \mathbb{E}(X)$ is sometimes referred to as "centering X". Then, if k is a non-negative integer, the kth *central moment* of X is the kth moment of $X - \mathbb{E}(X)$, assuming it is well-defined. In particular, the variance corresponds to the 2nd central moment. (Note that the 1st central moment is 0.)

Proposition 7.28. *For a random variable X with a 2nd moment,*

$$\mathrm{Var}(X) = \mathbb{E}(X^2) - \mathbb{E}(X)^2.$$

Proof   For the sake of clarity, set $\mu := \mathbb{E}(X)$. Then

$$(X - \mathbb{E}(X))^2 = (X - \mu)^2 = X^2 - 2\mu X + \mu^2.$$

Hence,

$$\begin{aligned}
\mathrm{Var}(X) &= \mathbb{E}(X^2 - 2\mu X + \mu^2) \\
&= \mathbb{E}(X^2) + \mathbb{E}(-2\mu X) + \mathbb{E}(\mu^2) \\
&= \mathbb{E}(X^2) - 2\mu\,\mathbb{E}(X) + \mu^2 \\
&= \mathbb{E}(X^2) - \mu^2.
\end{aligned}$$

We used Proposition 7.14, then (7.3), and the fact that the expectation of a constant is itself, and finally the definition of μ and some simplifying algebra.   □

Problem 7.29 Let X be random variable with a 2nd moment. Show that $\mathrm{Var}(X) = 0$ if and only if X is constant.

Problem 7.30 Prove that for a random variable X with a 2nd moment, and $a \in \mathbb{R}$,

$$\mathrm{Var}(aX) = a^2 \, \mathrm{Var}(X), \quad \mathrm{Var}(X + a) = \mathrm{Var}(X).$$

Problem 7.31 (Normal variance) Show that the normal distribution with parameters (μ, σ^2) has variance σ^2.

Proposition 7.32. *For two* independent *random variables X and Y with 2nd moments,*

$$\mathrm{Var}(X + Y) = \mathrm{Var}(X) + \mathrm{Var}(Y).$$

Compare with Proposition 7.14, which does not require independence.

Problem 7.33 Prove Proposition 7.32 using Proposition 7.14.

Problem 7.34 Extend Proposition 7.32 to more than two independent random variables. [There is a simple argument by induction.]

Problem 7.35 (Binomial variance) Show that the binomial distribution with parameters (n, p) has variance $np(1 - p)$.

7.4 Covariance and Correlation

In this section, all the random variables that we consider are assumed to have a 2nd moment.

Covariance

To generalize Proposition 7.32 to non-independent random variables requires the definition of the *covariance* of two random variables, given by

$$\mathrm{Cov}(X, Y) := \mathbb{E}\big((X - \mathbb{E}(X))(Y - \mathbb{E}(Y))\big).$$

Note that

$$\mathrm{Cov}(X, X) = \mathrm{Var}(X).$$

Problem 7.36 Prove that

$$\mathrm{Cov}(X, Y) = \mathbb{E}(XY) - \mathbb{E}(X)\,\mathbb{E}(Y).$$

Problem 7.37 Prove that

$$X \perp\!\!\!\perp Y \implies \mathrm{Cov}(X, Y) = 0.$$

Problem 7.38 For random variables X, Y, Z, and reals a, b, show that

$$\text{Cov}(aX, bY) = ab\,\text{Cov}(X, Y),$$

and

$$\text{Cov}(X + Y, Z) = \text{Cov}(X, Z) + \text{Cov}(Y, Z).$$

We are now ready to generalize Proposition 7.32.

Proposition 7.39. *For random variables X and Y with 2nd moment,*

$$\text{Var}(X + Y) = \text{Var}(X) + \text{Var}(Y) + 2\,\text{Cov}(X, Y).$$

Problem 7.40 More generally, prove (by induction) that for random variables $X_1, \ldots, X_r$,

$$\text{Var}\left(\sum_{i=1}^{r} X_i\right) = \sum_{i=1}^{r}\sum_{j=1}^{r} \text{Cov}(X_i, X_j)$$

$$= \sum_{i=1}^{r} \text{Var}(X_i) + 2\sum_{1 \le i < j \le r} \text{Cov}(X_i, X_j).$$

Problem 7.41 (Hypergeometric variance) Show that the hypergeometric distribution with parameters (n, r, b) has variance $np(1 - p)\frac{r+b-n}{r+b-1}$ where $p := r/(r + b)$. The easy way described in Problem 7.16 is recommended.

Correlation

The *correlation* of X and Y is defined as

$$\text{Corr}(X, Y) := \frac{\text{Cov}(X, Y)}{\sqrt{\text{Var}(X)\,\text{Var}(Y)}}. \tag{7.8}$$

Problem 7.42 Show that the correlation has no unit in the (usual) sense that it is invariant with respect to affine transformations, or in formula, that

$$\text{Corr}(aX + b, cY + d) = \text{Corr}(X, Y),$$
$$\text{for all } a, c > 0 \text{ and all } b, d \in \mathbb{R}.$$

Problem 7.43 Show that

$$\text{Corr}(X, Y) \in [-1, 1],$$

and equal to ± 1 if and only if there are $a, b \in \mathbb{R}$ such that $\mathbb{P}(X = aY + b) = 1$ or $\mathbb{P}(Y = aX + b) = 1$.

7.5 Conditional Expectation

Consider two random variables X and Y, with X having an expectation. Then conditionally on $Y = y$, X also has an expectation.

Problem 7.44 Prove this when both variables are discrete.

When the variables are both discrete, the *conditional expectation* of X given $Y = y$ can be expressed as follows:

$$\mathbb{E}(X \mid Y = y) = \sum_{x \in \mathbb{Z}} x f_{X \mid Y}(x \mid y).$$

When (X, Y) has a density, it can be expressed as

$$\mathbb{E}(X \mid Y = y) = \int_{-\infty}^{\infty} x f_{X \mid Y}(x \mid y) \mathrm{d}x.$$

Note that $\mathbb{E}(X \mid Y)$ is a random variable. In fact, $\mathbb{E}(X \mid Y) = g(Y)$ with $g(y) := \mathbb{E}(X \mid Y = y)$, which happens to be measurable.

Problem 7.45 (Law of Total Expectation) Show that

$$\mathbb{E}(\mathbb{E}(X \mid Y)) = \mathbb{E}(X).$$

Conditional Variance

If X has a 2nd moment, then this is also the case of $X \mid Y = y$, which therefore has a variance, called the *conditional variance* of X given $Y = y$ and denoted by $\mathrm{Var}(X \mid Y = y)$. Note that $\mathrm{Var}(X \mid Y)$ is a random variable.

Problem 7.46 (Law of Total Variance) Show that

$$\mathrm{Var}(X) = \mathbb{E}(\mathrm{Var}(X \mid Y)) + \mathrm{Var}(\mathbb{E}(X \mid Y)).$$

7.6 Moment Generating Function

The *moment generating function* of a random variable X is defined as

$$\zeta_X(t) := \mathbb{E}(\exp(tX)), \quad \text{for } t \in \mathbb{R}. \tag{7.9}$$

As a function taking values in $[0, \infty]$, it is indeed well-defined everywhere.

Problem 7.47 Show that $\{t : \zeta_X(t) < \infty\}$ is an interval (possibly a singleton). [Use Jensen's inequality.]

In the special case where X is supported on the non-negative integers,

$$\zeta_X(t) = \sum_{k \geq 0} f_X(k) \exp(tk).$$

The moment generating function derives its name from the following.

Proposition 7.48. *Assume that ζ_X is finite in an open interval containing 0.
Then ζ_X is infinitely differentiable (in fact, analytic) in that interval and*

$$\zeta_X^{(k)}(0) = \mathbb{E}(X^k), \quad \text{for all } k \geq 0.$$

Theorem 7.49. *Two distributions whose moment generating functions are
finite and coincide on an open interval containing zero must be equal.*

Remark 7.50 (Laplace transform) When X has a density, its moment
generating function may be expressed as

$$\zeta_X(t) = \int_{-\infty}^{\infty} f_X(x) \exp(tx) dx.$$

This coincides with the *Laplace transform* of f_X evaluated at $-t$, and a
standard proof of Theorem 7.49 relies on the fact that the Laplace transform
is invertible under the stated conditions.

7.7 Probability Generating Function

The *probability generating function* of a non-negative random variable X is
defined as

$$\gamma_X(z) := \mathbb{E}(z^X), \quad \text{for } z \in [-1, 1].$$

Note that

$$\zeta_X(t) = \gamma_X(e^t), \quad \text{for all } t \leq 0.$$

In the special case where X is supported on the non-negative integers,

$$\gamma_X(z) = \sum_{k \geq 0} f_X(k) z^k.$$

The probability generating function derives its name from the following.

Proposition 7.51. *Assume X is non-negative. Then γ_X is well-defined and
finite on $[-1, 1]$, and infinitely differentiable (in fact, analytic) in $(-1, 1)$.
Moreover, if X is supported on the non-negative integers,*

$$\gamma_X^{(k)}(0) = k! f_X(k), \quad \text{for all } k \geq 0.$$

Problem 7.52 Show that any distribution that is supported on the non-
negative integers is characterized by its probability generating function.

7.8 Characteristic Function

The *characteristic function* of a random variable X is defined as

$$\varphi_X(t) := \mathbb{E}(\exp(\imath t X)), \quad \text{for } t \in \mathbb{R},$$

where $\imath^2 = -1$. Compare with the definition of the moment generating function in (7.9). While the moment generating function may be infinite at any $t \neq 0$, the characteristic function is always well-defined for all $t \in \mathbb{R}$ as a complex-valued function.

Problem 7.53 Show that if X and Y are independent random variables then

$$\varphi_{X+Y}(t) = \varphi_X(t)\varphi_Y(t), \quad \text{for all } t \in \mathbb{R}. \tag{7.10}$$

The converse is not true, meaning that there are situations where (7.10) holds even though X and Y are not independent. Find an example of that.

The characteristic function owes its name to the following.

Theorem 7.54. *A distribution is characterized by it characteristic function. Furthermore, if X is supported on the non-negative integers, then*

$$f_X(x) = \frac{1}{2\pi} \int_0^{2\pi} \exp(-\imath t x)\varphi_X(t)\mathrm{d}t. \tag{7.11}$$

If instead X is absolutely continuous, and its characteristic function is absolutely integrable, then

$$f_X(x) = \frac{1}{2\pi} \int_{-\infty}^{\infty} \exp(-\imath t x)\varphi_X(t)\mathrm{d}t. \tag{7.12}$$

Problem 7.55 Prove (7.11).

Remark 7.56 (Fourier transform) When X has a density,

$$\varphi_X(t) = \int_{-\infty}^{\infty} \exp(\imath t x)f_X(x)\mathrm{d}x.$$

This coincides with the *Fourier transform* of f_X evaluated at $-t/2\pi$ and a standard proof of Theorem 7.54 relies on the fact that the Fourier transform is invertible.

Remark 7.57 It is possible to define the characteristic function of a random vector X. If X is r-dimensional, it is defined as

$$\varphi_X(t) := \mathbb{E}(\exp(\imath \langle t, X \rangle)), \quad \text{for } t \in \mathbb{R}^r,$$

where $\langle u, v \rangle$ denotes the inner product of $u, v \in \mathbb{R}^r$. We note that an analogue of Theorem 7.54 holds for random vectors.

7.9 Concentration Inequalities

An important question when examining a random variable is to know how far it strays away from its mean (which we assume is well-defined whenever needed). This is a probability statement, and we present some inequalities that bound the corresponding probability.

Proposition 7.58 (Markov's inequality[30]). *For a non-negative random variable X with expectation μ,*

$$\mathbb{P}(X \geq t\mu) \leq 1/t, \quad \textit{for all } t > 0.$$

For example, if X is non-negative with mean μ, then $X \geq 2\mu$ with at most 50% chance, while $X \geq 10\mu$ with at most 10% chance.

Proof We have

$$\{X \geq t\} = \{X/t \geq 1\} \leq X/t.$$

and we conclude by taking the expectation and using its monotonicity property (Problem 7.12).   □

Proposition 7.59 (Chebyshev's inequality[31]). *For a random variable X with expectation μ and standard deviation σ,*

$$\mathbb{P}(|X - \mu| \geq t\sigma) \leq 1/t^2, \quad \textit{for all } t > 0. \tag{7.13}$$

Moreover,

$$\mathbb{P}(X \geq \mu + t\sigma) \leq 1/(1 + t^2), \quad \textit{for all } t \geq 0, \tag{7.14}$$

and

$$\mathbb{P}(X \leq \mu - t\sigma) \leq 1/(1 + t^2), \quad \textit{for all } t \geq 0.$$

Problem 7.60 Prove these inequalities by applying Markov's inequality to carefully chosen random variables.

For example, if X has mean μ and standard deviation σ, then $|X - \mu| \geq 2\sigma$ with at most 25% chance and $|X - \mu| \geq 5\sigma$ with at most 4% chance.

Markov's and Chebyshev's inequalities are examples of *concentration inequalities*. These are inequalities that bound the probability that a random variable is away from its mean (or sometimes median) by a certain amount.

---

[30] Named after Andrey Markov (1856–1922).
[31] Named after Pafnuty Chebyshev (1821–1894).

Markov's inequality gives a concentration bound with a linear decay, while Chebyshev's inequality gives a concentration bound with a quadratic decay. Even stronger concentration is possible.

Problem 7.61 Consider a random variable Y with mean μ and such that $\alpha_s := \mathbb{E}(|Y - \mu|^s) < \infty$ for some $s > 0$ (not necessarily integer). Show that

$$\mathbb{P}(|Y - \mu| \geq y) \leq \alpha_s y^{-s}, \quad \text{for all } y > 0.$$

Proposition 7.62 (Chernoff's bound[32]). *Consider a random variable Y such that as $\zeta(\lambda) := \mathbb{E}(\exp(\lambda Y)) < \infty$ for some $\lambda > 0$. Then*

$$\mathbb{P}(Y \geq y) \leq \zeta(\lambda) \exp(-\lambda y), \quad \text{for all } y > 0.$$

Proof For any $y > 0$,

$$Y \geq y \implies \lambda Y - \lambda y \geq 0 \implies \exp(\lambda Y - \lambda y) \geq 1.$$

Thus,

$$\begin{aligned}
\mathbb{P}(Y \geq y) &\leq \mathbb{P}(\exp(\lambda Y - \lambda y) \geq 1) \\
&\leq \mathbb{E}(\exp(\lambda Y - \lambda y)) \\
&= \exp(-\lambda y)\zeta(\lambda) \\
&= \exp(-\lambda y + \log \zeta(\lambda)),
\end{aligned}$$

using Markov's inequality along the way. □

Binomial Distribution

Let $Y = X_1 + \cdots + X_n$, where the X_i are independent, each being Bernoulli with parameter p, so that Y is binomial with parameters (n, p).

Problem 7.63 Show that Y has moment generating function given by

$$\zeta(\lambda) = (1 - p + pe^\lambda)^n, \quad \text{for all } \lambda \in \mathbb{R}.$$

By Chernoff's bound, for any $\lambda \geq 0$,

$$\log \mathbb{P}(Y \geq y) \leq -\lambda y + \log \zeta(\lambda).$$

Since the left-hand side does not depend on λ, to sharpen the bound we minimize the right-hand side with respect to $\lambda \geq 0$, yielding

$$\log \mathbb{P}(Y \geq y) \leq \inf_{\lambda \geq 0} \left[-\lambda y + \log \zeta(\lambda) \right] = -\sup_{\lambda \geq 0} \left[\lambda y - \log \zeta(\lambda) \right].$$

---

[32] Named after Herman Chernoff (1923–), who attributes the result to a colleague of his, Herman Rubin.

We turn to maximizing $g(\lambda) := \lambda y - \log \zeta(\lambda)$ over $\lambda \geq 0$. Let $b = y/n$, so that $b \in [0, 1]$; however, since we are interested in deviations away from, and above the mean np, and because the case $b = 1$ requires a special (but trivial) treatment, we assume that $p < b < 1$.

Problem 7.64 Verify that g is infinitely differentiable and that it has a unique maximizer at

$$\lambda^* = \log \left(\frac{(1-p)b}{p(1-b)} \right),$$

and that $g(\lambda^*) = n\mathsf{H}_p(b)$, where

$$\mathsf{H}_p(b) := b \log \left(\frac{b}{p} \right) + (1-b) \log \left(\frac{1-b}{1-p} \right).$$

We thus arrived at the following.

Proposition 7.65 (Chernoff's bound for the binomial distribution). *For* $Y \sim Bin(n, p)$*, with* $0 < p < 1$*,*

$$\mathbb{P}(Y \geq nb) \leq \exp \left(-n\mathsf{H}_p(b) \right), \quad \text{for all } b \in [p, 1]. \tag{7.15}$$

Problem 7.66 Verify that the bound indeed applies to the cases that we left off, namely, when $b = p$ and when $b = 1$.

Remark 7.67 An upper bound for $\mathbb{P}(Y \leq nb)$ when $b \in [0, p]$ can be derived in a similar fashion, or using Property (4.5).

We end this section with a general exponential concentration inequality whose roots are also in Chernoff's inequality.

Theorem 7.68 (Bernstein's inequality). *Suppose that* $X_1, \ldots, X_n$ *are independent with zero mean and such that* $\max_i |X_i| \leq c$*. Define* $\sigma_i^2 = \text{Var}(X_i) = \mathbb{E}(X_i^2)$*. Then, for all* $y \geq 0$*,*

$$\mathbb{P} \left(\sum_{i=1}^{n} X_i \geq y \right) \leq \exp \left(-\frac{y^2/2}{\sum_{i=1}^{n} \sigma_i^2 + \frac{1}{3}cy} \right).$$

Problem 7.69 Apply Bernstein's inequality to get a concentration inequality for the binomial distribution. Compare the resulting bound with the one obtained from Chernoff's inequality in (7.15).

7.10 Further Topics

7.10.1 Random Sums of Random Variables

Suppose that $\{X_i : i \geq 1\}$ are independent with the same distribution, and independent of a random variable N supported on the non-negative integers. Together, these define the following *compound sum*

$$Y = \sum_{i=1}^{N} X_i.$$

Put differently, the distribution of Y given $N = n$ is that of $\sum_{i=1}^{n} X_i$. By convention, the sum is zero if $N = 0$.

Problem 7.70 Assume that the X_i have a 2nd moment. Derive the mean and variance of Y (showing in the process that Y has a 2nd moment).

Problem 7.71 Assume that the X_i are non-negative. Compute the probability generating function of Y.

When N has a Poisson distribution, the resulting distribution is called a *compound Poisson distribution.*

The negative binomial distribution is known to have a compound Poisson representation. In detail, first define the *logarithmic distribution* via its mass function

$$f_p(k) := \frac{1}{\log(\frac{1}{1-p})} \frac{p^k}{k}, \quad k \geq 1.$$

Problem 7.72 Show that, for $p \in (0,1)$, this defines a probability distribution on the positive integers.

Proposition 7.73. *Let* $(X_i : i \geq 1)$ *be independent with the logarithmic distribution with parameter* p, *and let* N *be Poisson with parameter* $m \log(\frac{1}{1-p})$. *Then* $\sum_{i=1}^{N} X_i$ *is negative binomial with parameters* (m, p).

Problem 7.74 Prove Proposition 7.73 using Problem 7.71 in combination with Theorem 7.49.

7.10.2 Estimation from Finite Samples

Consider an urn containing coins. Without additional information, to compute the average value of these coins, one would have to go through all coins and sum their values. But what if an approximation is sufficient – is it possible to do that without looking at all the coins? It turns out that the answer is yes, at least under some sampling schemes, and this is so because of concentration.

Let the coin values be $c_1, \ldots, c_N \in \mathbb{R}$. The goal is to approximate their average, $\mu := \frac{1}{N}(c_1 + \cdots + c_N)$. We assume we have the ability to sample uniformly at random with replacement from the urn n times. We do so and let $X_1, \ldots, X_n$ denote the resulting a sample, with corresponding average $\bar{X}_n = \frac{1}{n}(X_1 + \cdots + X_n)$.

Problem 7.75 Based on Chebyshev's inequality, show that, for any $t > 0$,

$$\left| \bar{X}_n - \mu \right| \le t\sigma / \sqrt{n} \tag{7.16}$$

with probability at least $1 - 1/t^2$, where $\sigma^2 := \frac{1}{N}(c_1^2 + \cdots + c_N^2) - \mu^2$.

The surprising fact in the approximation bound (7.16) is that it depends on the ticket values only through μ and σ, so that N could be infinite in principle.

The fact that we can 'learn' about the contents of a possibly infinite urn based a finite sample from it is at the core of Statistics. It also explains why a carefully designed and conducted poll of a few thousand individuals can yield reliable information on a population of hundreds of millions (Section 11.1).

Problem 7.76 Obtain an approximation bound based Chernoff's bound instead. Compare this bound with that obtained in (7.16) via Chebyshev's inequality.

Remark 7.77 (Estimation) This sort of approximation based on a sample is often referred to as *estimation*, and will be developed in Part III.

7.10.3 Saint Petersburg Paradox

Suppose a casino offers a gambler the opportunity to play the following fictitious game, attributed to Nicolas Bernoulli (1687–1759). The game starts with $2 on the table. At each round a fair coin is flipped: if it lands heads, the amount is doubled and the game continues; if it lands tails, the game ends and the player pockets whatever is on the table. The question is: how much should the gambler be willing to pay to play the game?

A paradox arises when the gambler aims at optimizing his expected return, defined as $X - c$, where X is the gain (the amount on the table at the end of the game) and c is the entry cost (the amount the gambler pays the casino to enter the game).

Problem 7.78 Show that the expected return is infinite regardless of the cost.

Thus, in principle, a rational gambler would be willing to pay any amount to enter the game. However, a gambler with common sense would only be willing to pay very little to enter the game, hence, the paradox. Indeed, although the expected return is infinite, the probability of a positive return can be quite small.

Problem 7.79 Suppose the gambler pays c dollars to enter the game. Compute his chances of ending with a positive return in terms of c.

This, and other similar considerations, have led some commentators to argue that the expected return is not what the gambler should be optimizing. Daniel Bernoulli (1700–1782) proposed as a solution in [11] (translated from Latin to English in [12]) to optimize the expected log return, i.e., $\mathbb{E}(\log(X/c))$. This, he argued, was more natural.

Problem 7.80 Find in closed form or numerically (in R) the amount the gambler should be willing to pay to enter the game if his goal is to optimize the expected log return.

Another possibility is to optimize the median instead of the mean.

Problem 7.81 Find in closed form or numerically (in R) the amount the gambler should be willing to pay to enter the game if his goal is to optimize the median return.

Remark 7.82 (Pascal's wager) The essential component of the Saint Petersburg Paradox arises from the extremely unlikely possibility of an enormous gain. This was considered by the philosopher Pascal in questions of faith in the existence of God (in the context of his Catholic faith). As he saw it, a person had to decide whether to believe in God or not. From his book *Pensées* (here translated by F. Trotter):

> Let us weigh the gain and the loss in wagering that God is. Let us estimate these two chances. If you gain, you gain all; if you lose, you lose nothing.

7.11 Additional Problems

Problem 7.83 In the context of Problem 6.29, compute the expectation of X. First, do so directly using the definition of expectation (7.1) based on the distribution of X found in that problem. (You may use R for that purpose.) Then do this using Proposition 7.14.

Problem 7.84 Using R, compute the first 10 moments of the random variable X defined in Problem 6.29. [Do this efficiently using vector/matrix manipulations.]

Problem 7.85 Consider a location-scale family as in Section 5.5, therefore, of the form

$$F_{a,b} := F((x-b)/a), \quad \text{for } a > 0, \ b \in \mathbb{R},$$

where F is some given distribution function. Assume that F has finite 2nd moment and is not constant. Show that there is exactly one distribution in this family with mean 0 and variance 1. Assuming F itself is that distribution, compute the mean and variance of $F_{a,b}$ in terms of (a, b).

Problem 7.86 Recall the setting of Section 4.6. Compute the mean and variance of T.

Problem 7.87 Let X be a random variable with a 2nd moment. Show that $a \mapsto \mathbb{E}((X-a)^2)$ is uniquely minimized at the mean of X.

Problem 7.88 Let X be a random variable with a 1st moment. Show that $a \mapsto \mathbb{E}(|X-a|)$ is minimized at any median of X and that any minimizer is a median of X.

Problem 7.89 Compute the mean and variance of the distribution of Problem 4.31.

Problem 7.90 For $X_1, \ldots, X_n$ iid from a distribution on $(0, \infty)$, show that

$$\mathbb{E}\left(\frac{\sum_{i=1}^{k} X_i}{\sum_{i=1}^{n} X_i}\right) = \frac{k}{n}, \quad \text{for all } 1 \le k \le n.$$

Problem 7.91 Compute the characteristic function of:

(i)   the uniform distribution on $\{1, \ldots, N\}$;
(ii)   the Poisson distribution with mean λ;
(iii)   the geometric distribution with parameter p.

Repeat with the moment generating function, specifying where it is finite.

Problem 7.92 Compute the characteristic function of:

(i)   the binomial distribution with parameters n and p;
(ii)   the negative binomial distribution with parameters m and p.

Repeat with the moment generating function, specifying where it is finite.

Problem 7.93 Compute the characteristic function of:

(i)   the uniform distribution on $[a, b]$;
(ii)   the exponential distribution with rate λ;

(iii) the normal distribution parameters (μ, σ^2).

Repeat with the moment generating function, specifying where it is finite.

Problem 7.94 Compute the characteristic function of the Poisson distribution with mean λ. Then combine Problem 7.53 and Theorem 7.54 to prove Proposition 4.15.

Problem 7.95 Compute the characteristic function of the normal distribution with mean μ and variance σ^2. Then combine Problem 7.53 and Theorem 7.54 to prove Proposition 5.29.

Problem 7.96 (Hoeffding's covariance formula) Let X and Y be two random variables with finite second moments. Show that

$$\text{Cov}(X, Y) = \int_{-\infty}^{\infty} \int_{-\infty}^{\infty} \left(F_{X,Y}(x, y) - F_X(x) F_Y(y) \right) dx dy.$$

(In a sense, this is an analogue for two variables of (7.2).)

Problem 7.97 Suppose that X is supported on the non-negative integers. Show that

$$F_X(x) = \frac{1}{2\pi} \int_0^{2\pi} \frac{\sin(t(x+1)/2)}{\sin(t/2)} e^{-itx/2} \varphi_X(t) dt.$$

Problem 7.98 (Paley–Zygmund inequality) For a non-negative random variable X with finite 2nd moment, and $t \in [0, 1]$, show that

$$\mathbb{P}(X \ge t \mathbb{E}(X)) \ge (1 - t)^2 \frac{\mathbb{E}(X)^2}{\mathbb{E}(X^2)}.$$

Problem 7.99 (Markov vs Chebyshev) Evaluate the accuracy of these two inequalities for the exponential distribution with rate $\lambda = 1$. One way to do so is to draw the survival function, the bound given by the Markov inequality, and the bound given by the Chebyshev inequality (7.14). Do so in R, and start at $x = 1$ (which is the mean in this case). Put all the graphs in the same plot, in different colors identified by a legend.

Problem 7.100 (Entropy)

(i) Let f be a mass function on some arbitrary set. Show that

$$H(f) := -\sum f(x) \log f(x) \ge 0,$$

with equality if and only if f is supported on a single element. (This quantity is called the *entropy* of f.)

(ii)   Let f be a density on some Euclidean space. Provide conditions under which

$$H(f) := -\int f(x) \log f(x) dx$$

is well-defined. Provide an example of a density on the real line for which this quantity is *not* well-defined. Is it always non-negative? (This quantity, when it is well-defined, is called the *differential entropy* of f.)

Problem 7.101 (Kullback–Leibler divergence)

(i)   Let f and g be two mass functions with exactly the same support set, denoted $\mathcal{X}$ and assumed to be finite. Show that

$$\sum_{x \in \mathcal{X}} f(x) \log(f(x)/g(x)) \geq 0,$$

with equality if and only if $f = g$.

(ii)   Let f and g be two densities on a compact interval I, where there are bounded away from 0 and ∞. Show that

$$\int_I f(x) \log(f(x)/g(x)) dx \geq 0,$$

with equality if and only if $f = g$.

(In both cases, the quantity is the *Kullback–Leibler divergence*, aka *relative entropy*, from g to f, and can be shown to be well-defined, and with the same properties, much more broadly, if it is allowed to take the value ∞.)

Problem 7.102 Write an R function that generates k independent numbers from the compound Poisson distribution obtained when N is Poisson with parameter λ and the X_i are Bernoulli with parameter p. Perform some simulations to better understand this distribution for various choices of parameter values.

Problem 7.103 (Passphrases) The article [109] advocates choosing a strong password by selecting seven words at random from a list of 7776 English words. It claims that an adversary able to try one trillion guesses per second would have to keep trying for about 27 million years before discovering the correct passphrase. (This is so even if the adversary knows how the passphrase was generated.) Perform some calculations to corroborate this claim.

Problem 7.104 (Two envelopes: randomized strategy) In the Two Envelopes Problem (Section 2.5.3), it turns out that it is possible to do better than random guessing. This is possible with even less information, in a setting were we are not told anything about the amounts inside the envelopes. Cover [32] offered the following strategy, which relies on the ability to draw a random number. Having chosen a distribution with support the positive real line, we draw a number from this distribution and, if the amount in the envelope we opened is less than that number, we switch, otherwise we keep the opened envelope. Show that this strategy beats pure random guessing.

Problem 7.105 (Two envelopes: model 1) A possible model for the Two Envelopes Problem is the following. Suppose X is supported on the positive integers. Given $X = x$, put x in Envelope A and either $x/2$ or $2x$ in Envelope B, each with probability $1/2$. We are shown the contents of Envelope A and we need to decide whether to keep the amount found there or switch for the (unknown) amount in Envelope B. Consider three strategies: *(i)* always keep A; *(ii)* always switch to B; *(iii)* random switch (50% chance of keeping A, regardless of the amount it contains). For each strategy, compute the expected gain. Then describe an optimal strategy assuming the distribution of X is known. [Consider the discrete case first, and then the absolutely continuous case.]

Problem 7.106 (Two envelopes: model 2) Another model for the Two Envelopes Problem is the following. Here, given $X = x$, let the contents of the envelopes (A, B) be $(x, 2x)$, $(2x, x)$, $(x, x/2)$, $(x/2, x)$, each with probability $1/4$. We are shown the contents of Envelope A (although it does not matter in this model). For each strategy described in Problem 7.105, compute the expected gain. Then derive an optimal strategy.

8

Convergence of Random Variables

The convergence of random variables plays an important role in Probability Theory. This is particularly true of the Law of Large Numbers, which underpins the frequentist notion of probability. Another famous convergence result is a refinement known as the Central Limit Theorem, which underpins much of large-sample statistical theory.

In the entire chapter, the framework is that of a probability space $(\Omega, \Sigma, \mathbb{P})$. In particular, all events are elements of the σ-algebra Σ, and all the random variables are defined on the measurable space (Ω, Σ).

8.1 Zero-One Laws

We start with the Borel–Cantelli lemmas[33]. These lemmas have to do with an infinite sequence of events and whether infinitely many events among these will happen or not.

To formalize our discussion, consider a sequence of events, $(\mathcal{A}_n : n \geq 1)$. The event that infinitely many among these events happen is the so-called *limit supremum* of these events, defined as

$$\bar{\mathcal{A}} := \bigcap_{m \geq 1} \bigcup_{n \geq m} \mathcal{A}_n.$$

Problem 8.1 (1st Borel–Cantelli lemma) Prove that

$$\sum_{n \geq 1} \mathbb{P}(\mathcal{A}_n) < \infty \implies \mathbb{P}(\bar{\mathcal{A}}) = 0.$$

The following converse requires independence.

---

[33] Named after Émile Borel (1871–1956) and Francesco Paolo Cantelli (1875–1966).

Problem 8.2 (2nd Borel–Cantelli lemma) Assuming in addition that the events are independent, prove that

$$\sum_{n \geq 1} \mathbb{P}(A_n) = \infty \implies \mathbb{P}(\bar{A}) = 1.$$

Combining these two lemmas, we arrive at the following.

Proposition 8.3 (Borel–Cantelli's zero-one law). *In the present context, and assuming in addition that the events are independent, we have $\mathbb{P}(\bar{A}) = 0$ or 1 according the whether $\sum_{n \geq 1} \mathbb{P}(A_n) < \infty$ or $= \infty$.*

Thus, in the context of this proposition, the situation is black or white: the limit supremum event has probability equal to 0 or 1. This is an example of a *zero-one law*. Another famous example is the following.

Theorem 8.4 (Kolmogorov's zero-one law). *Consider an infinite sequence of independent random variables. Any event determined by this sequence, but independent of any finite subsequence, has probability zero or one.*

Problem 8.5 Show that the assumption of independence is crucial for the result to hold in this generality, by providing a simple counter-example.

Problem 8.6 Use Kolmogorov's zero-one law to prove that, for any sequence of independent random variables, $(X_n : n \geq 1)$, $\limsup_n X_n$ is constant.

8.2 Convergence of Random Variables

Random variables being functions on the sample space, defining notions of convergence for random variables relies on similar notions for sequences of functions. We present the two main notions here.

8.2.1 Convergence in Probability

We say that a sequence of random variables $(X_n : n \geq 1)$ *converges in probability* towards a random variable X if, for any fixed $\varepsilon > 0$,

$$\mathbb{P}(|X_n - X| \geq \varepsilon) \longrightarrow 0, \quad \text{as } n \to \infty.$$

We will denote this convergence by $X_n \to_{\mathrm{P}} X$.

Example 8.7 For a simple example, let Y be a random variable, and let $g: \mathbb{R}^2 \to \mathbb{R}$ be continuous in the second variable, and define $X_n = g(Y, 1/n)$.

Then

$$X_n \xrightarrow{\text{P}} X := g(Y,0).$$

This example encompasses instances like $X_n = a_n Y + b_n$, where (a_n) and (b_n) are convergent deterministic sequences.

Problem 8.8 Show that

$$X_n \xrightarrow{\text{P}} X \iff X_n - X \xrightarrow{\text{P}} 0.$$

Problem 8.9 Show that if $X_n \geq 0$ for all n and $\mathbb{E}(X_n) \to 0$, then $X_n \to_P 0$.

Problem 8.10 Show that if $\mathbb{E}(X_n) = 0$ for all n and $\text{Var}(X_n) \to 0$, then $X_n \to_P 0$.

Proposition 8.11 (Dominated convergence). *Suppose that $X_n \to_P X$ and that $|X_n| \leq Y$ for all n, where Y has finite expectation. Then X_n, for all n, and X have an expectation, and $\mathbb{E}(X_n) \to \mathbb{E}(X)$ as $n \to \infty$.*

Problem 8.12 Prove Proposition 8.11, at least when Y is constant.

8.2.2 Convergence in Distribution

We say that a sequence of distribution functions $(\mathsf{F}_n : n \geq 1)$ *converges weakly* to a distribution function F if, for any point $x \in \mathbb{R}$ where F is continuous,

$$\mathsf{F}_n(x) \longrightarrow \mathsf{F}(x), \quad \text{as } n \to \infty.$$

We will denote this by $\mathsf{F}_n \to_{\mathcal{L}} \mathsf{F}$. Now, a sequence of random variables $(X_n : n \geq 1)$ *converges in distribution* to a random variable X if $\mathsf{F}_{X_n} \to_{\mathcal{L}} \mathsf{F}_X$. We will denote this by $X_n \to_{\mathcal{L}} X$.

Remark 8.13 The consideration of continuity points is important. As an illustration, take the simple example of constant variables, $X_n \equiv 1/n$. We anticipate that (X_n) converges weakly to $X \equiv 0$. Indeed, the distribution function of X_n is $\mathsf{F}_n(x) := \{x \leq 1/n\}$, while that of X is $\mathsf{F}(x) := \{x \leq 0\}$. Clearly, $\mathsf{F}_n(x) \to \mathsf{F}(x)$ for all $x \neq 0$, but not at 0 since $\mathsf{F}_n(0) = 0$ and $\mathsf{F}(0) = 1$. Fortunately, weak convergence only requires a pointwise convergence at the continuity points of F, which is the case here.

Problem 8.14 Show that convergence in distribution to a constant implies convergence in probability to that constant.

Problem 8.15 Prove that convergence in probability implies convergence in distribution, meaning that if $X_n \to_P X$ then $X_n \to_{\mathcal{L}} X$. Then prove that the converse is not true in general by providing a counter-example.

Even though convergence in distribution does not imply convergence in probability, these two notions of convergence are nonetheless intimately related.

Theorem 8.16 (Skorokhod's representation theorem[34]). *Suppose that* (F_n) *converges weakly to* F. *Then there exist* (Y_n) *and* Y, *random variables defined on the same probability space, with* Y_n *having distribution* F_n *and* Y *having distribution* F, *and such that* $Y_n \to_P Y$.

8.3 Law of Large Numbers

Consider Bernoulli trials where a fair coin (i.e., a p-coin with $p = 1/2$) is tossed repeatedly. Common sense would lead one to anticipate that, after a large number of tosses, the proportion of heads would be close to $1/2$. Thankfully, this is also the case within the theoretical framework built on Kolmogorov's axioms.

Remark 8.17 (iid sequences) Random variables that are independent and have the same marginal distribution are said to be *independent and identically distributed (iid)*.

Theorem 8.18 (Law of Large Numbers). *Let* (X_n) *be a sequence of iid random variables with expectation* μ. *Then*

$$\frac{1}{n} \sum_{i=1}^{n} X_i \xrightarrow{P} \mu, \quad as \ n \to \infty.$$

See Figure 8.1 for an illustration.

If the variables have a 2nd moment, then the result is easy to prove using Chebyshev's inequality.

Problem 8.19 Let $X_1, \ldots, X_n$ be random variables with the same mean μ and variances all bounded by σ^2, and assume their pairwise covariances are non-positive. Let $Y_n = \sum_{i=1}^{n} X_i$. Show that

$$\mathbb{P}\big(|Y_n/n - \mu| \geq \varepsilon\big) \leq \frac{\sigma^2}{n\varepsilon^2}, \quad \text{for all } \varepsilon > 0. \tag{8.1}$$

---

[34] Named after Anatoliy Skorokhod (1930–2011).

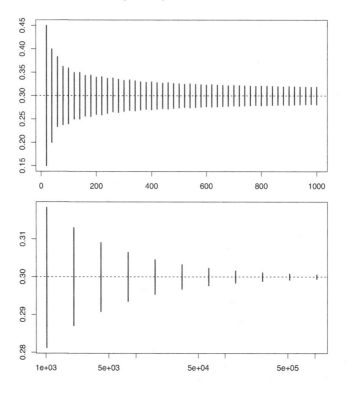

Figure 8.1 An illustration of the Law of Large Number in the context of Bernoulli trials. The horizontal axis represents the sample size n. The vertical segment corresponding to a sample size n is defined by the 0.005 and 0.995 quantiles of the distribution of the mean of n Bernoulli trials with parameter $p = 1/2$.

Problem 8.20 Apply Problem 8.19 to the number of heads in a sequence of n tosses of a p-coin. In particular, find ε such that the probability bound is 5%. Turn this into a statement about the "typical" number of heads in 100 tosses of a fair coin.

Problem 8.21 Repeat with the number of red balls drawn without replacement n times from an urn with r red balls and b blue balls. [The pairwise covariances were computed as part of Problem 7.41.] Make the statement about the "typical" number of red balls in 100 draws from an urn with 100 red and 100 blue balls. How does your statement change when there are 1000 red and 1000 blue balls instead?

Remark 8.22 Theorem 8.18 is in fact known as the Weak Law of Large Numbers. There is indeed a Strong Law of Large Numbers, and it says that, under the same conditions, with probability one,

$$\frac{1}{n}\sum_{i=1}^{n}X_i \longrightarrow \mu, \quad \text{as } n \to \infty.$$

8.4 Central Limit Theorems

The bound (8.1) can be rewritten as

$$\mathbb{P}\left(\frac{|Y_n - n\mu|}{\sigma\sqrt{n}} \geq t\right) \leq \frac{1}{t^2},$$

for all $t > 0$ and all $n \geq 1$ integer. In fact, under some additional conditions, it is possible to obtain the exact limit of the left-hand side as $n \to \infty$.

8.4.1 Classical Central Limit Theorem

Theorem 8.23 (Central Limit Theorem). *Let (X_n) be a sequence of iid random variables with mean μ and variance σ^2. Let $Y_n = \sum_{i=1}^{n} X_i$. Then $(Y_n - n\mu)/\sigma\sqrt{n}$ converges in distribution to the standard normal distribution, or equivalently, for all $t \in \mathbb{R}$,*

$$\mathbb{P}\left(\frac{Y_n - n\mu}{\sigma\sqrt{n}} \leq t\right) \longrightarrow \Phi(t), \quad \text{as } n \to \infty,$$

where Φ is the distribution function of the standard normal distribution, given in (5.1).

Importantly, $n\mu$ is the mean of Y_n and $\sigma\sqrt{n}$ is its standard deviation. So the Central Limit Theorem says that the standardized sum of iid random variables (with 2nd moment) converges to the standard normal distribution.

Problem 8.24 Show that the Central Limit Theorem encompasses the De Moivre–Laplace Theorem (Theorem 5.4) as a special case. In particular, if $(X_i : i \geq 1)$ is a sequence of iid Bernoulli random variables with parameter $p \in (0, 1)$, then

$$\frac{\sum_{i=1}^{n} X_i - np}{\sqrt{np(1-p)}} \xrightarrow{\mathcal{L}} \mathcal{N}(0, 1), \quad \text{as } n \to \infty.$$

A standard proof of Theorem 8.23 relies on the Fourier transform, and for that reason is rather sophisticated. So we only provide some pointers. We assume that $\mu = 0$ and $\sigma = 1$, which can be done without loss of generality.

We focus on the case where the X_i have density f and characteristic function φ that is integrable. In that case, we have the Fourier inversion formula (7.12).

Let f_n denote the density of Y_n. Based on (6.7), we know that f_n exists and furthermore that it is the nth convolution power of f. Let $Z_n = Y_n/\sqrt{n}$, which has density $g_n(z) := \sqrt{n}\, f_n(\sqrt{n}z)$. We want to show, or at least argue, that g_n converges to the standard normal density, denoted

$$\phi(z) := \frac{\exp(-z^2/2)}{\sqrt{2\pi}}, \quad \text{for } z \in \mathbb{R}.$$

Because of (7.10), Y_n has characteristic function φ^n.

Problem 8.25 Show that, for any positive integer $n \geq 1$, φ^n is integrable when φ is integrable.

We may therefore apply (7.12) to derive

$$g_n(z) = \frac{\sqrt{n}}{2\pi} \int_{-\infty}^{\infty} e^{-\iota t \sqrt{n}z} \varphi(t)^n dt$$

$$= \frac{1}{2\pi} \int_{-\infty}^{\infty} e^{-\iota s z} \varphi(s/\sqrt{n})^n ds,$$

using a simple change of variables in the 2nd line.

Problem 8.26 Recall that we assumed that f has zero mean and unit variance. Based on that, show that φ is twice continuously differentiable, with $\varphi(0) = 1$, $\varphi'(0) = 0$, and $\varphi''(0) = 1$. Deduce that

$$\varphi(s/\sqrt{n})^n = (1 - s^2/2n + o(1/n))^n \to e^{-s^2/2}, \quad \text{as } n \to \infty.$$

Thus, if passing to the limit under the integral is justified (⁙), we obtain

$$g_n(z) \longrightarrow \frac{1}{2\pi} \int_{-\infty}^{\infty} e^{-\iota s z} e^{-s^2/2} ds, \quad \text{as } n \to \infty.$$

Problem 8.27 Prove that the limit is $\phi(z)$, either directly, or using a combination of (7.12) and the fact that the standard normal characteristic function is $e^{-s^2/2}$ (Problem 7.93).

This completes the proof that $g_n \to \phi$ pointwise, modulo (⁙) above. Even then, this does not prove Theorem 8.23, which is a result on the distribution functions rather than the densities.

8.4.2 Other Central Limit Theorems

While Theorem 8.23 is the most classical version, there are other central limit theorems. Lindeberg's is a particularly useful version as it does not require the variables to be identically distributed. Lyapunov's is a simplified version of Lindeberg's.

Theorem 8.28 (Lindeberg's central limit theorem[35]). *Let $(X_i : i \geq 1)$ be independent random variables, with X_i having mean μ_i and variance σ_i^2. Define $s_n^2 = \sum_{i=1}^n \sigma_i^2$ and assume that, for any fixed $\varepsilon > 0$,*

$$\lim_{n \to \infty} \frac{1}{s_n^2} \sum_{i=1}^n \mathbb{E}\left((X_i - \mu_i)^2 \left\{ |X_i - \mu_i| > \varepsilon s_n \right\} \right) = 0. \tag{8.2}$$

Then $s_n^{-1} \sum_{i=1}^n (X_i - \mu_i)$ converges in distribution to the standard normal distribution.

Problem 8.29 Verify that Theorem 8.28 implies Theorem 8.23.

Problem 8.30 Consider independent Bernoulli variables, $X_i \sim \text{Ber}(p_i)$. Assume that $p_i \leq 1/2$ for all i. Show that (8.2) holds if and only if $\sum_{i=1}^n p_i \to \infty$ as $n \to \infty$.

Problem 8.31 (Lyapunov's central limit theorem[36]) Show that (8.2) holds when there is $\delta > 0$ such that

$$\lim_{n \to \infty} \frac{1}{s_n^{2+\delta}} \sum_{i=1}^n \mathbb{E}\left(|X_i - \mu_i|^{2+\delta} \right) = 0.$$

[Use Jensen's inequality.]

8.5 Extreme Value Theory

Extreme Value Theory is the branch of Probability that studies such things as the extrema of iid random variables. Its main results are 'universal' convergence results, the most famous of which is the following.

Theorem 8.32 (Extreme Value Theorem). *Let (X_n) be iid random variables. Let $Y_n = \max_{i \leq n} X_i$. Suppose that there are deterministic sequences (a_n) and (b_n) such that $a_n Y_n + b_n \to_{\mathcal{L}} Z$ where Z is not constant. Then Z has either a Weibull, a Gumbel, or a Fréchet distribution.*

---

[35] Named after Jarl Lindeberg (1876–1932).
[36] Named after Aleksandr Lyapunov (1857–1918).

The *Weibull family*[37] with shape parameter $\kappa > 0$ is the location-scale family generated by

$$\mathsf{G}_\kappa(z) := 1 - \exp(-z^\kappa), \quad z > 0. \tag{8.3}$$

(Thus, the entire Weibull family has three parameters.)

The *Gumbel family*[38] is the location-scale family generated by

$$\mathsf{G}(z) := 1 - \exp(-\exp(-z)), \quad z \in \mathbb{R}. \tag{8.4}$$

(Thus, the entire Gumbel family has two parameters.)

The *Fréchet family*[39] with shape parameter $\kappa > 0$ is the location-scale family generated by

$$\mathsf{G}_\kappa(z) := \exp(-z^{-\kappa}), \quad z > 0. \tag{8.5}$$

(Thus, the entire Fréchet family has three parameters.)

Problem 8.33 Verify that (8.3), (8.4), and (8.5) are bona fide distribution functions.

Problem 8.34 (Distributions with finite support) Suppose that the distribution generating the iid sequence has finite support, say, $\{c_1, \ldots, c_N\}$ with $c_1 < \cdots < c_N$. Show that

$$\mathbb{P}(Y_n = c_N) \longrightarrow 1, \quad \text{as } n \to \infty.$$

Deduce that the Extreme Value Theorem does not apply to this case. [Note that N is fixed in this problem.]

Rather than proving the theorem, we provide some examples, one for each case. We place ourselves in the context of the theorem.

Problem 8.35 Let F denote the distribution function of the X_i. Show that the distribution function of Y_n is F^n.

Problem 8.36 (Maximum of a uniform sample) Let $X_1, \ldots, X_n$ be iid uniform in $[0, 1]$. Show that, for any $z > 0$,

$$\mathbb{P}(n(1 - Y_n) \le z) \longrightarrow 1 - \exp(-z), \quad \text{as } n \to \infty.$$

Thus, the limiting distribution is in the Weibull family.

---

[37] Named after Waloddi Weibull (1887–1979).
[38] Named after Emil Julius Gumbel (1891–1966).
[39] Named after Maurice René Fréchet (1878–1973).

Problem 8.37 (Maximum of a normal sample) Let $X_1, \ldots, X_n$ be iid standard normal. Show that, for any $z \in \mathbb{R}$,

$$\mathbb{P}(a_n Y_n + b_n \le z) \longrightarrow \exp(-\exp(-z)), \quad \text{as } n \to \infty,$$

where

$$a_n := \sqrt{2 \log n}, \quad b_n = -2 \log n + \frac{1}{2} \log \log n + \frac{1}{2} \log(4\pi).$$

Thus, the limiting distribution is in the Gumbel family. [To prove the result, use the fact that, Φ denoting the standard normal distribution function,

$$1 - \Phi(x) \sim \frac{1}{\sqrt{2\pi} x} \exp(-x^2/2), \quad \text{as } x \to \infty,$$

which can be obtained via integration by parts.]

Problem 8.38 (Maximum of a Cauchy sample) Let $X_1, \ldots, X_n$ be iid from the Cauchy distribution. We saw the density in (5.4), and the corresponding distribution function is given by

$$F(x) := \frac{1}{\pi} \tan^{-1}(x) + \frac{1}{2}.$$

Show that, for any $z > 0$,

$$\mathbb{P}\left((\pi/n)Y_n \le z\right) \longrightarrow 1 - \exp(-1/z), \quad \text{as } n \to \infty.$$

Thus, the limiting distribution is in the Fréchet family. [To prove the result, use the fact that $1 - F(x) \sim 1/\pi x$ as $x \to \infty$.]

8.6 Further Topics

8.6.1 Continuous Mapping Theorem and the Delta Method

The following result says that applying a continuous function to a convergent sequence of random variables results in a convergent sequence of random variables, where the type of convergence remains the same. (The theorem applies to random vectors as well.)

Problem 8.39 (Continuous Mapping Theorem) Let (X_n) be a sequence of random variables and let $g: \mathbb{R} \to \mathbb{R}$ be continuous. Prove that, if (X_n) converges in probability (resp. in distribution) to X, then $(g(X_n))$ converges in probability (resp. in distribution) to $g(X)$.

The following is a simple corollary.

Problem 8.40 (Slutky's theorem) Show that, if $X_n \to_{\mathcal{L}} X$ while $A_n \to_P a$ and $B_n \to_P b$, where a and b are constants, then $A_n X_n + B_n \to_{\mathcal{L}} aX + b$.

The following is a refinement of the Continuous Mapping Theorem.

Problem 8.41 (Delta Method) Let (Y_n) be a sequence of random variables and (a_n) a sequence of real numbers such that $a_n \to \infty$ and $a_n Y_n \to_{\mathcal{L}} Z$, where Z is some random variable. Let $g \colon \mathbb{R} \to \mathbb{R}$ be differentiable at 0 and such that $g(0) = 0$. Prove that $a_n g(Y_n) \to_{\mathcal{L}} g'(0)Z$.

8.6.2 Exchangeable Random Variables

The random variables $X_1, \ldots, X_n$ are said to be *exchangeable* if their joint distribution is invariant with respect to permutations. This means that for any permutation $(\pi_1, \ldots, \pi_n)$ of $(1, \ldots, n)$, the random vectors $(X_1, \ldots, X_n)$ and $(X_{\pi_1}, \ldots, X_{\pi_n})$ have the same distribution.

Problem 8.42 Show that $X_1, \ldots, X_n$ are exchangeable if and only if their joint distribution function is invariant with respect to permutations. Show that the same is true of the mass function (if discrete) or density (if absolutely continuous).

Problem 8.43 Show that if $X_1, \ldots, X_n$ are exchangeable then they necessarily have the same marginal distribution.

Problem 8.44 Show that independent and identically distributed random variables are exchangeable. Show that the converse is *not* true.

Problem 8.45 Let $X_1, \ldots, X_n$ and Y be random variables such that, conditionally on Y the X_i are independent and identically distributed. Show that $X_1, \ldots, X_n$ are exchangeable.

Remark 8.46 The relation between exchangeability and conditional independence was perhaps first explored by Bruno de Finetti (1906–1985). For a more quantitative examination, see [44].

8.7 Additional Problems

Problem 8.47 (From discrete to continuous uniform) Let X_N be uniform on $\{\frac{1}{N+1}, \frac{2}{N+1}, \ldots, \frac{N}{N+1}\}$. Show that (X_N) converges in distribution to the uniform distribution on $[0, 1]$.

Problem 8.48 (From geometric to exponential) Let X_N be geometric with parameter p_N, where $N p_N \to \lambda > 0$. Show that (X_N/N) converges in distribution to the exponential distribution with rate λ.

Problem 8.49 Consider a sequence of distribution functions (F_n) that converges weakly to some distribution function F. Recall the definition of pseudo-inverse defined in (3.9) and show that $F_n^-(u) \to F^-(u)$ at any u where F is continuous and strictly increasing.

Problem 8.50 Recall the setting of Section 4.6. Show that the X_i are exchangeable but not independent.

Problem 8.51 Continuing with the setting of Section 4.6, we denote T by T_N and let $N \to \infty$. It is known since [58] that, for any $a \in \mathbb{R}$,

$$\mathbb{P}(T_N \leq N \log N + aN) \to \exp(-\exp(-a)), \quad N \to \infty.$$

Re-express this statement as a convergence in distribution. Note that the limiting distribution is in the Gumbel family. Using the function implemented in Problem 4.20, perform some simulations to confirm this mathematical result.

Problem 8.52 Continuing with the setting of the previous problem, for $q \in [0, 1]$, $T_{\lceil qN \rceil}$ is the number of trials needed to sample a fraction of at least q of the entire collection of N coupons. Show that, when $q \in (0, 1)$ is fixed while $N \to \infty$, the limiting distribution of $T_{\lceil qN \rceil}$ is in the normal family. [First, express $T_{\lceil qN \rceil}$ as the sum of certain W_i as in Problem 4.19. Then apply Lyapunov's central limit theorem (Problem 8.31).]

Problem 8.53 Suppose that $X_1, \ldots, X_n$ are exchangeable and define $Y = \#\{i : X_i \geq X_1\}$. First, assume that X_1 has a continuous distribution and show that Y has the uniform distribution on $\{1, 2, \ldots, n\}$. In any case, show that

$$\mathbb{P}(Y \leq y) \leq y/n, \quad \text{for all } y \geq 1.$$

Problem 8.54 Suppose the distribution of the random vector $(X_1, \ldots, X_n)$ is invariant with respect to some set of permutations $\mathcal{S}$, and that $\mathcal{S}$ is such that, for every pair of distinct $i, j \in \{1, \ldots, n\}$, there is $\sigma = (\sigma_1, \ldots, \sigma_n) \in \mathcal{S}$ such that $\sigma_i = j$. Show that the conclusions of Problem 8.53 apply to this more general situation.

Problem 8.55 Consider drawing from an urn with red and blue balls. Let $X_i = 1$ if the ith draw is red and $X_i = 0$ if it is blue. Show that, whether the sampling is without or with replacement (Section 2.4), or follows Pólya's scheme (Section 2.4.4), $X_1, \ldots, X_n$ are exchangeable.

Problem 8.56 Let $X_1, \ldots, X_n$ be exchangeable non-negative random variables. For any $k \leq n$, compute

$$\mathbb{E}\left(\frac{\sum_{i=1}^{k} X_i}{\sum_{i=1}^{n} X_i}\right).$$

Problem 8.57 Suppose that a box contains m balls. The goal is to estimate m given the ability to sample uniformly at random with replacement from the urn. We consider a protocol which consists in repeatedly sampling from the urn and marking the resulting ball with a unique symbol. (This can be seen as a form of *capture-recapture* sampling scheme used in Ecology, for example.) The process stops when the ball we draw has been previously marked. Let K denote the total number of draws in this process. Show that $K/\sqrt{m} \to \sqrt{\pi/2}$ in probability as $m \to \infty$.

Problem 8.58 (Tracy–Widom distribution) Let $\Lambda_{m,n}$ denote the square of the largest singular value of an m-by-n matrix with iid standard normal coefficients. Then there are deterministic sequences, $a_{m,n}$ and $b_{m,n}$ such that, as $m/n \to \gamma \in [0, \infty]$, $(\Lambda_{m,n} - a_{m,n})/b_{m,n}$ converges in distribution to the so-called *Tracy–Widom distribution* of order 1. In R, perform some numerical simulations to probe into this phenomenon. [Note that the amount of computation might be substantial.]

9

Stochastic Processes

Stochastic processes model experiments whose outcomes are collections of variables organized in some fashion. We focus here on Markov processes, which include random walks and branching processes. As this material is rather advanced, we only cover some fundamentals.

As usual, our foundation is a probability space $(\Omega, \Sigma, \mathbb{P})$.

9.1 Markov Chains

Some situations are poorly modeled by sequences of independent random variables. Think, for example, of the daily closing price of a stock, or the maximum daily temperature on successive days. Markov chains offer a simple way to model dependencies and may be more relevant models for such phenomena. We provide a very brief introduction, focusing on the case where the observations are in a discrete space. The topic is more extensively treated, for example, in the textbook [84].

9.1.1 Definition

Let $\mathcal{X}$ be a discrete space and let $f(\cdot | \cdot)$ denote a conditional mass function on $\mathcal{X} \times \mathcal{X}$, namely, for each $x_0 \in \mathcal{X}$, $x \mapsto f(x | x_0)$ is a mass function on $\mathcal{X}$. The corresponding chain, starting at $x_0 \in \mathcal{X}$, proceeds as follows:

(i) X_1 is drawn from $f(\cdot | x_0)$ resulting in x_1;
(ii)   for $t \geq 1$, given $X_t = x_t$, X_{t+1} is drawn from $f(\cdot | x_t)$ resulting in x_{t+1}.

The outcome of this experiment is $(x_1, x_2, \dots)$, or $(X_1, X_2, \dots)$ when left unspecified as a sequence of random variables.

Remark 9.1 If in actuality $f(x | x_0)$ does not depend on x_0, in which case we write it as $f(x)$, the process generates an iid sequence from f. Hence, an iid sequence is a Markov chain.

If the chain is at state $x \in \mathcal{X}$, that state is referred to as the present state. The next state, generated using $f(\cdot \mid x)$, is the state one (time) step into the future. That future state is generated without any reference to previous states except for the present state. In that sense, a Markov chain only 'remembers' the present state. See Section 9.1.6 for an extension.

9.1.2 Two-State Markov Chains

Let us consider the simple setting of a state space $\mathcal{X}$ with two elements, say, $\mathcal{X} = \{1, 2\}$. Because $f(\cdot \mid \cdot)$ is a conditional mass function it needs to satisfy

$$a := f(1 \mid 1) = 1 - f(2 \mid 1),$$
$$b := f(2 \mid 2) = 1 - f(1 \mid 2).$$

The parameters $a, b \in [0, 1]$ are free and define the Markov chain. The conditional probabilities above may be organized in a so-called *transition matrix*, which here takes the form

$$\text{present state} \left\{ \begin{array}{c} \overbrace{\begin{pmatrix} a & 1-a \\ 1-b & b \end{pmatrix}}^{\text{next state}} \end{array} \right. \tag{9.1}$$

Chains are sometimes depicted as an *automaton*. See Figure 9.1 for an illustration of the chain with transition matrix (9.1).

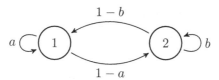

Figure 9.1 A representation of the chain given by (9.1) as an automaton. (Thanks to Phong Alain Chau for this figure.)

Problem 9.2 Starting at $x_0 = 1$, compute the probability of observing $(x_1, x_2, x_3) = (1, 2, 1)$ as a function of (a, b).

Problem 9.3 In R, write a function taking as input the parameters of the chain (a, b), the starting state $x_0 \in \{1, 2\}$, and the number of steps t, and returning a realization, $(x_1, \ldots, x_t)$, from the corresponding process.

9.1.3 General Markov Chains

Since we assume the state space $\mathcal{X}$ to be discrete, we may take it to be the positive integers, $\mathcal{X} = \{1, 2, \dots\}$, without loss of generality. For $i, j \in \mathcal{X}$, let $\theta_{ij} = f(j|i)$, which is the probability of transitioning from i to j. These are organized into a transition matrix

$$
\text{present state} \left\{ \overbrace{\begin{pmatrix} \theta_{11} & \theta_{12} & \theta_{13} & \cdots \\ \theta_{21} & \theta_{22} & \theta_{33} & \cdots \\ \theta_{31} & \theta_{32} & \theta_{33} & \cdots \\ \vdots & \vdots & \vdots & \vdots \end{pmatrix}}^{\text{next state}} \right.
$$

Note that the matrix can be infinite in principle. This general case includes the case where the state space is finite. Denote the transition matrix by

$$
\Theta := (\theta_{ij} : i, j \geq 1).
$$

The transition matrix defines the chain, and together with the initial state, defines the distribution of the random process.

Problem 9.4 Show that, for any $t \geq 1$,

$$
\mathbb{P}(X_t = j \mid X_0 = i) = \theta_{ij}^{(t)},
$$

if we denote the coefficients of the tth power of Θ by $\Theta^t = (\theta_{ij}^{(t)} : i, j \geq 1)$.

9.1.4 Long-Term Behavior

In the study of a Markov chain, quantities of interest include the long-term behavior of the chain, the average time (number of steps) it takes to visit a given state (or set of states) starting from a given state (or set of states), the average number of such visits, and more. We focus on the limiting marginal distribution.

We say that a mass function on $\mathcal{X}$, $\mathsf{q} := (q_1, q_2, \dots)$, is a *stationary distribution* of the chain with transition matrix $\Theta := (\theta_{ij} : i, j \geq 1)$ if $X_0 \sim \mathsf{q}$ and $X \mid X_0 \sim f(\cdot \mid X_0)$ yields $X \sim \mathsf{q}$.

Problem 9.5 Show that q is a stationary distribution of Θ if and only if $\mathsf{q}\Theta = \mathsf{q}$, when interpreting q as a row vector. (Note that the multiplication is on the left.)

The chain Θ is said to be *irreducible* if for any $i, j \geq 1$ there is some $t \geq 1$ such that $\theta_{ij}^{(t)} > 0$. This means that, starting at any state, the chain can eventually reach any other state with positive probability.

The state i is said to be *aperiodic* if

$$\gcd(t \geq 1 : \theta_{ii}^{(t)} > 0) = 1,$$

where $\gcd$ is short for 'greatest common divisor'. To understand this, suppose that state i is such that

$$\gcd(t \geq 1 : \theta_{ii}^{(t)} > 0) = 2.$$

Then this would imply that the chain starting at i cannot be at i after an odd number of steps.

A chain is aperiodic if all its states are aperiodic.

Proposition 9.6. *Suppose that the chain is irreducible. If one state is aperiodic then all states are aperiodic.*

State i is *positive recurrent* if, starting at i, the expected time it takes the chain to return to i is finite. A chain is positive recurrent if all its states are positive recurrent.

Proposition 9.7. *A finite irreducible chain is positive recurrent.*

(A finite chain is a chain over a finite state space.)

Theorem 9.8. *An irreducible, aperiodic, and positively recurrent chain has a unique stationary distribution. Moreover, the chain converges weakly to the stationary distribution regardless of the initial state, meaning that, if X_t denotes the state the chain is at time t, and if $\mathsf{q} = (q_1, q_2, \dots)$ denotes the stationary distribution, then*

$$\lim_{t \to \infty} \mathbb{P}(X_t = j \mid X_0 = i) = q_j, \quad \text{for any states } i \text{ and } j.$$

Problem 9.9 Verify that a finite chain whose transition probabilities are all positive satisfies the requirements of the theorem.

Problem 9.10 Provide necessary and sufficient conditions for a two-state Markov chain to satisfy the requirements of the theorem. When these are satisfied, derive the limiting distribution.

9.1.5 Reversible Markov Chains

Running a Chain Backward

Consider a chain with transition probabilities (θ_{ij}) with unique stationary distribution $\mathsf{q} = (q_i)$. Assuming the state is i at time $t = 0$, a step forward is taken according to

$$\mathbb{P}(X_1 = j \mid X_0 = i) = \theta_{ij}, \quad \text{for all } i, j.$$

In contrast, a step backward is taken according to

$$\mathbb{P}(X_{-1} = j \mid X_0 = i) = \frac{\theta_{ji} q_j}{q_i}, \quad \text{for all } i, j. \tag{9.2}$$

Problem 9.11 To support (9.2), show that $\mathbb{P}(X_0 = j \mid X_1 = i) = \theta_{ji} q_j / q_i$, for all i, j.

Reversible Chains

In essence, a chain is reversible if running it forward is equivalent (in terms of distribution) to running it backward. More generally, a chain with transition probabilities (θ_{ij}) is said to be *reversible* if there is a probability mass function $\mathsf{q} = (q_i)$ such that

$$q_i \theta_{ij} = q_j \theta_{ji}, \quad \text{for all } i, j. \tag{9.3}$$

This means that, if we draw a state from q and then run the chain for one step, the probability of obtaining (i, j) is the same as that of obtaining (j, i).

Problem 9.12 Show that a distribution q satisfying (9.3) is stationary for the chain.

Problem 9.13 Show that if X_0 is sampled according to q and we run the chain for t steps resulting in $X_1, \ldots, X_t$, the distribution of $(X_0, \ldots, X_t)$ is the same as that of $(X_t, \ldots, X_0)$. [Note that this does *not* imply that these variables are exchangeable, only that we can reverse the order. See Problem 9.32.]

Example 9.14 (Simple random walk on a graph) A *graph* is a set of *nodes* and *edges* between some pairs of nodes. Two nodes connected by an edge are said to be *neighbors*. Assume that each node has a finite number of neighbors. Consider the following process: starting at some node, at each step choose a neighbor uniformly at random among those of the current node. Then the resulting chain is reversible.

9.1.6 Extensions

We have discussed the simplest variant of Markov chain: it is called a discrete time, discrete space, time homogeneous Markov process. The definition of a continuous time Markov process requires technicalities that we will avoid. But we can elaborate on the other aspects.

General State Space

The state space $\mathcal{X}$ does not need to be discrete. Indeed, suppose the state space is equipped with a σ-algebra Γ. What are needed are transition probabilities, $\{P(\cdot \mid x) : x \in \mathcal{X}\}$, on Γ. Then, given a present state $x \in \mathcal{X}$, the next state is drawn from $P(\cdot \mid x)$.

Problem 9.15 Suppose that $(W_t : t \geq 1)$ are iid with distribution P on $(\mathbb{R}, \mathcal{B})$. Starting at $X_0 = x_0 \in \mathbb{R}$, successively define $X_t = X_{t-1} + W_t$. (Equivalently, $X_t = x_0 + W_1 + \cdots + W_t$.) What are the transition probabilities in this case?

More Memory

A Markov chain only remembers the present. However, with little effort, it is possible to have it remember some of its past as well.

The number of states it remembers is the *order* of the chain. A Markov chain of order m is such that

$$\mathbb{P}(X_t = i_t \mid X_{t-m} = i_{t-m}, \ldots, X_{t-1} = i_{t-1})$$
$$= \theta(i_{t-m}, \ldots, i_t), \quad \text{for all } i_{t-m}, \ldots, i_t,$$

where the $(m+1)$-dimensional array

$$(\theta(i_0, i_1, \ldots, i_m) : i_0, i_1, \ldots, i_m \geq 1)$$

now defines the chain.

In fact, a finite-order chain can be seen as an order 1 chain in an enlarged state space. Indeed, consider a chain of order m on $\mathcal{X}$, and let X_t denote its state at time t. Based on this, define

$$Y_t := (X_{t-m+1}, \ldots, X_{t-1}, X_t).$$

Then $(Y_m, Y_{m+1}, \ldots)$ forms a Markov chain of order 1 on $\mathcal{X}^m$, with transition probabilities

$$\mathbb{P}(Y_t = (i_{t-m+1}, \ldots, i_t) \mid Y_{t-1} = (i_{t-m}, \ldots, i_{t-1}))$$
$$= \theta(i_{t-m}, \ldots, i_t), \quad \text{for all } i_{t-m}, \ldots, i_t,$$

and all other possible transitions given the probability 0.

Time-Varying Transitions

The transition probabilities may depend on time, meaning

$$\mathbb{P}(X_{t+1} = j \mid X_t = i) = \theta_{ij}(t), \quad \text{for all } i, j,$$

where a sequence of transition matrices, $\Theta(t) := (\theta_{ij}(t))$, defines the chain.

9.2 Simple Random Walks

Let $(X_i : i \geq 1)$ be iid with

$$\mathbb{P}(X_i = 1) = p, \quad \mathbb{P}(X_i = -1) = 1 - p,$$

and define $S_n = \sum_{i=1}^n X_i$. Note that $(X_i + 1)/2 \sim \mathrm{Ber}(p)$. Then $(S_n : n \geq 0)$, with $S_0 = 0$ by default, is a *simple random walk*[40]. See Figure 9.2.

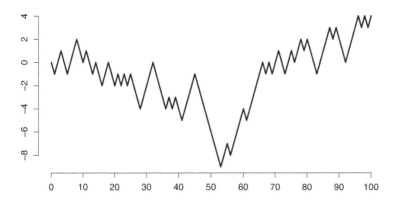

Figure 9.2 A realization of a symmetric ($p = 1/2$) simple random walk, specifically, a plot of a linear interpolation of a realization of $\{(n, S_n) : n = 0, \ldots, 100\}$.

This is a special case of Example 9.14, where the graph is the one-dimensional lattice: if the process is at $k \in \mathbb{Z}$, it moves to $k+1$ with probability p, or $k - 1$ with probability $1 - p$. When $p = 1/2$, the walk is *symmetric*.

Problem 9.16 Show that a simple random walk is a Markov chain with state space $\mathbb{Z}$. Display the transition matrix. Show that the chain is irreducible and periodic. (What is the period?)

Proposition 9.17. *The simple random walk is positive recurrent if and only if it is symmetric.*

---

[40] The simple random walk is one of the most well-studied stochastic processes. We refer the reader to Feller's classic textbook [61] for a thorough yet accessible exposition.

9.2.1 Gambler's Ruin

Consider a gambler than bets one dollar at every trial and doubles or looses that dollar, each with probability $1/2$ independently of the other trials. Suppose the gambler starts with s dollars and that he stops gambling if that amount reaches 0 (loss) or some prescribed amount $w > s$ (win).

Problem 9.18 Leaving w implicit, let γ_s denote the probability that the gambler loses. Show that

$$\gamma_s = p\gamma_{s+1} + (1-p)\gamma_{s-1}, \quad \text{for all } s \in \{1,\ldots,w-1\}.$$

Deduce that

$$\gamma_s = 1 - s/w, \quad \text{if } p = 1/2,$$

while

$$\gamma_s = \frac{1 - (\frac{p}{1-p})^{w-s}}{1 - (\frac{p}{1-p})^{w}}, \quad \text{if } p \neq 1/2.$$

The result applies to $w = \infty$, meaning to the setting where the gambler keeps on playing as long as he has money. In that case, we see that the gambler loses with probability 1 if $p \leq 1/2$ and with probability $(1/p - 1)^s$ if $p > 1/2$. In particular, if $p > 1/2$, with probability $1 - (1/p - 1)^s$, the gambler's fortune increases without bound.

9.2.2 Fluctuations

A realization of a simple random walk can be surprising. Indeed, if asked to have a guess at a realization, most (untrained) people would have the tendency to make the walk fluctuate much more than it typically does.

Problem 9.19 In R, write a function which plots $\{(k, S_k) : k = 0,\ldots,n\}$, where $S_0, S_1,\ldots,S_n$ is a realization of a simple random walk with parameter p initialized at the origin. Try your function on several choices for (n, p).

We say that a sign change occurs at step n if $S_{n-1}S_{n+1} < 0$. Note that this implies that $S_n = 0$.

Proposition 9.20 (Sign changes). *The number of sign changes in the first $2n + 1$ steps of a symmetric simple random walk equals k with probability $2^{-2n}\binom{2n+1}{2k+1}$.*

9.2.3 Maximum

There is a beautifully simple argument, called the *reflection principle*, which leads to the following clear description of how the maximum of the random walk up to step n behaves

Proposition 9.21. *For a symmetric simple random walk, for all $n \geq 1$ and all $r \geq 0$,*

$$\mathbb{P}\left(\max_{k \leq n} S_k = r \right) = \mathbb{P}(S_n = r) + \mathbb{P}(S_n = r + 1).$$

Assume the walk is symmetric. Since S_n has mean 0 and standard deviation $\sqrt{n}$, it is natural to study the normalized random walk given by $(S_n / \sqrt{n} : n \geq 1)$.

Theorem 9.22 (Erdös and Rényi [37]). *Suppose that $(X_i : i \geq 1)$ are iid with mean 0 and variance 1. Define $S_n = \sum_{i \leq n} X_i$, as well as $a_n = \sqrt{2 \log \log n}$ and $b_n = \frac{1}{2} \log \log \log n$. Then for any $t \in \mathbb{R}$,*

$$\lim_{n \to \infty} \mathbb{P}\left(\max_{k \leq n} \frac{S_k}{\sqrt{k}} \leq a_n + \frac{b_n}{a_n} + \frac{t}{a_n} \right) = \exp\left(-e^{-t} / 2\sqrt{\pi} \right).$$

Thus, the maximum of a simple random walk, properly normalized, converges to a Gumbel distribution.

Problem 9.23 In the context of this theorem, show that

$$\sqrt{2 \log \log n} \leq \max_{k \leq n} \frac{S_k}{\sqrt{k}} \leq \sqrt{2 \log \log n} + 1,$$

with probability tending to 1 as n increases.

9.3 Galton–Watson Processes

Francis Galton (1822–1911) and Henry William Watson (1827–1903) were interested in the extinction of family names. Their model assumes that the family name is passed from father to son and that each male has a number of male descendants, with the number of male descendants being independent and identically distributed across siblings and generations.

More formally, suppose that we start with one male with a certain family name. Let $X_0 = 1$. The male has $\xi_{0,1}$ male descendants. If $\xi_{0,1} = 0$, the family name dies. Otherwise, the first male descendant has $\xi_{1,1}$ male descendants, the second has $\xi_{1,2}$ male descendants, etc. In general, let $\xi_{n,j}$ be the number of male descendants of the jth male in the nth generation. The order within each generation is arbitrary and only used for identification purposes. The central assumption is that the $\xi_{n,j}$ are iid. Let f denote their common mass function. See Figure 9.3 for an illustration.

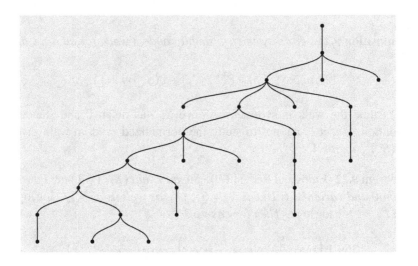

Figure 9.3 A realization of a Galton–Watson branching process with progeny distribution the Poisson distribution with mean 1 (thus at criticality). In this particular realization, the family name died with the 8th generation (meaning that none of the individuals at the 8th generation had any descendants).

The number of male individuals in the nth generation is thus

$$X_n = \sum_{j=1}^{X_{n-1}} \xi_{n-1,j}.$$

(This is an example of a compound sum as seen in Section 7.10.1.)

Problem 9.24 Show that $(X_1, X_2, \dots)$ forms a Markov chain and give the transition probabilities in terms of f.

Problem 9.25 Provide sufficient conditions on f under which, as a Markov chain, the process is irreducible and aperiodic.

Note that if $X_n = 0$, then the family name dies and, in particular, $X_m = 0$ for all $m \geq n$. Of interest, therefore, is

$$d_n := \mathbb{P}(X_n = 0).$$

Clearly, (d_n) is increasing, and being in $[0, 1]$, converges to a limit d_∞, which is the probability that the male line dies out eventually. A basic problem is to determine d_∞.

Remark 9.26 In Markov chain parlance, the state 0 is *absorbing* in the sense that, once reached, the chain stays there forever after.

Problem 9.27 Show that $d_\infty > 0$ if and only if $f(0) > 0$.

Problem 9.28 Show that an irreducible chain with an absorbing state cannot be (positive) recurrent. [You can start with a two-state chain, with one state being absorbing, and then generalize from there.]

Problem 9.29 (Trivial setting) The setting is trivial when f has support in $\{0, 1\}$. Compute d_n in that case and show that $d_\infty = 1$ when $f(0) > 0$.

We assume henceforth that we are not in the trivial setting, meaning that $f(0) + f(1) < 1$.

Problem 9.30 In that case the state space cannot be taken to be finite.

First-Step Analysis

A *first-step analysis* consists in conditioning on the value of X_1. Suppose that $X_1 = k$, thus out of the first generation are born k lines. The whole line dies by the nth generation if and only if all these k lines die by the nth generation. But the nth generation in the whole line corresponds to the $(n-1)$th generation for these lines because they started at generation 1 instead of generation 0. By the Markov property, each of these lines has the same distribution as the whole line, and therefore dies by its $(n-1)$th generation with probability d_{n-1}. Thus, by independence, they all die by their $(n-1)$th generation with probability d_{n-1}^k, establishing that

$$\mathbb{P}(X_n = 0 \mid X_1 = k) = d_{n-1}^k.$$

By the Law of Total Probability, this yields

$$\begin{aligned} d_n &= \mathbb{P}(X_n = 0 \mid X_0 = 1) \\ &= \sum_{k \geq 0} \mathbb{P}(X_n = 0 \mid X_1 = k)\, \mathbb{P}(X_1 = k \mid X_0 = 1) \\ &= \sum_{k \geq 0} d_{n-1}^k f(k). \end{aligned}$$

Thus, letting γ denote the probability generating function of f as defined in Section 7.7, we have that

$$d_n = \gamma(d_{n-1}), \quad \text{for all } n \geq 1.$$

Note that $d_0 = 0$ since $X_0 = 1$. Thus, by the continuity of γ on $[0, 1]$, d_∞ is necessarily a fixed point of γ, meaning that $\gamma(d_\infty) = d_\infty$. There are some standard ways of dealing with this situation and we refer the reader to the textbook [84] for details.

Let μ be the (possibly infinite) mean of f. This is the mean number of male descendants of a given male. The mean plays a special role, in particular because $\gamma'(1) = \mu$.

Theorem 9.31 (Probability of extinction). *The line becomes extinct with probability one (meaning $d_\infty = 1$) if and only if $\mu \leq 1$. If $\mu > 1$, d_∞ is the unique fixed point of γ in $(0, 1)$.*

Many more things are known about this process, such as the average growth of the population and features of the population tree.

9.4 Additional Problems

Problem 9.32 Consider a Markov chain and a distribution q on the state space such that, if the chain is started at $X_0 \sim$ q, then $(X_0, X_1, X_2, \dots)$ are exchangeable, meaning that $X_0, \dots, X_t$ are exchangeable for any $t \geq 0$. Show that, necessarily, the sequence is iid with distribution q.

Problem 9.33 (Three-state Markov chains) Repeat Problem 9.10 with a three-state Markov chain.

Problem 9.34 Consider a Markov chain with a symmetric transition matrix. Show that the chain is reversible and that the uniform distribution is stationary.

Problem 9.35 (Random walks) A sequence of iid random variables, (X_i), defines a *random walk* by taking the partial sums, $S_n := X_1 + \dots + X_n$ for $n \geq 1$. The X_i are called the *increments*. It turns out that random walks with increments having zero mean and finite 2nd moment behave similarly. Perform some numerical experiments to ascertain this claim.

Part II

Practical Considerations

10

Sampling and Simulation

In this chapter, we introduce some tools for sampling from a distribution. We also explain how to use computer simulations to approximate probabilities and, more generally, expectations, which can allow one to circumvent complicated mathematical derivations.

10.1 Monte Carlo Simulation

Suppose, in the context of a probability space $(\Omega, \Sigma, \mathbb{P})$, we want to compute the probability $\mathbb{P}(\mathcal{A})$ of a given event $\mathcal{A} \in \Sigma$. We start by contrasting two very different avenues for doing that.

Analytic Calculations

In some situations, it might be possible to compute this probability (or at least approximate it) directly by calculations 'with pen and paper' (or via the use of a computer to perform symbolic calculations), and possibly a simple calculator to numerically evaluate the final expression. In the pre-computer age, researchers with sophisticated mathematical skills spent a lot of effort on such problems and, almost universally, had to rely on some form of approximation to arrive at a useful result.

Numerical Simulations

In some situations, it might be possible to generate independent realizations from $\mathbb{P}$. By the Law of Large Numbers (Theorem 8.18), if $\omega_1, \omega_2, \ldots$ are sampled iid from $\mathbb{P}$, then

$$Q_m := \frac{1}{m} \sum_{i=1}^{m} \{\omega_i \in \mathcal{A}\} \xrightarrow{\text{P}} \mathbb{P}(\mathcal{A}), \quad \text{as } m \to \infty.$$

This leads to the idea of choosing a large integer m, generate $\omega_1, \ldots, \omega_m$ iid from $\mathbb{P}$, and then output Q_m as an approximation to $\mathbb{P}(\mathcal{A})$. The larger

m, the better the approximation, and a priori the only reason to settle for a particular m are the available computational resources. This approach is often called *Monte Carlo simulation*.[41]

Applying Chebyshev's inequality (7.13), we derive

$$|Q_m - \mathbb{P}(\mathcal{A})| \le \frac{t}{2\sqrt{m}}, \qquad (10.1)$$

with probability at least $1 - 1/t^2$. For example, choosing $t = 10$ we have that $|Q_m - \mathbb{P}(\mathcal{A})| \le 5/\sqrt{m}$ with probability at least 99%. Therefore, an approximation based on m Monte Carlo draws is accurate to within order $1/\sqrt{m}$.

Problem 10.1 Verify the assertions made here.

Problem 10.2 Apply Chernoff's bound for the binomial distribution (7.15) to obtain a sharper bound on the probability of (10.1).

Problem 10.3 Consider the following problem.[42] A gardener plants three maple trees, four oaks, and five birch trees in a row; they are planted in random order, each arrangement being equally likely. What is the probability that no two birch trees are next to one another? Compute this probability analytically. Then, using R, approximate this probability by Monte Carlo simulation.

Problem 10.4 (Monty Hall by simulation) In [192], Andrew Vazsonyi tells us that even Paul Erdös, a prominent figure in Probability Theory and Combinatorics, was challenged by the Monty Hall problem (Example 1.33). Vazsonyi performed some computer simulations to demonstrate to Erdös that the solution was indeed $1/3$ (no switch) and $2/3$ (switch). Do the same in R. First, simulate the process when there is no switch. Do that many times (say $m = 10^6$) and record the fraction of successes. Repeat, this time when there is a switch.

10.2 Monte Carlo Integration

Monte Carlo integration applies the principle underlying Monte Carlo simulation to the computation of expectations.

Suppose we want to integrate a function h on $[0, 1]^d$. Typical numerical integration methods work by evaluating h on a grid of points and

---

[41] See [56] for an account of early developments at Los Alamos National Laboratory in the context of research in nuclear fission.

[42] It appeared in the *American Invitational Mathematics Examination* (1984 ed.).

approximating the integral with a linear combination of these values. The simplest scheme of this sort is based on the definition of the Riemann integral. For example, in dimension $d = 1$,

$$\int_0^1 h(x)dx \approx \frac{1}{m}\sum_{i=1}^m h(x_i),$$

where $x_i := i/m$. This is based on a piecewise constant approximation to h. If h is smoother, a higher-order approximation would yield a better approximation for the same value of m.

R corner The function integrate in R uses a quadratic approximation on an adaptive grid.

Such methods work well in any dimension, but only in theory. Indeed, in practice, a grid in dimension d, even for moderately large d, is too large, even for modern computers. For instance, suppose that we want to sample the function every $1/10$ along each coordinate. In dimension d, this requires a grid of size 10^d. If $d \geq 10$, this number is quite large already, and if $d \geq 100$, it is beyond hope for any computer.

Monte Carlo integration provides a way to approximate the integral of h at a rate of order $1/\sqrt{m}$ regardless of the dimension d. The simplest scheme uses randomness and is motivated as before by the Law of Large Numbers. Indeed, let $X_1, \ldots, X_m$ be iid uniform in $[0,1]^d$. Then $h(X_1), \ldots, h(X_m)$ are iid with mean

$$I_h := \int_{[0,1]^d} h(x)dx,$$

which is the quantity of interest.

Problem 10.5 Show that $h(X_1), \ldots, h(X_m)$ have finite 2nd moment if and only if h is square-integrable (meaning that h^2 is integrable) over $[0,1]^d$. Then compute their variance (denoted σ_h^2 in what follows).

Applying Chebyshev's inequality to

$$I_m := \frac{1}{m}\sum_{i=1}^m h(X_i),$$

we derive

$$|I_m - I_h| \leq \frac{t\sigma_h}{\sqrt{m}},$$

with probability at least $1 - 1/t^2$.

Although computing σ_h is likely as hard, or harder, than computing I_h itself, it can be easily bounded when an upper bound on h is known. Indeed, if it is known that $|h| \le b$ over $[0,1]^d$, then $\sigma_h \le b$. (Note that numerically verifying that $|h| \le b$ over $[0,1]^d$ can be a challenge in high dimensions.)

Problem 10.6 Verify these assertions.

Problem 10.7 Compute $\int_0^1 \sqrt{1 - x^2}\, \mathrm{d}x$ in three ways:

(i)   Analytically.
(ii)   Numerically, using the function integrate in R.
(iii) By Monte Carlo integration, also in R.

10.3 Rejection Sampling

Suppose we want to sample from the uniform distribution with support $\mathcal{A}$, a compact set in $\mathbb{R}^d$. Translating and scaling $\mathcal{A}$ as needed, we may assume without loss of generality that $\mathcal{A} \subset [0,1]^d$. Then consider the following procedure: repeatedly sample a point from $\mathrm{Unif}([0,1]^d)$ until the point belongs to $\mathcal{A}$, and return that last point. It turns out that the resulting point has the uniform distribution on $\mathcal{A}$.

Problem 10.8 Let $\mathcal{A} \subset \mathcal{B}$, where both $\mathcal{A}$ and $\mathcal{B}$ are compact subsets of $\mathbb{R}^d$. Let $X \sim \mathrm{Unif}(\mathcal{B})$. Show that, conditional on $X \in \mathcal{A}$, X is uniform in $\mathcal{A}$.

Problem 10.9 In R, implement this procedure for sampling from the lozenge in the plane with vertices $(1,0), (0,1), (-1,0), (0,-1)$.

This is arguably the most basic example of *rejection sampling*. The name comes from the fact that draws are rejected unless a prescribed condition is met.

Problem 10.10 In R, implement a rejection sampling algorithm for 'estimating' the number π based on the fact that the unit disc (centered at the origin and of radius one) has surface area equal to π. How many samples should be generated to estimate π with this method to within precision ε with probability at least $1 - \delta$? Here $\varepsilon > 0$ and $\delta > 0$ are given. In particular, how many samples would be needed to confirm that the 2nd decimal of π is 4 to within 99% confidence?

In general, suppose that we want to sample from a distribution with density f on $\mathbb{R}^d$. Let f_0 be another density on $\mathbb{R}^d$ such that

$$f(x) \le c f_0(x), \quad \text{for all } x \text{ in the support of } f, \tag{10.2}$$

for some known constant $c \ge 1$. The density f_0 plays the role of *proposal distribution*. Besides (10.2), the other requirement is that we need to be able to sample from f_0. Assuming this is the case, consider Algorithm 10.1.

Algorithm 10.1	Basic rejection sampling

Input: target density f, proposal density f_0, constant c satisfying (10.2).
Output: one realization from f

Repeat: generate y from f_0 and u from Unif($[0,1]$), independently
Until $u \le f(y)/cf_0(y)$
Return the last y

To see that Algorithm 10.1 outputs a realization from f, let $Y \sim f_0$ and $U \sim$ Unif$(0,1)$ be independent, and define the event $\mathcal{V} := \{U \le f(Y)/cf_0(Y)\}$. If X denotes the output of Algorithm 10.1, then X has the distribution of $Y \mid \mathcal{V}$. Thus, for any Borel set $\mathcal{A}$,

$$
\begin{aligned}
\mathbb{P}(X \in \mathcal{A}) &= \mathbb{P}(Y \in \mathcal{A} \mid \mathcal{V}) \\
&= \frac{\mathbb{P}(Y \in \mathcal{A} \text{ and } \mathcal{V})}{\mathbb{P}(\mathcal{V})},
\end{aligned}
$$

with

$$
\begin{aligned}
\mathbb{P}(Y \in \mathcal{A} \text{ and } \mathcal{V}) &= \int_{\mathcal{A}} \mathbb{P}(\mathcal{V} \mid Y = y) f_0(y) dy \\
&= \int_{\mathcal{A}} \frac{f(y)}{cf_0(y)} f_0(y) dy \\
&= \frac{1}{c} \int_{\mathcal{A}} f(y) dy,
\end{aligned}
$$

where in the 2nd line we used the fact that

$$
\begin{aligned}
\mathbb{P}(\mathcal{V} \mid Y = y) &= \mathbb{P}(U \le f(y)/cf_0(y) \mid Y = y) \\
&= \mathbb{P}(U \le f(y)/cf_0(y)) \\
&= f(y)/cf_0(y),
\end{aligned}
$$

since U is independent of Y and uniform in $[0,1]$, and (10.2) holds. By taking $\mathcal{A} = \mathbb{R}^d$, this gives

$$
\mathbb{P}(\mathcal{V}) = \mathbb{P}(Y \in \mathbb{R}^d \text{ and } \mathcal{V}) = \frac{1}{c} \int_{\mathbb{R}} f(y) dy = \frac{1}{c}.
$$

Hence,

$$
\mathbb{P}(X \in \mathcal{A}) = \int_{\mathcal{A}} f(y) dy,
$$

and this being valid for any Borel set $\mathcal{A}$, we have established that X has density f, as desired.

Problem 10.11 Let S be the number of samples generated by Algorithm 10.1. Show that $\mathbb{E}(S) = c$. What is the distribution of S?

Problem 10.12 From the previous problem, we see that the algorithm is more efficient the smaller c is. Show that $c \geq 1$ with equality if and only if f and f_0 are densities for the same distribution.

The *ratio of uniforms* is another rejection sampling method proposed by Kinderman and Monahan [107]. It is based on the following.

Problem 10.13 Suppose that g is non-negative and integrable over the real line with integral b, and define $\mathcal{A} := \left\{ (u, v) : 0 < v < \sqrt{g(u/v)} \right\}$. Assuming that $\mathcal{A}$ has finite area, show that if (U, V) is uniform in $\mathcal{A}$, then $X := U/V$ has distribution $f := g/b$.

Problem 10.14 Implement the method in R for the special case where g is supported on $[0, 1]$.

10.4 Markov Chain Monte Carlo (MCMC)

Markov chains can be used to sample from a distribution when doing so 'directly' is not available. In discrete settings, this may be the case because the space is too large and there is no simple way of enumerating the elements in the space. We consider such a setting in what follows, in particular since we only discussed Markov chains over discrete state spaces. Let $q = (q_i)$ be a mass function on a discrete space from which we want to sample. The idea is to construct a chain, meaning devise a transition matrix $\Theta = (\theta_{ij})$, such that the reversibility condition (9.3) holds. If, in addition, the chain satisfies the requirements of Theorem 9.8, then the chain converges in distribution to q. Thus, a possible method for generating an observation from q, at least approximately, is as in Algorithm 10.2.

Algorithm 10.2   Basic MCMC sampling

Input: chain Θ, initial distribution q_0, total number of steps t
Output: one state

Initialize: draw a state according to q_0
Run the chain Θ for t steps
Return the last state

Remark 10.15 Obviously, we need to be able to sample from the distribution q_0. In the present context, choosing q_0 equal to q is, therefore, not an option. However, if q_0 were taken equal to q, the method would then be exact since q is stationary for the chain. More generally, the closer q_0 is to q, the more accurate the method is (for a given number of steps t).

10.4.1 Binary Matrices with Given Row and Column Sums

We are tasked with sampling uniformly at random from the set of $m \times n$ matrices with entries in $\{0, 1\}$ with given row sums, $r_1, \ldots, r_m$, and given column sums, $c_1, \ldots, c_n$. Let that set be denoted by $\mathcal{M}(r, c)$, where $r := (r_1, \ldots, r_m)$ and $c := (c_1, \ldots, c_n)$. Importantly, we assume that we already have in our possession one such matrix, which has the added benefit of guaranteeing that this set is non-empty. This setting is motivated by applications in Psychometry and Ecology (Section 22.1).

The space $\mathcal{M}(r, c)$ is often gigantic and there is no known way to enumerate it to enable drawing from the uniform distribution directly. (In fact, merely computing the cardinality of $\mathcal{M}(r, c)$ is difficult enough [45].) However, an MCMC approach is viable. The following is based on the work of Besag and Clifford [14].

The chain is defined as follows. At each step, choose two rows and two columns uniformly at random. If the resulting submatrix is of the form

$$\begin{pmatrix} 1 & 0 \\ 0 & 1 \end{pmatrix} \quad \text{or} \quad \begin{pmatrix} 0 & 1 \\ 1 & 0 \end{pmatrix}$$

then switch one for the other. If the resulting submatrix is not of this form, then stay put.

Problem 10.16 Show that this chain is indeed a chain on $\mathcal{M}(r, c)$, and a reversible one, and that the uniform distribution is stationary. To complete the picture, show that the chain satisfies the requirements of Theorem 9.8. [The only real difficulty is proving irreducibility.]

Problem 10.17 Staying put may seem, at first, unnecessary, and it may be tempting Show that this necessary, however, in that without doing this, the uniform distribution may not be stationary anymore. For example, examine the case of 3-by-3 binary matrices with row and column sums equal to $(2, 1, 1)$. [The space of such matrices is very manageable.] Perform simulations in R to compare the two chains: the one that stays put and the one that keeps sampling until a witch is made.

10.4.2 Generating a Sample

Most typically, there is a need to generate not one but several independent samples from a given distribution q. Doing so (approximately) using MCMC is possible if we already have a Markov chain with q as limiting distribution. An obvious procedure is to repeat the process, say Algorithm 10.2, the desired number of times n to obtain an iid sample of size n from a distribution that approximates the target distribution q.

This process is deemed wasteful in situations where the chain converges slowly to its stationary distribution. Indeed, in such circumstances, the number of steps t in Algorithm 10.2 to generate a single draw can be quite large, and if $(x_0, \ldots, x_t)$ represents a realization of the chain, then $x_0, \ldots, x_{t-1}$ are discarded and only x_t is returned by the algorithm. This would be repeated n times, thus generating $n(t+1)$ states to only keep n of them.

Various methods and heuristics exist to attempt to make better use of the states computed along the way. The main issue is that states generated in sequence by a single run of the chain are dependent. Nevertheless, the following generalization of the Law of Large Numbers is true[43].

Theorem 10.18 (Ergodic Theorem). *Consider a Markov chain on a discrete state space $\mathcal{X}$. Assume the chain is irreducible and positive recurrent, and with stationary distribution q. Let X_0 have any distribution on $\mathcal{X}$ and start the chain at that state, resulting in $X_1, X_2, \ldots$. Then, for any bounded function h,*

$$\frac{1}{t} \sum_{s=1}^{t} h(X_s) \xrightarrow{P} \sum_{x \in \mathcal{X}} q(x) h(x), \quad as\ t \to \infty.$$

10.5 Metropolis–Hastings Algorithm

This algorithm offers a method for constructing a reversible Markov chain for MCMC sampling. It is closely related to rejection sampling, as we shall see. We consider the discrete case, although the same procedure applies more generally almost verbatim.

Suppose we want to sample from a distribution with mass function q. The algorithm seeks to express the transition probability $p(\cdot \mid \cdot)$ as follows:

$$p(x \mid x_0) = p_0(x \mid x_0) a(x \mid x_0), \tag{10.3}$$

---

[43] A central limit theorem also holds under some additional conditions.

where $p_0(\cdot\,|\,\cdot)$ is the 'proposal' conditional mass function and $a(\cdot\,|\,\cdot)$ is the 'acceptance' probability function. The transition probability $p(\cdot\,|\,\cdot)$ is reversible with stationary distribution q if

$$p(x\,|\,x_0)q(x_0) = p(x_0\,|\,x)q(x), \quad \text{for all } x, x_0.$$

When p is as in (10.3), this condition is equivalent to

$$\frac{a(x\,|\,x_0)}{a(x_0\,|\,x)} = \frac{p_0(x_0\,|\,x)q(x)}{p_0(x\,|\,x_0)q(x_0)}, \quad \text{for all } x, x_0.$$

Problem 10.19 Prove that

$$a(x\,|\,x_0) := 1 \wedge \frac{p_0(x_0\,|\,x)q(x)}{p_0(x\,|\,x_0)q(x_0)} \tag{10.4}$$

satisfies this condition.

The *Metropolis–Hastings Algorithm*[44] is an MCMC algorithm with Markov chain of the form (10.3), with p_0 chosen by the user and a as in (10.4). A detailed description is given in Algorithm 10.3.

Algorithm 10.3   Metropolis–Hastings sampling

Input: target q, proposal p_0, initial distribution q_0, number of steps t
Output: one state

Initialize: draw x_0 from q_0
For $s = 1, \ldots, t$
    draw x from $p_0(\cdot\,|\,x_{s-1})$
    draw u from $\text{Unif}(0, 1)$
    set $x_s = x$ if $u \leq a(x\,|\,x_{t-1})$ and $x_s = x_{s-1}$ otherwise
Return the last state x_t

Remark 10.20 We only need to be able to compute $q(x)/q(x_0)$ for two states x_0, x. This makes the method applicable in settings where $q = c\tilde{q}$ with c being a normalizing constant that is hard to compute and $\tilde{q}$ being a function that is relatively easy to evaluate.

Example 10.21 (Ising model[45]) The *Ising model* is a model of ferromagnetism where the (iron) atoms are organized in a regular lattice and each atom has a spin which is either $-$ or $+$. We consider such a model in dimension two. Let $x_{ij} \in \{-1, +1\}$ denote the spin of the atom at position

---

[44] Named after Nicholas Metropolis (1915–1999) and Wilfred K. Hastings (1930–2016).
[45] Named after Ernst Ising (1900–1998).

(i, j) in the m-by-n rectangular lattice $\{1, \ldots, m\} \times \{1, \ldots, n\}$. In its simplest form, the Ising model presumes that the set of random variables $X = (X_{ij})$ has a distribution of the form

$$\mathbb{P}(X = x) = C(u, v) \exp(u\xi(x) + v\zeta(x)),$$

where $u, v \in \mathbb{R}$ are parameters, $x = (x_{ij})$ with $x_{ij} \in \{-1, +1\}$, and

$$\xi(x) := \sum_{(i,j)} x_{ij},$$

and

$$\zeta(x) := \frac{1}{2} \sum_{(i,j) \leftrightarrow (k,l)} x_{ij} x_{kl},$$

with $(i, j) \leftrightarrow (k, l)$ if and only if $i = k$ and $|j - l| = 1$ or $|i - k| = 1$ and $j = l$.

The normalization constant $C(u,v)$ may be difficult to compute in general, as in principle it involves summing over the whole state space, which is of size 2^{mn}. However, the functions ξ and ζ are rather easy to evaluate, which makes a Metropolis–Hastings approach particularly attractive.

We present a simple variant. We say that x and x' are neighbors, denoted $x \leftrightarrow x'$, if they differ in exactly one entry. We choose as $p_0(\cdot \,|\, x')$ the uniform distribution over the neighbors of x'. Then the acceptance probability takes the following simple form

$$a(x \,|\, x') = 1 \wedge \frac{q(x)}{q(x')}$$

$$= 1 \wedge \exp\left[u(\xi(x) - \xi(x')) + v(\zeta(x) - \zeta(x'))\right].$$

Note that, if x and x' differ at (i, j), meaning $x_{ij} \neq x'_{ij}$, then

$$\xi(x) - \xi(x') = x_{ij} - x'_{ij},$$

while

$$\zeta(x) - \zeta(x') = \sum_{(k,l) \leftrightarrow (i,j)} \left(x_{ij} x_{kl} - x'_{ij} x'_{kl}\right),$$

where the last sum is only over the neighbors of (i, j) (and there are, at most, four of them).

Problem 10.22 In R, simulate realizations of such an Ising model using the Metropolis–Hastings algorithm just described. Do so for $m = 100$ and $n = 200$, and various choices of parameters a and b, chosen carefully to exhibit different regimes. [The realizations can be visualized using the function image.]

10.6 Pseudo-Random Numbers

We have assumed in several places that we have the ability to generate random numbers, at least from simple distributions, such as the uniform distribution on $[0, 1]$. Doing so, in fact, presents quite a conundrum since the computer is a deterministic machine. The conundrum is solved by the use of a *pseudo-random number generator*, which is a program that outputs a sequence of numbers that are not random but designed to behave as if they were random.

Linear Congruential Generators

These generators produce sequences (x_n) of the form

$$x_n = (ax_{n-1} + c) \mod m,$$

where a, c, m are given integers chosen appropriately. The starting value x_0 needs to be provided and is called the *seed*.

The sequence (x_n) is in $\{0, \ldots, m - 1\}$ and designed to behave like an iid sequence from the uniform distribution on that set.

R corner By default, the pseudo-random number generator in R is the Mersenne–Twister algorithm [126]. We refer the reader to [54] for more details, as well as a comprehensive discussion of pseudo-random number generators in R.

11

Data Collection

Statistics is the science of data collection and data analysis. In this chapter, we provide a brief introduction to principles and techniques for data collection, traditionally divided into Survey Sampling and Experimental Design – each the subject of a rich literature.

While most of the concepts and methods presented in this book are grounded in mathematical theory, the collection of data is, by nature, much more practical and almost always requires domain-specific knowledge.

Example 11.1 (Collection of data in ESP experiments) In [149], magician and paranormal investigator James 'The Amazing' Randi relates the story of how scientists at the Stanford Research Institute (SRI) were investigating a person claiming to have psychic abilities. The scientists were apparently fooled by relatively standard magic tricks into believing that this person was indeed psychic. This has led Randi, and others such as Persi Diaconis [43], to strongly recommend that a person competent in magic or deception be present during an ESP experiment or be consulted during the planning phase of the experiment.

Even though we will spend much more time on data analysis (Part III), careful data collection is of paramount importance.

> Data that were improperly collected can be completely useless and unsalvageable by any technique of analysis.

And it is worth keeping in mind that the collection phase is typically much more expensive that the analysis phase that ensues (e.g., clinical trials, car crash tests, etc.). Thus, the collection of data should be carefully planned according to well-established protocols, and preferably with expert advice.

11.1 Survey Sampling

Survey Sampling is the field that focuses on developing methods for sampling a population to determine characteristics of that population. The type of surveys that we will consider are those that involve sampling only a (usually very small) fraction of the population.

Remark 11.2 (Census) A census aims for an exhaustive survey of the population. Any statistical analysis of census data is necessarily descriptive since (at least in principle) the entire population is revealed. Some adjustments, based on complex statistical modeling, may be performed to attempt to palliate some deficiencies having to do with the undercounting of some subpopulations. See [71] for a relatively nontechnical introduction to the census as conducted by the US Census Bureau and a critique of such adjustments.

We present in this section some essentials and refer the reader to Chapter 19 in [66] or Chapter 4 in [191] for more comprehensive, yet gentle introductions.

11.1.1 Survey Sampling as an Urn Model

Consider polling a given population. Suppose the poll consists of one multiple-choice question with s possible choices. Let's say that n people are polled and asked to answer the question (with a single choice). Then the survey can be modeled as sampling from an urn (the population) made of balls with s possible colors.

Note that the possible choices usually include one or several options like "I do not know", "I am not aware of the issue", etc., for individuals that do not have an opinion on the topic, or are unaware of the issue, or are undecided in other ways. See Table 11.1 for an example. Special care may be warranted to deal with nonrespondents and people that did not properly fill the questionnaire.

Table 11.1 *New York Times – CBS News poll of 1022 adults in the US (March 11–15, 2016). "Do you think re-establishing relations between the US and Cuba will be mostly good for the US or mostly bad for the US?"*

Mostly good	Mostly bad	Unsure/No answer
62%	24%	15%

11.1.2 Simple Random Sampling

When sampling uniformly at random from the population is possible, the experiment can be simply modeled as sampling from an urn where at each stage each ball has the same probability of being drawn. We studied the resulting probability model in Section 2.4.2. (The sampling is, indeed, typically done without replacement. This is the case in standard polls, where a person is only interviewed once.)

It turns out that sampling uniformly at random from a population of interest is rather difficult to do in practice. Modern sampling of human populations is often done by phone based on a method for sampling phone numbers that has been designed with great care.

11.1.3 Bias in Survey Sampling

When simple random sampling is desired but the actual survey results in a different sampling distribution, it is said that the sampling is *biased*.

There are a number of factors that could lead to a biased sample, including the following:

- *Self-selection bias*   This may occur when people can volunteer to take the poll.
- *Non-response bias*   This occurs when the nonrespondents differ in opinion on the question of interest from the respondents.
- *Response bias*   This may occur, for example, when the way the question is presented has an unintended (and often unanticipated) influence on the response.

Self-selection bias and non-response bias are closely related. See Section 11.1.4 for an example where they may have played out. The following provides an example where response bias might have influence the outcome of a US presidential election.

Example 11.3 (2000 US presidential election) In 2000, George W. Bush won the US presidential election by an extremely small margin against his main opponent, Al Gore. Indeed, Bush won the deciding state, Florida, by a mere 537 votes.[46] It has been argued that this very slim advantage might have been reversed if not for some difficulties with some ballot designs used in the state, in particular, the "butterfly ballot" used in the county of Palm

---

[46] This margin was in fact so small that it required a recount (by state law). However, in a controversial (and 5:4 split) decision, the US Supreme Court halted the recount, in the process overruling the Florida Supreme Court.

Beach, which may have led some voters to vote for another candidate, Pat Buchanan, instead of Gore [1, 199, 172].

The following types of sampling are generally known to generate biased samples:

- *Quota sampling*   In this scheme, each interviewer is required to interview a certain number of people with certain characteristics (e.g., socioeconomic). Within these quotas, the interviewer is left to decide which persons to interview. The natural inclination (typically unconscious) is to reach out to people that are more accessible and seem more likely to agree to be interviewed. This generates a bias.
- *Sampling volunteers*   This is when individuals are given the opportunity to volunteer to take the poll. A prototypical example is a television poll where viewers are asked a question and are given the opportunity to answer that question by calling or texting a given phone number. This sampling scheme leads, almost by definition, to self-selection bias.
- *Convenience sampling*   The last two schemes above are examples of convenience sampling. A prototypical example is an individual asking his friends who they will vote for in an upcoming election. Almost by construction, there is no hope that this will result in a sample that is representative of the population of interest (presumably, all eligible voters).

Remark 11.4 (Coverage error) Bias typically leads to some members of the population being sampled with a higher probability than other members of the population. This is problematic if the intent is to sample the population uniformly at random. On the other hand, this is fine if the resulting sampling distribution is known, as there are ways to deal with non-uniform distributions. See Remark 11.6.

11.1.4 A Large Biased Sample is Still Biased

A typical sample is much smaller than the entire population it is drawn from. Even then, as long as the sampling was done appropriately, a sample of moderate size can be representative of the population. By contrast, a biased sample is, by definition, not representative of the population, and this is so even if the sample is very large in size. We can thus state the following:

> Large samples offer no protection against bias.

Literary Digest Poll

An emblematic example of this is the *Literary Digest* poll on the occasion of the 1936 US presidential election. The main contenders that year were Franklin Roosevelt (D) and Alf Landon (R). The *Literary Digest* (a US magazine at the time) mailed 10 million questionnaires resulting in 2.3 million returned. The poll predicted an overwhelming victory for Landon, with about 57% of the respondents in his favor. On election day, however, Roosevelt won by a landslide with 62% of the vote.[47]

Problem 11.5 That year, about 44.5 million people voted. Suppose, for a moment, that the sample collected by the *Literary Digest* was unbiased. If so, as in Problem 8.21, derive a typical range for the proportion favoring Roosevelt in such a poll.

What happened? For one thing, the response rate (about 23%) was rather small, so that any non-response bias could be substantial. Also, and perhaps more importantly, the sampling favored more affluent people. Indeed, the list of recipients was compiled from a variety of sources, including car and telephone owners, club memberships, and the magazine's readers, and, in the 1930s, a person on that list would likely be more affluent (and therefore more supportive of the Republican candidate) than the typical voter.

By comparison, George Gallup – who went on to found the renown Gallup polling company – accurately predicted the result of the election with a sample of size 50000. In fact, modern polls are typically even smaller, based on samples of size in the few thousands.

The moral of this story is that a smaller, but less biased sample may be preferable to a larger, but more biased sample. (This statement assumes that there is no possibility of correcting for the bias.) The story itself is worth reading in more detail, for example, in [66, 191, 176].

11.1.5 Examples of Sampling Plans

The following sampling designs are typically used for reasons of efficiency, cost, or (limited) resources.

- *Systematic sampling*   An example of such a sampling plan would be, in the context of an election poll conducted in a residential suburb, to interview every tenth household. Here one relies implicitly on the assumption that the residents are distributed – within the suburb – in a way that is independent of the question of interest.

---

[47] The numbers are taken from [66]. They are a little bit different in [176].

- *Cluster sampling*   An example of such a sampling plan would be, in the same context, to interview every household on several blocks chosen in some random fashion among all blocks in the suburb. For a *two-stage* variant, households could be sampled at random within each selected block. (This is an example of *multi-stage sampling*.)
- *Network sampling*   An example of that would be, in the context of surveying a population of drug addicts, to ask a person all the addicts he knows, which are then interviewed in turn, or otherwise observed or counted. This type of sampling is indeed popular when surveying hard-to-reach populations. There are many variants, known under various names such as *chain-referral sampling*, *respondent-driven sampling*, or *snowball sampling*. These are related to *web crawls* performed in the World Wide Web. Some of these designs may be considered to be of convenience, particularly when they do not involve randomization.

The following sampling design is meant to improve on simple random sampling.

- *Stratified sampling*   An example would be, in the context of estimating the average household income in a city, to divide the city into socio-economic neighborhoods (which play the role of strata here and would have to be known in advance) and then sample at random housing units in each neighborhood. In general, stratified sampling improves on simple random sampling when the strata are more homogeneous in terms of the quantity or characteristic of interest.

Remark 11.6 (Post-stratification)   A *stratification* is sometimes done *after* collecting the sample. This can be used, in some circumstances, to correct a possible bias in the sample.

Example 11.7 (Polling gamers) For an example of post-stratification, consider the polling scheme described in [200]. Quoting from the text, this was "an opt-in poll which was available continuously on the Xbox gaming platform during the 45 days preceding the 2012 US presidential election. Each day, three to five questions were posted, one of which gauged voter intention. [...] The respondents were allowed to answer at most once per day. The first time they participated in an Xbox poll, respondents were also asked to provide basic demographic information about themselves, including their sex, race, age, education, state [of residence], party [affiliation], political ideology, and who they voted for in the 2008 presidential election".

The survey resulted in a very large sample. However, as the authors warn, "the pool of Xbox respondents is far from being representative of the voting population". An (apparently successful) attempt to correct for this bias was made using post-stratification based on the side information collected on each respondent. In the authors' own words, "the central idea is to partition the data into thousands of demographic cells, estimate voter intent at the cell level [...], and finally aggregate the cell-level estimates in accordance with the target population's demographic composition".

11.2 Experimental Design

> To consult the statistician after an experiment is finished is often merely to ask him to conduct a post mortem examination. He can perhaps say what the experiment died of.
>
> Ronald A. Fisher

An experiment is designed with a particular purpose in mind. An example is that of comparing treatments for a certain medical condition. Another example is that of comparing NPK fertilizer mixtures to optimize the yield of a certain crop.

In this section, we present some essential elements of Experimental Design and describe some classical designs. For a book-length exposition, we recommend the book by Oehlert [139].

A good design follows some fundamental principles proven to lead to correct analyses (avoiding systematic bias) with good power (due to increased precision). Some of these principles include

- *Randomization* To avoid systematic bias.
- *Replication* To increase power.
- *Blocking* To better control variation.

Importantly, the design needs to be chosen with care *before* the collection of data starts.

11.2.1 Setting

The setting of an experiment consists of experimental units, a set of treatments assigned to the experimental units, and the response(s) measured on the experimental units. For example, in a medical experiment comparing two or more treatments for a certain disease, the experimental units are the human subjects and the response is a measure of the severity of the symptoms associated with the disease. (Such an experiment is often referred to as a *clinical trial*.) In an agricultural experiment where several fertilizer mixtures are compared in terms of yield for a certain crop, the experimental units may be the plots of land, the treatments are the fertilizer mixtures, and the response is the yield. (The plants may be referred to as measurement units.)

The way the experimental units are chosen, the way the treatments are assigned to the experimental units, and the way the response is measured, are all part of the design.

There may be several (real-valued) response measurements that are collected in a single experiment. Even then, there is typically one primary response, the other responses being secondary, and the primary response needs to be specified at the design stage, before the collection of data starts, to steer the subsequent analysis clear of a *Texas Sharpshooter Fallacy*. We will focus on the primary response (henceforth simply called 'response').

Although it is not part of the design itself, the method of analysis should also be decided beforehand, as failure to do so may result in a biased analysis [169, 76].

11.2.2 Enrollment/Sampling

The inference drawn from the experiment (e.g., 'treatment A shows a statistically significant improvement over treatment B') applies, strictly speaking, to the experimental units themselves. For the inference to apply more broadly – which is typically desired – additional conditions on the design need to be fulfilled in addition to randomization (Section 11.2.4).

For example, in biomedical research, a clinical trial is typically set up to compare the efficacy of two or more treatments for a particular disease or set of symptoms. Human subjects are enrolled in the trial and their response to one (or several) of the treatment(s) is recorded. Based on this limited information, the goal is often to draw conclusions on the relative effectiveness of the treatments when given to members of a certain population. (This is invariably true of clinical trials for drug development.) For this to be possible, as we saw in Section 11.1, the sample of subjects in the trial needs to be representative of the population.

Example 11.8 (Psychology experiments on campuses) Psychology experiments carried out on academic campuses have been criticized on that basis, due to the fact that the experiments are often conducted on volunteer, and often paid, students while the conclusions are, at least implicitly, supposed to generalize to a much larger population under different circumstances [164, 93]. Undeniably, such samples are used for convenience, and generalization to a larger population, although potentially valid, is far from automatic.

11.2.3 Power Calculations

In addition to a protocol for enrolling subjects that will (hopefully) yield a sample that is representative of the population of interest, the sample size needs to be large enough that the experiment, properly conducted and analyzed, would detect an effect (typically the difference in efficacy between treatments) of a certain prescribed size if such an effect is present. The target effect size is typically chosen based on scientific, societal, or commercial importance. For instance, for an experiment comparing two treatments, the investigators might want to calculate a minimum sample size n that would enable them to detect a potential 10% difference in response between the two treatments.

Such *power calculations* are important. Indeed, simply put, if the sample size is too small to detect what would be considered an interesting or important effect, then what is the point of conducting the experiment? The issue is exacerbated by the fact that conducting an experiment typically requires substantial resources. Not only could an underpowered study lead to waste [26], it could effectively be unethical if it involves human or animal subjects [87]. In addition, the multiplicity of small, and thus typically underpowered studies can lead to misinformation [17, 127].

11.2.4 Randomization

Randomization consists in assigning treatments to experimental units following a pre-specified random mechanism. This process is typically performed on a computer. We have seen the central role that randomness plays in survey sampling. The same is true in experimental design, where randomization helps avoid systematic bias.

Confounding

Systematic bias may be due to *confounding*, which happens when the effect of a factor (i.e., a characteristic of the experimental units) is related both to the received treatment and to the measured response. This is particularly problematic when the confounding factor is not measured as part of the data being collected.

Example 11.9 (Prostate cancer) It is routine for surgical patients with prostate cancer to have their lymph nodes dissected for inspection. (This is often called *surgical staging*.) If the nodes are found to be cancerous, it is accepted that the disease has spread over the body and a prostatectomy (the removal of the prostate) is not performed. Such staging is not part of other approaches to prostate cancer such as radiotherapy and, as argued in [119], can lead to a comparison that unfairly favors surgery. In such comparisons, the survival time is often the response, and in the context of comparing the effectiveness of surgery with, say, radiotherapy in prostate cancer, the grade (i.e., severity) of the cancer is a likely confounder.

Randomization offers some protection against confounding, and for this reason, an experiment where randomization is employed may enable causal inference (Section 22.2).

Blinding

Just as in survey sampling where little or no freedom of decision is given to the surveyor, randomization is done using a computer to prevent an experimenter (a human administering the treatments) from introducing bias. However, experimenters have been known to bias the results in other, sometimes quite subtle ways. To minimize bias of any kind, the experimenter is often blinded to the treatments.

In addition to that, in particular in experiments involving human subjects, the subjects are blinded to the treatment they are receiving. This is done, for example, to minimize non-compliance.

A clinical trial where both the experimenter and the subjects are blind to the administered treatment is said to be *double-blind*. This is the gold standard and has motivated the routine use of a *placebo* when no other control is available.[48] See [39] for a historical perspective.

Example 11.10 (Placebo effect) As a treatment is often costly and may induce undesirable side effects, it is important that it perform better than a placebo. As defined in [105], "placebo effects are improvements in patients' symptoms that are attributable to their participation in the therapeutic

---

[48] In fact, placebos are often the only control, even when a competing treatment is available. Indeed, according to [177]: "The FDA requires developers of new treatments to demonstrate that they are safe and effective in order to receive approval for market entry, but the agency demands proof of superiority to existing products only when it is patently unethical to withhold treatment from study patients, as in the cases of therapies for AIDS and cancer. Many new drugs are approved on the basis of demonstrated superiority to placebo. Even less is required for many new medical devices."

encounter, with its rituals, symbols, and interactions". These effects can be substantial. For example, as argued in [198], "for psychological disorders, particularly depression, it has been shown that pill placebos are nearly as effective as active medications".

Example 11.11 (Sham surgery) A double-blind design is not always possible. This is particularly true in surgery as, almost invariably, the operating surgeon knows whether he is performing a real or a sham procedure. For example, in [167], arthroscopic partial meniscectomy (which amounts to trimming the meniscus) is compared with sham surgery (which serves as placebo) to relieve symptoms attributed to meniscal tear. (Incidentally, this is another example where the placebo is shown to be as effective as the active procedure.)

To further avoid bias, this time at the level of the analysis, it is sometimes recommended that the analyst also be blinded to the detailed results of the experiments, for example, by anonymizing the treatments. This blinding is, in principle, unnecessary if the analysis is decided before the experiment starts.

Remark 11.12 (Keeping a trial blind) Humans are inquisitive creatures, and in a long clinical trial – some lasting a decade or more – it can become very tempting to guess what treatment is given to whom and which treatment is most effective. It turns out that keeping a clinical trial blind is a nontrivial aspect of its design [73, Ch 7].

11.2.5 Some Classical Designs

Completely Randomized Designs

A completely randomized design simply assigns the treatments at random with the only constraint being on the treatment group sizes, which are chosen beforehand. In detail, assume there are n experimental units and g treatments. Suppose we decide that Treatment j is to be assigned to a total of n_j units, where $n_1 + \cdots + n_g = n$. Then n_1 units are chosen uniformly at random and given Treatment 1; n_2 units are chosen uniformly at random and given Treatment 2; etc. (The sampling is without replacement.)

Example 11.13 (Back pain) In [2], the efficacy of a device for managing back pain is examined. In this randomized, double-blind trial, the treatment consisted of six sessions of transcutaneous electrical nerve stimulation (TENS) combined with mechanical pressure, while the placebo excluded the electrical stimulation.

Remark 11.14 (Balanced designs) To optimize the power of the subsequent analysis, it is usually preferable that the design be *balanced* as much as possible, meaning that the treatment group sizes be as close to equal as possible. In a perfectly balanced design, $r := n/g$ is an integer representing the number of units per treatment group.

Remark 11.15 (Sequential designs) Sequential designs are employed in clinical trials where subjects enter the trial over time. A simple variant consists in assigning Treatment j with probability p_j to a subject entering the trial, where $p_1, \ldots, p_g$ are chosen beforehand. (*A/B testing* is the term most often used in the tech industry: think of the administrator of some website that wants to try different layout configurations to maximize revenue.)

Complete Block Designs

Blocking is used to reduce the variance. Blocks are defined with the goal of forming more homogeneous groups of units. In its simplest variant, called a *randomized complete block design*, each block is randomized as in a completely randomized design, and the randomization is independent from block to block. When there is at least one unit for each treatment within each block, the design is complete.

In a balanced design, if there is exactly one unit within each block assigned to each treatment, we have $n = gb$, where b denotes the number of blocks. Further replication can be realized when, for example, each experimental unit contains several measurement units.

Blocking is often done based on one or several characteristics (aka factors) of the units. In clinical trials, common blocking variables include gender and age group.

Example 11.16 (Carbon sequestration) A randomized complete block design is used in [122] to study the effects of different cultural practices on carbon sequestration in soil planted with switchgrass.

Incomplete Block Designs

These are designs that involve blocking but in which the blocks have fewer units than treatments. The simplest variant is the *balanced incomplete block design*. Suppose, as before, that there are g treatments and that n experimental units are available for the experiments, and let b denote the number of blocks. Such a design is structured so that each block has the same number of units (say k) and each treatment is assigned to the same number of units (say r). This is referred to as *first-order balance* and requires $n = kb = gr$, while *second-order balance* refers to each pair of treatments appearing together in the same number of blocks (say λ) and requires that $\lambda := r(k-1)/(g-1)$. See Table 11.2 for an illustration.

Table 11.2 *Example of a balanced incomplete block design with*
g = 5 treatments (labeled A, B, C, D, E), b = 5 blocks each with
k = 4 units, r = 4 replicates per treatment, resulting in each pair
of treatments appearing together in λ = 3 blocks.

Block 1	C	A	D	B
Block 2	A	E	B	C
Block 3	B	D	E	C
Block 4	E	C	A	D
Block 5	B	D	E	A

Split Plots

As the name indicates, this design comes from agricultural experiments and is useful in situations where a factor is 'hard to vary'. To paraphrase an example given in [139], suppose that some land is available for an experiment meant to compare the productivity of several varieties of rice. We want to control the effect of irrigation and consider 2 levels of irrigation, say 'high' and 'low'. Irrigation may be hard and/or costly to control spatially. An option is to consider plots (called *whole plots*) that are sufficiently separated so that their irrigation can be done independently of one another. These plots are then subdivided into subplots (called *split plots*), each planted with one of the varieties of rice under consideration. In such a design, irrigation is randomized at the whole plot level, while seed variety is randomized at the split plot level within each whole plot.

Group Testing

Robert Dorfman [51] proposed an experimental design during World War II to minimize the number of blood tests required to test for syphilis in soldiers. His idea was to test blood samples from a number of soldiers simultaneously, repeating the process a few times. Group testing has been used in many other settings including quality control in product manufacturing and cDNA library screening.

More generally, consider a setting where we are testing n individuals for a disease based on blood samples. We consider the case where the design is set beforehand, as opposed to sequential. In that case, the design can be represented by an n-by-t binary matrix $X = (x_{i,j})$ where $x_{i,j} = 1$ if blood from the ith individual is in the jth testing pool, and $x_{i,j} = 0$ otherwise. In particular, t denotes the total number of testing pools. The result of each test is either positive $\oplus$ or negative $\ominus$. We assume for simplicity that the test is perfectly accurate in that it comes up positive when applied to a testing pool if and only if the pool includes the blood sample of at least one affected individual.

The design is *d-disjunct* if the sum of any of its d rows does not contain any other row [53], where in the present context we say that a row vector $u = (u_j)$ contains a row vector $v = (v_j)$ if $u_j \geq v_j$ for all j.

Problem 11.17 If the design is d-disjunct, and there are at most d diseased individuals in total, then each non-diseased individual will appear in at least one pool with no diseased individual. Deduce from this property a simple procedure for identifying the diseased individuals.

Thus, a design that is d-disjunct allows the experimenter to identify the diseased individuals as long as there are at most d of them. Note that this property is sufficient for that, but not necessary, although it offers the advantage of a particularly simple identification procedure.

A number of ways have been proposed for constructing disjunct designs, the goal being to achieve a d-disjunct design with a minimum number of testing pools t for a given number of subjects n. (There is a parallel with the construction of codes in the area of Information Theory.) We content ourselves with constructions that rely on randomness. In the simplest such construction, the elements of the design matrix, meaning the $x_{i,j}$, are the realization of iid $\mathrm{Ber}(p)$ random variables.

Proposition 11.18. *The probability that a Bernoulli n-by-t design with parameter p is d-disjunct is at least*

$$1 - (d+1)\binom{n}{d+1}\left[1 - p(1-p)^d\right]^t.$$

The proof is elementary and relies on the union bound [53, Thm 8.1.3].

Problem 11.19 For which value of p is the design most likely to be d-disjunct? For that value of p, deduce that there is a numeric constant C_0 such that this random group design is d-disjunct with probability at least $1 - \delta$ when

$$t \geq C_0\left[d^2 \log(n/d) + d \log(d/\delta)\right].$$

Repeated Measures

This is a type of design that is commonly used in longitudinal studies. For example, some human subjects are 'followed' over time to better understand how a certain condition evolves depending on a number of factors.

Example 11.20 (Neck pain) In [21], 191 patients with chronic neck pain were randomized to one of 3 treatments: (A) spinal manipulation combined with rehabilitative neck exercise, (B) MedX rehabilitative neck exercise, or (C) spinal manipulation alone. After 20 sessions, the patients were assessed 3, 6, and 12 months afterward for self-rated neck pain, neck disability, functional health status, global improvement, satisfaction with care, and medication use, as well as range of motion, muscle strength, and muscle endurance.

Such a design looks like a split plot design, with subjects as whole plots and the successive evaluations as split plots. The main difference is that there is no randomization at the split plot level.

Crossover Design

In a clinical trial with a crossover design, each subject is given each one of the treatments that are being compared. There is generally a *washout* period between treatments in an attempt to isolate the effect of each treatment or, said differently, to minimize the *residual effect* (aka *carryover effect*) of the preceding treatment. In addition, the order in which a subject receives the treatments needs to be randomized to avoid any systematic bias. First-order balance further imposes that the number of subjects that receive treatment X as their jth treatment not depend on X or j. When this is the case, the design is called a *Latin square design*. See Table 11.3 for an illustration. Second-order balance further imposes that each treatment follows every other treatment an equal number of times.

Table 11.3 *Example of a crossover design with 4 subjects, each receiving 4 treatments in sequence based on a Latin square design, with each treatment order appearing only once.*

Subject 1	A	B	C	D
Subject 2	C	D	A	B
Subject 3	B	D	E	C
Subject 4	D	C	B	A

Problem 11.21 Determine whether the design in Table 11.3 is second-order balanced. If not, find such a design.

Example 11.22 (Medical cannabis) The study [57] evaluates the potential of cannabis for pain management in 34 HIV patients suffering from neuropathic pain in a crossover trial where the treatments being compared are placebo (without THC) and active (with THC) cannabis cigarettes.

Example 11.23 (Gluten intolerance) The study [42] is on gluten intolerance. It involves 61 adult subjects without celiac disease or any other (formally diagnosed) wheat allergy, who nevertheless believe that ingesting gluten causes them some digestive problems. The design is a crossover design comparing rice starch (which serves as placebo) and actual gluten.

Matched-Pairs Design

In a study that adopts this design to compare 2 treatments, subjects that are similar in key attributes (i.e., possible confounders) are matched and the randomization to treatment happens within each pair. (Thus, this is a special case of a complete block design where each pair forms a block.)

Example 11.24 (Cognitive therapy for PTSD) In the study [123], which took place in Dresden, Germany, 42 motor vehicle accident survivors with post-traumatic syndrome were recruited to be enrolled in a study designed to examine the efficacy of cognitive behavioral treatment (CBT) protocols and methods. Subjects were matched after an initial assessment and then randomized to either CBT or control (which consisted of no treatment).

11.3 Observational Studies

Let us start with an example.

Example 11.25 (Role of anxiety in postoperative pain) In [104], 53 women who underwent an hysterectomy where assessed for anxiety, coping style, and perceived stress 2 weeks prior to the intervention. This was repeated several times during the recovery period. Analgesic consumption and level of perceived pain were also measured.

This study shares some of the attributes of a repeated measures design or a crossover design, yet there was no randomization involved. Broadly speaking, the ability to implement randomization is what distinguishes a controlled experiment from an observational study.

11.3.1 Why Observational Studies?

There is wide agreement that randomization should be a central part of a study whenever possible, because it offers some protection against confounding. As discussed in [171, 206, 145], there are quite a few examples of observational studies that were later refuted by randomized trials.

However, there are situations where researchers resort to an observational study. We follow Nick Black [16], who argues that experiments and observational studies are complementary. Although conceding that clinical trials are the gold standard, he says that "observational methods are needed [when] experimentation [is] unnecessary, inappropriate, impossible, or inadequate".

- *Experimentation may be unnecessary* when the effect size is very large. (Black cites, as an example, the effectiveness of penicillin for treating bacterial infections.)

- *Experimentation may be inappropriate* in situations where the sample size needed to detect a rare side effect of a drug, for example, far exceeds the size of a feasible clinical trial. Another example is the detection of long-term adverse effects, as this would require a study that exceeds in length that of most typical clinical trials.

- *Experimentation may be impossible* for a number of reasons. One reason could be *ethics*. For example, randomizing pregnant women to smoking and non-smoking groups is not ethical in large part because we know that smoking carries substantial negative effects for both the mother and the fetus/baby. Another reason could be lack of *control*. For example, randomizing US cities to a minimum wage of, say, $15 per hour versus no minimum wage is difficult;[49] similarly, increasing the levels of carbon dioxide in the atmosphere to better understand the resulting effects on climate is not an option.

- *Experimentation may be inadequate* when it comes to generalizing the findings to a broader population and to how medicine is practiced in that population on a day-to-day basis. While observational studies, almost by definition, take stock of how medicine is practiced in real life, clinical trials occur in more controlled settings, often take place in university hospitals or clinics, and may attract particular types of subjects.

---

[49] An example of large-scale experimentation with policy includes Finland's Design for Government project, and an experiment with the universal income in Canada [22]. In fact, the 2019 Nobel Prize in Economics was awarded jointly to Abhijit Banerjee, Esther Duflo, and Michael Kremer, "for their experimental approach to alleviating global poverty".

Example 11.26 (Effects of smoking) In the study of how smoking affects lung cancer and other ailments, for ethical reasons, researchers have had to rely on observational studies and experiments on animals. These have provided very strong circumstantial evidence that tobacco consumption is associated with the onset and development of various pulmonary and cardiovascular diseases. In the meantime, the Tobacco Industry has insisted that these are only associations and that no causation has been established [129]. (The story might be repeating itself with the consumption of sugar [121].)

Example 11.27 (Human role in climate change) There is a parallel in the question of climate change, which is another area where experimentation is hardly possible. To determine the role of human activity in climate change, scientists have had to rely on indirect evidence such as the known warming effects of carbon dioxide, methane, and other 'greenhouse' gases, combined with the fact that the increased presence of these gases in the atmosphere is due to human activity. Scientists have also been able to rely computer models. Here too, the sum total evidence is overwhelming, and scientific consensus is essentially unanimous, yet the Fossil Fuel Industry still claims that all this does not prove that human activity is a substantial contributor to climate change [140].

11.3.2 Types of Observational Studies

Problem 11.28 In the examples given below, identify as many obstacles to randomization as you can among the ones listed above, and possibly others.

Cohort Study

A cohort in this context is a group of individuals sharing one or several characteristics of interest. For example, a birth cohort is made of subjects that were born at about the same time.

Example 11.29 (Obesity in children) In the context of a large longitudinal study, the Avon Longitudinal Study of Parents and Children, the goal of researchers in [151] was to "identify risk factors in early life (up to 3 years of age) for obesity in children (by age 7) in the United Kingdom".

Another example would be people that smoke, that are then followed to understand the implications of smoking, in which case another cohort of non-smokers may be used to serve as control. In general, such a study follows subjects with a certain condition with the goal of drawing associations with particular outcomes.

Example 11.30 (Head injuries) The study [188] followed almost 3000 individuals that were admitted to one of a number of hospitals in the Glasgow area, Scotland, after a head injury. The stated goal was to "determine the frequency of disability in young people and adults admitted to hospital with a head injury and to estimate the annual incidence in the community".

A matched-pairs cohort study[50] is a variant where subjects are matched and then followed as a cohort.

Example 11.31 (Helmet use in motorcycle crashes) The study [138] is based on the "Fatality Analysis Reporting System data, from 1980 through 1998, for motorcycles that crashed with two riders and either the driver or the passenger, or both, died". Matching was by motorcycle. To quote the authors of the study, "by estimating effects among naturally matched-pairs on the same motorcycle, one can account for potential confounding by motorcycle characteristics, crash characteristics, and other factors that may influence the outcome".

Case-Control Study

In a case-control study, subjects with a certain condition (typically a disease) of interest are identified and included in the study to serve as cases. At the same time, subjects not experiencing that condition (without the disease or with the disease but of lower severity) are identified and included in the study to serve as controls.

Example 11.32 (Lipoprotein(a) and coronary heart disease) [153] reports on a study that involves a sample of men aged 50 at the start of the study (therefore, a birth cohort) from Gothenburg, Sweden. At baseline, a blood sample was taken from each subject and frozen. After six years, the concentration of lipoprotein(a) was measured in men having suffered a myocardial infarction or died of coronary heart disease. For each of these men – which represent the cases in this study – four men were sampled at random among the remaining ones to serve as controls and their blood concentration of lipoprotein(a) was measured. The goal was to "examine the association between the serum lipoprotein(a) concentration and subsequent coronary heart disease".

Example 11.33 (Venous thromboembolism and hormone replacement therapy) Based on the General Practice Research Database (United Kingdom), in [85], "a cohort of 347253 women aged 50 to 79 without major risk factors for venous thromboembolism was identified. The cases

---

[50] Compare with the randomized matched-pairs design.

were the 292 women admitted to hospital for a first episode of pulmonary embolism or deep venous thrombosis. The controls were 10000 women randomly selected from the source cohort." (The cohort here is a birth cohort and not based on a particular risk factor.) The goal was to "evaluate the association between use of hormone replacement therapy and the risk of idiopathic venous thromboembolism".

Remark 11.34 (Cohort vs case-control) A cohort study starts with a possible risk factor (e.g., smoking) and aims at discovering the diseases it is associated with. A case-control study, on the other hand, starts with the disease and aims at discovering risk factors associated with it.[51]

Problem 11.35 (Rare diseases) A case-control study is often more suitable when studying a rare disease, which would otherwise require following a very large cohort. Consider a disease affecting 1 out of 100000 people in a certain large population (many millions). How large would a sample need to be in order to include 10 cases with probability at least 99%?

Cross-Sectional Study

While a cohort study and a xcase-control study both follow a certain sample of subjects over time, a cross-sectional study examines a sample of individuals at a specific point in time. In particular, associations are comparatively harder to interpret in cross-sectional studies. The main advantage is simply cost, as such studies can be based on data collected for other purposes.

Example 11.36 (Green tea and heart disease) In [99], "1371 men aged over 40 years residing in Yoshimi [were] surveyed on their living habits including daily consumption of green tea. Their peripheral blood samples were subjected to several biochemical assays." The goal was to "investigate the association between consumption of green tea and various serum markers in a Japanese population, with special reference to preventive effects of green tea against cardiovascular disease and disorders of the liver."

11.3.3 Matching

While in an observational study randomization cannot be purposely implemented, a number of techniques have been proposed to at least minimize the influence of other factors [30].

---

[51] For an illustration, compare Figures 3.5 and 3.6 in [18].

Matching is an umbrella name for a variety of techniques that aim at matching cases with controls in order to have a treatment group and a control group that look alike in important aspects. These important attributes are typically chosen for their potential to affect the response under consideration, that is, they are believed to be risk factors.

Example 11.37 (Effects of coaching on SAT) The study [144] attempted to measure the effect of attending an SAT coaching program on the resulting test score. (The paper focused on *SAT I: Reasoning Test Scores*.) These programs are offered by commercial test preparation companies that claim a certain level of success. The data were based on a survey of about 4000 SAT takers in 1995–96, where about 12% had attended a coaching program. Besides the response (the SAT score itself), some 27 variables were measured on each test taker, including coaching status (the main factor of interest), racial and socioeconomic indicators (e.g., father's education), various measures of academic preparation (e.g., math grade), etc. The intention, of course, was to isolate the effect of coaching from these other factors, some of which are undoubtedly important. The authors applied a number of techniques including a variant of matching. Other variants are applied to the same data in [88].

The intention behind matching is to control for confounding by balancing possible confounding factors with the intention of canceling out their confounding effect. We formalize this in Section 22.2.3, where we show that matching works under some conditions, the most important one being that there are no unmeasured variables that confound the effect. The beauty (and usefulness) of randomization is that it achieves this balancing automatically (although only on average) without a priori knowledge of any confounding variable. This is shown formally in Section 22.2.2.

11.3.4 Natural Experiments

As we mentioned above, and as will be detailed in Section 22.2, a proper use of randomization allows for causal inference. However, randomization is only possible in controlled experiments. In observational studies, matching can allow for causal inference, but only if one can guarantee that there are no confounders.

In general, the issue of drawing causal inferences from observational studies is complex, and in fact remains controversial, so we will keep the discussion simple and focused on the widely accepted view that causal inference is possible in the context of a *natural experiment* [33, 112].

In an observational study, treatment assignment is, by definition, out of the control of the experimenter; yet, if it can be argued that it was nonetheless done (by 'Nature') independently of the response to treatment, then the conditions resemble those of a controlled experiment. Recognizing such a natural experiment requires domain-specific expertise. (David Freedman speaks of a 'shoe leather' approach to data analysis [68].)

Example 11.38 (Snow's discovery of cholera) A classic example of a natural experiment is the one uncovered and leveraged by John Snow in nineteenth-century London to identify the source of a cholera epidemic [173]. (In the process, Snow discovered that cholera could be transmitted by the consumption of contaminated water – a fact that may have been unknown at the time.) The account is worth reading in more detail, but in a nutshell, Snow suspected that the consumption of water had something to do with the spread of cholera, and to confirm his hypothesis, he examined the houses in London served by one of two water companies, Southwark & Vauxhall and Lambeth, and found that the death rate in the houses served by the former was several times higher. He then explained this by the fact that, although both companies sourced their water from the River Thames, Lambeth was taping the river upstream of the main sewage discharge points, while Southwark & Vauxhall was getting its water downstream. Although this is an observational study, a case for 'natural' randomization can be argued, as Snow did: "Each company supplies both rich and poor, both large houses and small; there is no difference either in the condition or occupation of the persons receiving the water of the different Companies."

Example 11.39 (The Oregon Experiment on Medicaid[52]) In 2008, the state of Oregon in the US decided to expand a joint federal and state health insurance program for people with low-income known as Medicaid. As described in [20]: "Correctly anticipating excess demand for the available new enrollment slots, the state conducted a lottery, randomly selecting individuals from a list of those who signed up in early 2008. Lottery winners and members of their households were able to apply for Medicaid. Applicants who met the eligibility requirements were then enrolled in [the] Oregon Health Plan Standard." In the end, 29834 individuals won the lottery out of 74922 individuals who participated. The participants, both winners and losers of this lottery, have nevertheless been followed and compared on various metrics (e.g., health care utilization).

---

[52] https://www.nber.org/oregon/

Natural experiments are relatively rare, but some situations have been identified where they arise routinely. For example, in Genetics, pairs of twins are sought after and examined to better understand how much a particular behavior is due to 'nature' versus 'nurture'.

Regression Discontinuity Design

This is an area of statistics specializing in situations where an intervention is applied or not based on some sort of score being above or below some threshold. If the threshold is arbitrary to some degree, it may be justifiable to compare, in terms of outcome, cases with a score right above the threshold with cases with a score right below the threshold.

Example 11.40 (Medicare) In the US, most people become eligible at age 65 to enroll in a federal health insurance program called Medicare. This has led researchers, as in [24], to examine the effects of access to this program on various health outcomes.

Example 11.41 (Elite secondary schools in Kenya) Students from elite schools tend to perform better, but is this due to elite schools being truly superior to other schools, or simply a result of attracting the best and brightest students? This question is examined in [120] in the context of secondary schools in Kenya, employing a regression discontinuity design approach that takes advantage of "the random variation generated by the centralized school admissions process".

Remark 11.42 In a highly cited paper [95], Bradford Hill proposes nine aspects to pay attention to when attempting to draw causal inferences from observational studies. One of the main aspects is *specificity*, which is akin to identifying a natural experiment. Another important aspect, when present, is that of *experimental* evidence, which is akin to identifying a discontinuity.

Part III

Elements of Statistical Inference

12

Models, Estimators, and Tests

A prototypical (although somewhat idealized) workflow in any scientific investigation starts with the design of the experiment to probe a question or hypothesis of interest. The experiment is modeled using several plausible mechanisms. The experiment is conducted and the data are collected. These data are finally analyzed to identify the most adequate mechanism – the one among those considered that best 'explains' the data.

Although an experiment is supposed to be repeatable, this is not always possible, particularly if the system under study is chaotic or random in nature. When this is the case, the mechanisms above are expressed as probability distributions. We then talk about *probabilistic modeling*, albeit here there are several probability distributions under consideration. It is as if we contemplate several probability experiments (in the sense of Chapter 1), and the goal of *statistical inference* is to decide on the most plausible in view of the collected data.

To illustrate the various concepts introduced in this chapter, we use Bernoulli trials as our running example of an experiment. This includes, as special case, an experiment where we sample with replacement from an urn (Remark 2.17). Although basic, this model is relevant in a number of important real-life situations (e.g., election polls, clinical trials, etc.). We will study Bernoulli trials in more detail in Section 14.1.

12.1 Statistical Models

A *statistical model* for a given experiment is of the form $(\Omega, \Sigma, \mathcal{P})$, where Ω is the sample space containing all possible outcomes, Σ is the class of events of interest, and $\mathcal{P}$ is a *family* of distributions on Σ.[53] Modeling the

[53] Compare with a probability space, which only includes one distribution. A statistical model includes several distributions to model situations where the mechanism driving the experimental results is not perfectly known.

experiment with $(\Omega, \Sigma, \mathcal{P})$ postulates that the outcome of the experiment $\omega \in \Omega$ was generated from a distribution in $\mathcal{P}$. The goal, then, is to determine which distribution in $\mathcal{P}$ is most congruent with the observed data.

We follow the tradition of parameterizing the family $\mathcal{P}$. This is natural in some contexts and can always be done without loss of generality (since any set can be parameterized by itself). By default, the *parameter* will be denoted by θ and the *parameter space* (where θ belongs) by Θ, so that

$$\mathcal{P} = \{\mathbb{P}_\theta : \theta \in \Theta\}. \tag{12.1}$$

Remark 12.1 (Identifiability) Unless otherwise specified, we will assume that the model (12.1) is *identifiable*, meaning that the parameterization is one-to-one, or in formula, $\mathbb{P}_\theta \neq \mathbb{P}_{\theta'}$ when $\theta \neq \theta'$.

Example 12.2 (Binomial experiment) Suppose we model an experiment as a sequence of Bernoulli trials with probability parameter θ and predetermined length n. This assumes that each trial results in one of two possible values. Labeling these values as н and т, which we can always do at least formally, the sample space is the set of all sequences of length n with values in $\{$н, т$\}$, or in formula

$$\Omega = \{$н, т$\}^n.$$

(We already saw this in Example 1.8.) The statistical model also assumes that the observed sequence was generated by one of the Bernoulli distributions, namely

$$\mathbb{P}_\theta(\{\omega\}) = \theta^{Y(\omega)}(1 - \theta)^{n - Y(\omega)}, \tag{12.2}$$

where $Y(\omega)$ is the number of heads in ω. (We already saw that in (2.5).) Therefore, the family of distributions under consideration is

$$\mathcal{P} = \{\mathbb{P}_\theta : \theta \in \Theta\}, \quad \text{where } \Theta := [0, 1].$$

Note that the dependency in n is left implicit as n is given and not a parameter of the family. We call this a *binomial experiment* because of the central role that Y plays and we know that Y has the binomial distribution with parameters (n, θ).

Remark 12.3 (Correctness of the model) As is the custom, we will proceed assuming that the model is correct, that indeed one of the distributions in the family $\mathcal{P}$ generated the data. This depends, in large part, on how the data were generated or collected (Chapter 11). For example, a (binary) poll that successfully implements simple random sampling from a very large population can be accurately modeled as a binomial experiment. When the model is correct, we will sometimes use θ_* to denote the true value of the parameter. In practice, a model is rarely strictly correct. When the model is only approximate, the resulting inference will necessarily be approximate also.

12.2 Statistics and Estimators

A *statistic* is a random variable on the sample space Ω. It is meant to summarize the data in a way that is useful for the purpose of drawing inferences from the data.

Let φ be a function defined on the parameter space Θ representing a feature of interest (e.g., the mean). Note that φ is often used to denote $\varphi(\theta)$ (a clear abuse of notation) when confusion is unlikely. We will adopt this common practice. It is often the case that θ itself is the feature of interest, in which case $\varphi(\theta) = \theta$.

An *estimator* for $\varphi(\theta)$ is a statistic, say S, chosen for the purpose of approximating it. Note that while φ is defined on the parameter space (i.e., Θ), S is defined on the sample space (i.e., Ω).

Remark 12.4 The problem of estimating $\varphi(\theta)$ is well-defined if $\mathbb{P}_\theta = \mathbb{P}_{\theta'}$ implies that $\varphi(\theta) = \varphi(\theta')$, which is always the case if the model is identifiable.

Remark 12.5 (Estimators and estimates) By definition, an estimator is a statistic. The value that an estimator takes in a given experiment is often called an *estimate*. For example, if S is an estimator, and ω denotes the data, then $S(\omega)$ is an estimate.

Desiderata A good estimator is one which returns an estimate that is 'close' to the quantity of interest.

12.2.1 Measures of Performance

Quantifying closeness is not completely trivial as we are talking about estimators, whose output is random by definition. (An estimator is a function of the data and the data are modeled as random.) We detail the situation where we want to estimate a real-valued parameter $\varphi(\theta)$ in the context of a parametric model $\{\mathbb{P}_\theta : \theta \in \Theta\}$.

Mean Squared Error

A popular measure of performance for an estimator S is the *mean squared error* (MSE), defined as

$$\mathrm{mse}_\theta(S) := \mathbb{E}_\theta\left[(S - \varphi(\theta))^2\right],$$

where $\mathbb{E}_\theta$ denotes the expectation with respect to $\mathbb{P}_\theta$.

The MSE of S is closely related to its *bias*, defined as

$$\mathbb{E}_\theta(S) - \varphi(\theta), \tag{12.3}$$

and its variance, $\mathrm{Var}_\theta(S)$.

Problem 12.6 (Bias-variance decomposition) Show that

$$\mathrm{mse}_\theta(S) = \underbrace{(\mathbb{E}_\theta(S) - \varphi(\theta))^2}_{\text{squared bias}} + \underbrace{\mathrm{Var}_\theta(S)}_{\text{variance}}. \tag{12.4}$$

Mean Absolute Error

Another popular measure of performance for an estimator S is the *mean absolute error* (MAE), defined as

$$\mathrm{mae}_\theta(S) := \mathbb{E}_\theta\left[|S - \varphi(\theta)|\right].$$

Other Loss Functions

In general, let $\mathcal{L}(s, \theta)$ be a function measuring the discrepancy between s and $\varphi(\theta)$. This is called a *loss function* as it is meant to quantify the loss incurred when the true parameter is $\varphi(\theta)$ and our estimate is s. A popular choice is

$$\mathcal{L}(s, \theta) = |s - \varphi(\theta)|^\gamma,$$

for some pre-specified $\gamma > 0$. This includes the MSE and the MAE as special cases.

Risk

Having chosen a loss function we define the *risk* of an estimator as its expected loss, which for a loss function $\mathcal{L}$ and an estimator S may be expressed as

$$\mathcal{R}_\theta(S) := \mathbb{E}_\theta\left[\mathcal{L}(S, \theta)\right]. \tag{12.5}$$

Remark 12.7 (Frequentist interpretation) Let θ denote the true value of the parameter and let S denote an estimator. Suppose the experiment is repeated independently m times and let ω_j denote the outcome of the jth experiment.

Compute the average loss over these m experiments, meaning

$$L_m := \frac{1}{m} \sum_{j=1}^{m} \mathscr{L}(S(\omega_j), \theta).$$

By the Law of Large Numbers (which, as we saw, is at the core of the frequentist interpretation of probability),

$$L_m \xrightarrow{P} \mathscr{R}_\theta(S), \quad \text{as } m \to \infty.$$

12.2.2 Maximum Likelihood Estimator

As the name indicates, this estimator returns a distribution that maximizes, among those in the family, the chances that the experiment would result in the observed data. Assume that the statistical model is discrete in the sense that the sample space is discrete. See Section 12.5.1 for other situations.

Denoting the data by $\omega \in \Omega$ as before, the *likelihood function* is defined as

$$\text{lik}(\theta) = \mathbb{P}_\theta(\{\omega\}).$$

(Note that this function also depends on the data, but this dependency is traditionally left implicit.) Assuming that, for all possible outcomes, this function has a unique maximizer, the *maximum likelihood estimator* (MLE) is defined as that maximizer. If the resulting estimate for θ is denoted $\hat{\theta}$, then the corresponding estimate for $\varphi(\theta)$ is simply $\hat{\varphi} := \varphi(\hat{\theta})$.

Remark 12.8 When the likelihood admits several maximizers, one of them can be chosen according to some criterion.

Example 12.9 (Binomial experiment) In the setting of Example 12.2, the MLE is found by maximizing the likelihood (12.2) with respect to $\theta \in [0,1]$. To simplify the expression a little bit, let $y = Y(\omega)$, which is the number of heads in the sequence ω. We then have the following expression for the likelihood

$$\text{lik}(\theta) := \theta^y (1-\theta)^{n-y}.$$

This is a polynomial in θ and therefore differentiable. To maximize the likelihood we thus look for critical points. First assume that $1 \leq y \leq n-1$. In that case, setting the derivative to 0, we obtain the equation

$$\theta^y (1-\theta)^{n-y} \big(y(1-\theta) - (n-y)\theta\big) = 0.$$

The solutions are $\theta = 0$, $\theta = 1$, and $\theta = y/n$. Since the likelihood is zero at $\theta = 0$ or $\theta = 1$, and strictly positive at $\theta = y/n$, we conclude that the maximizer is unique and equal to y/n. If $y = 0$, the likelihood is easily seen to have a unique maximizer at $\theta = 0$. If $y = n$, the likelihood is easily seen to have a unique maximizer at $\theta = 1$. Thus, in any case, the maximizer is unique and given by y/n. We conclude that the MLE is well-defined and equal to Y/n, that is, the proportion of heads in the data.

12.3 Confidence Intervals/Regions

An estimator, as presented above, provides a way to obtain an informed guess (the estimate) for the true value of the parameter. In addition to that, it is often quite useful to know how far we can expect the estimate to be from the true value of the parameter.

When the parameter is real-valued, this is typically done via an interval whose bounds are random variables. We say that such an interval, denoted $I(\omega)$, is a $(1 - \alpha)$-level *confidence interval* for $\varphi(\theta)$ if

$$\mathbb{P}_\theta(\varphi(\theta) \in I) \geq 1 - \alpha, \quad \text{for all } \theta \in \Theta. \tag{12.6}$$

For example, $\alpha = 0.10$ gives a 90% confidence interval. See Figure 12.1.

In some cases, in particular when the parameter of interest is multivariate, the use of an interval may be too constraining or even inappropriate. In that case, we talk of a *confidence region*. In particular, a $(1 - \alpha)$-level confidence region for θ is a set-valued (measurable) function R such that

$$\mathbb{P}_\theta(\theta \in R) \geq 1 - \alpha, \quad \text{for all } \theta \in \Theta.$$

Although one can often directly construct a confidence interval (or region) for $\varphi(\theta)$, it can also be derived from a confidence region for θ. Indeed, suppose that R is a $(1 - \alpha)$-level confidence region for θ. Then $I := \varphi(R)$ $(1 - \alpha)$-level confidence region for $\varphi(\theta)$.

Desiderata A good confidence interval is one that has the prescribed level of confidence and is relatively short compared to other confidence intervals having the same level of confidence.

Example 12.10 (Binomial experiment) Confidence intervals are often constructed based on an estimator. We illustrate this in the context of Example 12.2, where we construct a confidence interval for θ based on the MLE, or equivalently, on the number of heads, controlling its deviations using Chebyshev's inequality (7.13). Indeed, because S has mean θ and variance $\theta(1 - \theta)/n$ under θ, the inequality gives

$$\mathbb{P}_\theta\left(|S - \theta| < c\sqrt{\theta(1 - \theta)/n}\right) \geq 1 - 1/c^2,$$

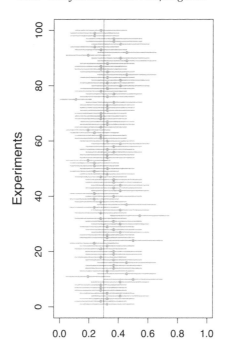

Figure 12.1 An illustration of the concept of confidence interval. We consider a binomial experiment with parameters $n = 10$ and $\theta = 0.3$. We repeat the experiment 100 times, each time computing the Clopper–Pearson two-sided 90% confidence interval for θ specified in (14.4). The vertical line is at the true value of θ.

for any $c > 0$ and any $\theta \in \Theta$. We will see in Problem 14.5 that the set of θ such that $|S - \theta| < c\sqrt{\theta(1 - \theta)/n}$ is an interval for any given value of S (in $[0, 1]$) and any $c > 0x$. And if we choose $c = 1/\sqrt{\alpha}$, so that $1 - 1/c^2 = 1 - \alpha$, this (random) interval is a $(1 - \alpha)$-level confidence interval for θ. Trading accuracy for simplicity – so as to provide a more concrete construction – we use the fact that $\theta(1 - \theta) \leq 1/4$ for all $\theta \in [0, 1]$ to derive the inequality

$$\mathbb{P}_\theta\big(|S - \theta| < c/(2\sqrt{n})\big) \geq 1 - 1/c^2,$$

from which we conclude that

$$\left[S - \frac{1}{2\sqrt{\alpha n}}, S + \frac{1}{2\sqrt{\alpha n}}\right] \tag{12.7}$$

is a $(1 - \alpha)$-level confidence interval for θ.

12.4 Testing Statistical Hypotheses

Suppose we do not need to estimate a function of the parameter, but rather only need to know whether the parameter value satisfies a given property. In what follows, we take the default hypothesis, called the *null hypothesis* and often denoted $\mathcal{H}_0$, to be that the parameter value satisfies the property. Deciding whether the data are congruent with this hypothesis is called *testing* the hypothesis.

Let $\Theta_0 \subset \Theta$ denote the subset of parameter values that satisfy the property. We will call Θ_0 the *null set*. The null hypothesis can be expressed as

$$\mathcal{H}_0 : \theta_* \in \Theta_0.$$

(Recall that θ_* denotes the true value of the parameter, assuming the statistical model is correct.) The complement of Θ_0 in Θ,

$$\Theta_1 := \Theta \smallsetminus \Theta_0, \qquad (12.8)$$

is often called the *alternative set*, and

$$\mathcal{H}_1 : \theta_* \in \Theta_1$$

is often called the *alternative hypothesis*.

Remark 12.11 A null hypothesis of the form $\varphi_* := \varphi(\theta_*) \in \Phi_0$ can be equivalently written as $\theta_* \in \Theta_0$, where $\Theta_0 := \{\theta : \varphi(\theta) \in \Phi_0\}$ — although the former null hypothesis can often be tackled directly without such a reformulation.

Example (Binomial experiment) In the setting of Example 12.2, we may want to know whether the parameter value does not exceed some given θ_0. This corresponds to testing the null hypothesis

$$\mathcal{H}_0 : \theta_* \leq \theta_0, \qquad (12.9)$$

which corresponds to the following null set

$$\Theta_0 = \{\theta \in [0, 1] : \theta \leq \theta_0\} = [0, \theta_0].$$

12.4.1 Test Statistics

A *test statistic* is used to decide whether the null hypothesis is reasonable. If, based on a chosen test statistic, the hypothesis is found to be 'substantially incompatible' with the data, it is *rejected*.

Although the goal may not be that of estimation, good estimators typically make good test statistics. Indeed, given an estimator S for θ, one could think of rejecting $\mathcal{H}_0$ when $S(\omega) \notin \Theta_0$. Tempting as it is, this is typically too harsh, as it does not properly account for the randomness in the estimator. Instead, it is generally better to reject $\mathcal{H}_0$ if $S(\omega)$ is 'far enough' from Θ_0.

Desiderata A good test statistic is one that behaves differently according to whether the null hypothesis is true or not.

Example (Binomial experiment) Consider the problem of testing the hypothesis $\mathcal{H}_0$ of (12.9). As a test statistic, let us use the maximum likelihood estimator, $S := Y/n$. In that case, it is tempting to reject $\mathcal{H}_0$ when $S > \theta_0$. However, doing so would lead us to reject by mistake quite often if θ_* is in the null set yet close to the alternative set: in the most extreme case where $\theta_* = \theta_0$, the probability of rejection approaches $1/2$ as the sample size n increases.

Problem 12.12 Prove the last claim using the Central Limit Theorem. Then examine the situation where $\theta_* < \theta_0$.

Remark 12.13 (Equivalent test statistics) We say that two test statistics, S and T, are equivalent if there is a strictly monotone function g such that $T = g(S)$. Clearly, equivalent statistics provide the same amount of evidence against the null hypothesis, since we can recover one from the other.

12.4.2 Likelihood Ratio

The *likelihood ratio* (LR) is to hypothesis testing what the MLE is to parameter estimation. It presents a general procedure for deriving a test statistic. In the setting of Section 12.2.2, having observed ω, the LR is defined as[54]

$$\frac{\max_{\theta \in \Theta_1} \mathsf{lik}(\theta)}{\max_{\theta \in \Theta_0} \mathsf{lik}(\theta)}.$$

By construction, a large value of that statistic provides evidence against the null hypothesis.

Remark 12.14 (Variants) The LR is sometimes defined differently, for example,

$$\frac{\max_{\theta \in \Theta} \mathsf{lik}(\theta)}{\max_{\theta \in \Theta_0} \mathsf{lik}(\theta)}. \tag{12.10}$$

---

[54] The same statistic is sometimes referred to as the *generalized likelihood ratio* (GLR).

or its inverse (in which case small values of the statistic weigh against the null hypothesis). However, all these variants are strictly monotonic functions of each other and are therefore equivalent for testing purposes (Remark 12.13).

Example (Binomial experiment) Consider the problem of testing the hypothesis $\mathcal{H}_0$ of (12.9). Recall the MLE is $S := Y/n$. With $y = Y(\omega)$ and $\hat{\theta} := y/n$ (which is the estimate), we compute

$$\max_{\theta \leq \theta_0} \text{lik}(\theta) = \max_{\theta \leq \theta_0} \theta^y (1 - \theta)^{n-y}$$

$$= \min(\hat{\theta}, \theta_0)^y (1 - \min(\hat{\theta}, \theta_0))^{n-y},$$

and

$$\max_{\theta \in [0,1]} \text{lik}(\theta) = \max_{\theta \in [0,1]} \theta^y (1 - \theta)^{n-y}$$

$$= \hat{\theta}^y (1 - \hat{\theta})^{n-y}.$$

Hence, the variant (12.10) of the LR is equal to 1 (the minimum possible value) if $\hat{\theta} \leq \theta_0$, and

$$\left(\frac{\hat{\theta}}{\theta_0}\right)^y \left(\frac{1 - \hat{\theta}}{1 - \theta_0}\right)^{n-y},$$

otherwise. Taking the logarithm and multiplying by $1/n$ yields an equivalent statistic equal to 0 if $\hat{\theta} \leq \theta_0$, and

$$\hat{\theta} \log\left(\frac{\hat{\theta}}{\theta_0}\right) + (1 - \hat{\theta}) \log\left(\frac{1 - \hat{\theta}}{1 - \theta_0}\right),$$

otherwise. Thus, the LR is a function of the MLE. However, this function is monotone, but not strictly monotone, and therefore the MLE and the LR are not equivalent test statistics. That said, they yield the same inference in most cases of interest (Problem 12.17).

12.4.3 P-Values

Given a test statistic, we need to decide what values of the statistic provide evidence against the null hypothesis. In other words, we need to decide what values of the statistic are 'unusual' or 'extreme' under the null, in the sense of being unlikely if the null (hypothesis) were true.

Suppose we decide that large values of a test statistic S are evidence that the null is not true. Let ω denote the observed data and let $s = S(\omega)$ denote the observed value of the statistic. In this context, we define the *p-value* as

$$\mathsf{pv}_S(s) := \sup_{\theta \in \Theta_0} \mathbb{P}_\theta(S \geq s). \tag{12.11}$$

To be sure, $\mathbb{P}_\theta(S \geq s)$ is shorthand for $\mathbb{P}_\theta\big(\{\omega' : S(\omega') \geq S(\omega)\}\big)$.

In words, the right-hand side in (12.11) is the supremum probability under any null distribution of observing a value of the test statistic as extreme as the observed value. Thus a a small p-value is evidence that the null hypothesis is false.

Note that a p-value is associated with a particular test statistic: different test statistics lead to different p-values, in general.

Remark 12.15 (Replication interpretation of the p-value) The definition itself leads us to believe that replications are needed to compute the p-value. Such replications may be out of the question, however. In many cases, the experiment has been performed and the data have been collected, and inference needs to be performed based on these data alone. This is where assuming a model is crucial. Indeed, if we are able to derive (or approximate) the distribution of the test statistic under any null distribution, then we can compute (or approximate) the p-value, at least in principle, without having to repeat the experiment.

Proposition 12.16 (Monotone transformation). *Consider a test statistic S such that large values of S provide evidence against the null hypothesis. Let g be strictly increasing (resp. decreasing). Then large (resp. small) values of $g(S)$ provide evidence against the null hypothesis and the resulting p-value is equal to that based on S.*

Proof Suppose, for instance, that g is strictly increasing and let $T = g(S)$. Suppose that the experiment resulted in ω. Then S is observed to be $s := S(\omega)$ and T is observed to be $t := T(\omega)$. Noting that $t = g(s)$, for any possible outcome ω', we have

$$T(\omega') \geq t \;\Leftrightarrow\; g(S(\omega')) \geq g(s) \;\Leftrightarrow\; S(\omega') \geq s.$$

From this we obtain $\mathsf{pv}_T(t) = \mathsf{pv}_S(s)$, which proves the result. $\quad\square$

Example (Binomial experiment) Consider testing the hypothesis $\mathcal{H}_0$ of (12.9). As a test statistic, we use the MLE, or equivalently, Y, with large values of Y weighing against $\mathcal{H}_0$. If the data are ω, $y := Y(\omega)$ is the observed value of the test statistic, and the resulting p-value is $\sup_{\theta \leq \theta_0} \mathbb{P}_\theta(Y \geq y)$. By Problem 4.24, this supremum is equal to $\mathbb{P}_{\theta_0}(Y \geq y)$, which can be computed by reference to $\mathrm{Bin}(n, \theta_0)$.

Problem 12.17 (Near equivalence of the MLE and LR) In Example 12.4.2, we saw that the MLE and the LR were not, strictly speaking, equivalent. Although they may yield different p-values, show that these coincide when they are smaller than $\mathbb{P}_{\theta_0}(S \geq \theta_0)$ (which is close to 0.5).

The p-value associated with a test statistic S is itself a statistic when seen as the random variable $\omega \mapsto \mathrm{pv}_S(S(\omega))$. As such, it satisfies the following important property.

Proposition 12.18. $P := \mathrm{pv}_S(S)$ *takes values in* $[0, 1]$ *and satisfies*

$$\sup_{\theta \in \Theta_0} \mathbb{P}_\theta (P \leq \alpha) \leq \alpha, \quad \text{for all } \alpha \in [0, 1]. \tag{12.12}$$

Proof Assume that $\alpha < 1$, for otherwise there is nothing to prove, and consider $u \in (\alpha, 1)$. Let F_θ and F_θ^- denote the distribution function and quantile function of S under θ, and let $\widetilde{\mathsf{F}}_\theta$ be defined as in Problem 3.15. Note that $P = \sup_{\theta \in \Theta_0} \widetilde{\mathsf{F}}_\theta(S)$. Fix any θ in Θ_0. Starting with the fact that $P \geq \widetilde{\mathsf{F}}_\theta(S)$, we have

$$\begin{aligned}
\mathbb{P}_\theta(P \leq \alpha) \leq \mathbb{P}_\theta(P < u) &\leq \mathbb{P}_\theta(\widetilde{\mathsf{F}}_\theta(S) < u) \\
&\leq \mathbb{P}_\theta(S > \mathsf{F}_\theta^-(1 - u)) \\
&= 1 - \mathsf{F}_\theta(\mathsf{F}_\theta^-(1 - u)) \\
&\leq 1 - (1 - u) = u.
\end{aligned}$$

In the second line we used the conclusions of Problem 3.15, and in the fourth we used the conclusions of Problem 3.14. Since $u > \alpha$ and $\theta \in \Theta_0$ are arbitrary, the proof is complete.   □

In general, we say that a random variable P is a *valid p-value* if Proposition 12.18 applies.

12.4.4 Tests

A test statistic yields a p-value that is used to quantify the amount of evidence in the data against the postulated null hypothesis. Sometimes, though, the end goal is actual decision: 'reject or not reject' is the question. Tests formalize such decisions.

Formally, a *test* is a statistic with values in $\{0, 1\}$, returning 1 if the null hypothesis is to be rejected and 0 otherwise. If large values of a test statistic S provide evidence against the null, then a test based on S will be of the

form
$$\phi(\omega) = \{S(\omega) > c\}.$$

The event $\{S > c\}$ is called the *rejection region* of the test. The threshold c is typically referred to as the *critical value*.

Test Errors

When applying a test to data, two types of error are possible:

- A *Type I error*, or *false positive*, happens when the test rejects even though the null hypothesis is true.
- A *Type II error*, or *false negative*, happens when the test does not reject even though the null hypothesis is false.

Table 12.1 illustrates the situation.

Table 12.1 *Types of error that a test can make.*

	Null is true	Null is false
Rejection	type I	correct
No rejection	correct	type II

For a test that rejects for large values of a statistic, the choice of critical value drives the probabilities of Type I and Type II errors. Qualitatively speaking, increasing the critical value makes the test reject less often, which decreases the probability of Type I error and increases the probability of Type II error. Of course, decreasing the critical value has the opposite effect.

12.4.5 Level

The *size* of a test ϕ is the maximum probability of Type I error,

$$\text{size}(\phi) := \sup_{\theta \in \Theta_0} \mathbb{P}_\theta(\phi = 1).$$

Let $\alpha \in [0, 1]$ denote the desired control on the probability of Type I error, called the *significance level*. A test ϕ is said to have level α if its size is bounded by α,

$$\text{size}(\phi) \leq \alpha.$$

Given a test statistic S whose large values weigh against the null hypothesis, the corresponding test has level α if the critical value c satisfies

$$\sup_{\theta \in \Theta_0} \mathbb{P}_\theta(S > c) \leq \alpha.$$

In order to minimize the probability of Type II error, we want to choose the smallest c that satisfies this requirement. This smallest c exist due to the fact that, for any θ, $c \mapsto \mathbb{P}_\theta(S > c)$ is lower semi-continuous (Problem 3.13) and the supremum of lower semi-continuous functions is itself lower semi-continuous. Let c_α denote that value, so that the resulting test has rejection region $\{S > c_\alpha\}$. This rejection region may be equivalently expressed directly in terms of the associated p-value.

Problem 12.19 Show that $\{S > c_\alpha\} = \{\mathsf{pv}_S(S) \leq \alpha\}$.

In any case, if P is a valid p-value, then in view of Proposition 12.18, the test with rejection region $\{P \leq \alpha\}$ has level α.

Example (Binomial experiment) Continuing with the same setting, and recalling that we use as test statistic the number of heads, Y, and that we reject for large values of that statistic, the critical value at level α is given by

$$
\begin{aligned}
c_\alpha &= \min\left\{c : \sup_{\theta \leq \theta_0} \mathbb{P}_\theta(Y > c) \leq \alpha\right\} \\
&= \min\left\{c : \inf_{\theta \leq \theta_0} \mathbb{P}_\theta(Y \leq c) \geq 1 - \alpha\right\} \\
&= \min\left\{c : \mathbb{P}_{\theta_0}(Y \leq c) \geq 1 - \alpha\right\},
\end{aligned}
$$

using the conclusions of Problem 4.24 at the end. Therefore, in view of (3.9), c_α is the $(1 - \alpha)$-quantile of $\mathrm{Bin}(n, \theta_0)$.

Controlling the level is equivalent to controlling the number of false alarms, which is crucial in applications.

Example 12.20 (Security at Y-12) We learn in [162] that, in 2012, three activists (including an 82-year-old nun) broke into the Y-12 National Security Complex in Oak Ridge, Tennessee. "Y-12 is the only industrial complex in the United States devoted to the fabrication and storage of weapons-grade uranium. Every nuclear warhead and bomb in the American arsenal contains uranium from Y-12." This is a highly guarded complex and the activists did set up an alarm, but there were several hundred false alarms per month. ([207] claims there were upward of 2000 alarms per day.) This reminds one of the allegory of The Boy Who Cried Wolf.

12.4.6 Power

The *power* of a test ϕ at $\theta \in \Theta$ is

$$\mathsf{pwr}_\phi(\theta) := \mathbb{P}_\theta(\phi = 1).$$

If the test is of the form $\phi = \{S > c\}$, then

$$\mathsf{pwr}_\phi(\theta) = \mathbb{P}_\theta(S > c).$$

Remark 12.21 The test ϕ has level α if and only if

$$\mathsf{pwr}_\phi(\theta) \le \alpha, \quad \text{for all } \theta \in \Theta_0.$$

Problem 12.22 (Binomial experiment) Consider the problem of testing the hypothesis $\mathscr{H}_0$ of (12.9) and continue to use the test derived from Y. Set the level at $\alpha = 0.01$ and, in R, plot the power as a function of $\theta \in [0, 1]$. Do this for $n \in \{10, 20, 50, 100, 200, 500, 1000\}$. Repeat, now setting the level at $\alpha = 0.10$ instead.

Desiderata A good test has large power against alternatives of interest when compared to other tests at the same significance level.

12.4.7 From Confidence Regions to Tests and Back

There is an important equivalence between confidence regions and tests of hypotheses. We focus our attention on the parameter θ itself.

A Confidence Region gives a Test

Suppose that I is a $(1 - \alpha)$-confidence region for θ. Define $\phi = \{\Theta_0 \cap I = \varnothing\}$, which is clearly a test. Moreover, it has level α. To see this, take $\theta \in \Theta_0$ and derive

$$\phi = 1 \iff \Theta_0 \cap I = \varnothing \implies \theta \notin I,$$

so that

$$\mathbb{P}_\theta(\phi = 1) \le \mathbb{P}_\theta(\theta \notin I) \le \alpha,$$

where the last inequality is due to the fact that I is a $(1 - \alpha)$-confidence region for θ.

A Family of Confidence Regions gives a P-Value

Consider a family of confidence regions for θ, which are denoted $\{I_\gamma : \gamma \in [0, 1]\}$, such that I_γ has confidence level γ and $I_\gamma \subset I_{\gamma'}$ when $\gamma \le \gamma'$. Then define

$$P := \sup \{\alpha : \Theta_0 \cap I_{1-\alpha} \neq \varnothing\}.$$

Note that this is a random variable with values in $[0, 1]$.

Proposition 12.23. *This quantity is a valid p-value in the sense that it satisfies* (12.12).

Proof Take any $\theta \in \Theta_0$ and any $\alpha \in [0, 1]$. We need to prove that

$$\mathbb{P}_\theta(P \leq \alpha) \leq \alpha. \tag{12.13}$$

By definition of P, we have $\Theta_0 \cap I_{1-u} = \varnothing$ for any $u > P$. In particular, for any $u > \alpha$,

$$\begin{aligned} \mathbb{P}_\theta(P \leq \alpha) &\leq \mathbb{P}_\theta(\Theta_0 \cap I_{1-u} = \varnothing) \\ &\leq \mathbb{P}_\theta(\theta \notin I_{1-u}) \\ &\leq 1 - (1 - u) = u, \end{aligned}$$

using at the very end the fact that I_{1-u} is a $(1 - u)$-confidence region for θ. The above being true for all $u > \alpha$, we obtain (12.13). □

A Family of Tests gives a Confidence Region

For each $\theta \in \Theta$, suppose we have available a level-α test denoted ϕ_θ, for the null hypothesis that $\theta_* = \theta$. Thus, ϕ_θ is testing the null hypothesis that the true value of the parameter is θ. Define

$$I(\omega) = \{\theta : \phi_\theta(\omega) = 0\}, \tag{12.14}$$

which is the set of θ whose associated test does not reject. Then I is a $(1 - \alpha)$-confidence region for θ, meaning it satisfies (12.6). To see that, take any $\theta \in \Theta$. By definition of I and the fact that each ϕ_θ has level α, we get

$$\mathbb{P}_\theta(\theta \notin I) = \mathbb{P}_\theta(\phi_\theta = 1) \leq \alpha.$$

Example (Binomial experiment) We continue to use Y as the test statistic. In line with how we tested (12.9), let ϕ_θ denote the level-α test based on Y for testing the null hypothesis that $\theta_* \leq \theta$. Note that ϕ_θ is also level α for testing $\theta_* = \theta$. Let F_θ and F_θ^- denote the distribution and quantile functions of $\text{Bin}(n, \theta)$. We saw that $\phi_\theta = \{Y > F_\theta^-(1 - \alpha)\}$, which implies that $I = \{\theta : Y \leq F_\theta^-(1 - \alpha)\}$. Define

$$\underline{S} := \inf\{\theta : Y \leq F_\theta^-(1 - \alpha)\}.$$

Note that $\underline{S}$ is a random variable through Y. Using the conclusions of Problem 4.26, we conclude that

$$I = (\underline{S}, 1]. \tag{12.15}$$

Problem 12.24 (Binomial one-sided confidence interval) Write an R function that computes this interval based on y and α as inputs. A simple grid search works: at each θ in a grid of values, one checks whether $y \le F_\theta^-(1-\alpha)$. However, taking advantage of the monotonicity established in Problem 4.26, a bisection search is applicable, which is much more efficient.

12.5 Further Topics

12.5.1 Likelihood Methods when the Model is Continuous

In our exposition of likelihood methods – ML estimation and likelihood ratio testing – we have assumed that the statistical model was discrete in the sense that the sample space Ω was discrete.

Consider the common situation where the experiment results in a d-dimensional random vector $X(\omega)$ and that our inference is based on that random vector. Assume that X has an absolute continuous distribution and let f_θ denote its density under $\mathbb{P}_\theta$.

The likelihood methods are defined as before with the density replacing the mass function. This is (at least morally) justified by the fact that, in a continuous setting, a density plays the role of a mass function. In more detail, letting $x = X(\omega)$, the *likelihood function* is now defined as

$$\mathrm{lik}(\theta) := f_\theta(x).$$

Based on this new definition of the likelihood, the MLE and the LR are defined as they were before.

12.5.2 Two-Sided P-Value

It is sometimes the case that large and small values of a test statistic provide evidence against the null hypothesis. For example, in the binomial experiment of Example 12.2, consider testing

$$\mathcal{H}_0 : \theta_* = \theta_0.$$

Problem 12.25 Show that the LR is an increasing function of

$$\hat{\theta} \log\left(\frac{\hat{\theta}}{\theta_0}\right) + (1 - \hat{\theta}) \log\left(\frac{1 - \hat{\theta}}{1 - \theta_0}\right),$$

where $\hat{\theta}$ denotes the maximum likelihood estimate as in Example 12.4.2.

Thus, the application of the LR procedure is as straightforward as it was in the one-sided situation considered earlier. In particular, the corresponding test rejects for large values of the LR and the associated p-value is computed based on the distribution of the LR under $\mathbb{P}_{\theta_0}$.

However, let's look directly at the MLE. A situation in which $\hat{\theta}$ is quite large, or if it is quite small, compared to θ_0, provides evidence against the null hypothesis. In such a situation, the p-value can be defined in a number of ways. A choice is twice the minimum of the two one-sided p-values, namely

$$2 \min \left\{ \mathbb{P}_{\theta_0}(Y \geq y), \mathbb{P}_{\theta_0}(Y \leq y) \right\},$$

where Y is the total number of heads in the sequence and $y = Y(\omega)$ is the observed value of Y, as before.

Problem 12.26 Compare this p-value with the p-value resulting from using the LR. [They are distinct, in general.]

12.6 Additional Problems

Problem 12.27 (German Tank Problem) Suppose we have an iid sample from the uniform distribution on $\{1, \ldots, \theta\}$, where $\theta \in \mathbb{N}$ is unknown. Derive the MLE for θ. [The name of the problem comes from World War II, where the Western Allies wanted to estimate the total number of German tanks in operation (θ here) based on the serial numbers of captured or destroyed German tanks. In such a setting, is the assumption of iid-ness reasonable?]

Problem 12.28 (Gosset's experiment) Fit a Poisson model to the data of Table 4.1 by maximum likelihood. Add a row to the table to display the corresponding expected counts so as to easily compare them with the actual counts. In R, do this visually by drawing side-by-side bar plots with different colors and a legend.

Problem 12.29 (Rutherford's experiment) Same as in Problem 12.28, but with the data of Table 4.2.

13

Properties of Estimators and Tests

In this chapter, we introduce and briefly discuss some properties of estimators and tests. This includes different notions of optimality, meaning, different ways comparing methods addressing the same statistical problem.

13.1 Sufficiency

We have referred to the setting of Example 12.2 as a binomial experiment. The main reason is that the binomial distribution is at the very center of the resulting statistical inference. This was a consequence of us relying on the number of heads, Y, which has the binomial distribution with parameters (n, θ). Surely, we could have based our inference on a different statistic. However, there is a fundamental reason that inference should be based on Y: because Y contains all the information about θ that we can extract from the experiment. This is rather intuitive because, in going from the sequence of tosses ω to the number of heads $Y(\omega)$, all that is lost is the position of the heads in the sequence, that is, the order. But, because the tosses are assumed iid, the order cannot provide any information on the parameter θ.

More formally, assume a statistical model $\{\mathbb{P}_\theta : \theta \in \Theta\}$. We say that a k-dimensional vector-valued statistic $\boldsymbol{Y}$ is *sufficient* for this family if

$$\text{For any event } \mathcal{A} \text{ and any } \boldsymbol{y} \in \mathbb{R}^k,$$
$$\mathbb{P}_\theta(\mathcal{A} \mid \boldsymbol{Y} = \boldsymbol{y}) \text{ does not depend on } \theta.$$

If $\boldsymbol{Y} = (Y_1, \ldots, Y_k)$, we say that the statistics $Y_1, \ldots, Y_k$ are *jointly sufficient*.

Thus, intuitively, a statistic $\boldsymbol{Y}$ is sufficient if the randomness left after conditioning on $\boldsymbol{Y}$ does not depend on the value of θ, so that this leftover randomness cannot be used to improve the inference.

The following result provides what is arguably the most direct way to establish that a statistic is sufficient.

Theorem 13.1 (Factorization criterion). *Consider a family of either mass functions or densities, $\{f_\theta : \theta \in \Theta\}$, over a sample space Ω. Then a statistic Y is sufficient for this model if and only if there are functions $\{g_\theta : \theta \in \Theta\}$ and h such that, for all $\theta \in \Theta$,*

$$f_\theta(\omega) = g_\theta(Y(\omega))h(\omega), \quad \text{for all } \omega \in \Omega.$$

Problem 13.2 In the binomial experiment (Example 12.2), show that the number of heads is sufficient.

Problem 13.3 In the German tank problem (Problem 12.27), show that the maximum of the observed (serial) numbers is sufficient.

Problem 13.4 More generally, in the context of this theorem, show that the maximum likelihood estimator for θ, assuming it exists, is a function of Y.

13.2 Consistency

The notion of consistency is best understood in an asymptotic model where the sample size becomes large. This can be confusing, as in a given experiment we only have access to a finite sample, which is fixed. We adopt here a rather formal stance for the sake of clarity.

Consider an experiment which consists of multiple, otherwise identical trials. If $(\Omega_0, \Sigma_0, \mathcal{P}_0)$ models the result of a single trial, $\mathcal{P}_0 = \{\mathbb{P}_{0,\theta} : \theta \in \Theta\}$, then an experiment consisting of n independent such trials is modeled by $(\Omega_n, \Sigma_n, \mathcal{P}_n)$, where $\Omega_n = \Omega_0^n$, where Σ_n is the algebra generated by sets of the form $\mathcal{A}_1 \times \cdots \times \mathcal{A}_n$ with $\mathcal{A}_i \in \Sigma_0$, and where $\mathcal{P}_n$ is the family of distributions of the form $\mathbb{P}_{n,\theta} := \mathbb{P}_{0,\theta}^{\otimes n}$, which is the *product distribution* uniquely defined by

$$\mathbb{P}_{0,\theta}^{\otimes n}(\mathcal{A}_1 \times \cdots \times \mathcal{A}_n) = \mathbb{P}_{0,\theta}(\mathcal{A}_1) \times \cdots \times \mathbb{P}_{0,\theta}(\mathcal{A}_n), \quad \text{for all } \mathcal{A}_i \in \Sigma.$$

Such an experiment results in an outcome of the form $(\omega_1, \ldots, \omega_n)$ with $\omega_i \in \Omega_0$, that is, a sample of size n drawn iid from some $\mathbb{P}_{0,\theta}$. A prototypical example of this is the binomial experiment of Example 12.2.

A *statistical procedure* is a sequence of statistics, therefore of the form $S = (S_n)$ with S_n being a statistic defined on Ω_n. An example of procedure in the context of estimation is the maximum likelihood method. An example of procedure in the context of testing is the likelihood ratio method.

13.2.1 Consistent Estimators

We say that an procedure $S = (S_n)$ is *consistent* for estimating $\varphi(\theta)$ if, for any $\varepsilon > 0$,

$$\mathbb{P}_{n,\theta}(|S_n - \varphi(\theta)| \geq \varepsilon) \to 0, \quad \text{as } n \to \infty.$$

Problem 13.5 (Binomial experiment) In Example 12.2, consider the task of estimating the parameter $\varphi(\theta) = \theta$. Show that the maximum likelihood method yields an estimation procedure that is consistent. [This is a simple consequence of the Law of Large Numbers.]

In fact, the maximum likelihood estimator (MLE) is consistent under fairly broad assumptions. Note that if the MLE for θ is well-defined and consistent, the same must be true for the MLE for $\varphi(\theta)$ if φ is continuous.

Problem 13.6 Consider a multiple trial experiment where a single trial is modeled by an identifiable family of mass functions $\{f_\theta : \theta \in \Theta\}$ on a finite sample space Ω. Assume that Θ is a compact subset of some Euclidean space; that $f_\theta(\omega) > 0$ and that $\theta \mapsto f_\theta(\omega)$ is continuous, for all $\omega \in \Omega$. Prove that the MLE for θ is well-defined and consistent.

Problem 13.7 (Population genetics) Consider a diploid population under Hardy–Weinberg equilibrium. Thus, at any given locus with two possible alleles denoted A and a with respective frequencies p and q, under random mating the frequencies of the possible allele pairs AA, aa, and Aa are p^2, q^2, and pq, respectively. Suppose that n individuals are sampled uniformly at random from the population. For each one of these individuals the pair of alleles at a given locus of interest is determined. Derive the MLE for the proportions of AA, aa, and Aa in the population. [Take the population to be infinite.]

Problem 13.8 (Beyond iid) We have focused on the iid setting, but the question of consistency can be formalized in other settings. For an example that resembles a binomial experiment but where the trials are not independent, consider a two-state Markov chain as in Section 9.1.2. The experiment consists in running such a chain for n steps starting at State 1. Derive the MLE for the parameters defining the chain, and show that it is consistent.

13.2.2 Consistent Tests

We say that a procedure $S = (S_n)$ is *consistent* for testing $\mathcal{H}_0 : \theta \in \Theta_0$ if there is a sequence of critical values (c_n) such that

$$\lim_{n \to \infty} \mathbb{P}_{n,\theta}(S_n \geq c_n) = 0, \quad \text{for all } \theta \in \Theta_0,$$

$$\lim_{n \to \infty} \mathbb{P}_{n,\theta}(S_n \geq c_n) = 1, \quad \text{for all } \theta \notin \Theta_0.$$

(We have implicitly assumed that large values of S_n weigh against the null hypothesis. This can be done without loss of generality.)

Problem 13.9 (Binomial experiment) In Example 12.2, consider the task of testing $\mathcal{H}_0 : \theta_* \leq \theta_0$. Show that the likelihood ratio (LR) method yields a procedure that is consistent for $\mathcal{H}_0$.

The likelihood ratio test (LRT) is also consistent under fairly broad assumptions.

Problem 13.10 State and prove a consistency result for the LRT under conditions similar to those in Problem 13.6.

Problem 13.11 In the context of Problem 13.8, derive the LRT for testing the null hypothesis that the sequence of trials is iid, and show that the test is consistent.

13.3 Notions of Optimality for Estimators

Given a loss function $\mathcal{L}$, the risk of an estimator S for θ is defined in (12.5). Obviously, the smaller the risk the better the estimator. However, this presents a difficulty since the risk is a function of θ rather than just a number.

13.3.1 Maximum Risk and Average Risk

We present two ways of reducing the risk function to a number.

Maximum Risk

The first avenue is to consider the *maximum risk*, namely the maximum of the risk function,

$$\mathcal{R}_{\max}(S) := \sup_{\theta \in \Theta} \mathcal{R}_\theta(S).$$

An estimator that minimizes the maximum risk, if one exists, is said to be *minimax*, and its risk is called the *minimax risk*, denoted $\mathcal{R}_{\max}^*$.

Average Risk

The second avenue is to consider the *average risk*. To do so, we need to choose a distribution on the parameter space. (A distribution on the parameter space is often called a *prior*.) Assuming that Θ is a subset of some Euclidean space, let λ be a density supported on Θ. We can then consider the average risk with respect to λ,

$$\mathcal{R}_\lambda(S) := \int_\Theta \mathcal{R}_\theta(S)\lambda(\theta)\mathrm{d}\theta.$$

An estimator that minimizes this average risk is called a *Bayes estimator* (with respect to λ) and its risk is called the *Bayes risk*, denoted $\mathcal{R}_\lambda^*$.

A Bayes estimator is often derived as follows. For simplicity, assume a family of densities $\{f_\theta : \theta \in \Theta\}$ on some Euclidean sample space Ω and that we are estimating θ itself. Applying the Fubini–Tonelli Theorem, we have

$$\mathcal{R}_\lambda(S) = \int_\Theta \int_\Omega \mathcal{L}(S(\omega), \theta) f_\theta(\omega)\mathrm{d}\omega\lambda(\theta)\mathrm{d}\theta$$
$$= \int_\Omega \int_\Theta \mathcal{L}(S(\omega), \theta) f_\theta(\omega)\lambda(\theta)\mathrm{d}\theta\mathrm{d}\omega.$$

Thus, if the following is well-defined,

$$S_\lambda(\omega) := \arg\min_s \int_\Theta \mathcal{L}(s, \theta) f_\theta(\omega)\lambda(\theta)\mathrm{d}\theta,$$

it is readily seen to minimize the λ-average risk.

Problem 13.12 (Binomial experiment) Consider a binomial experiment as in Example 12.9 and the estimation of θ under squared error loss.

(i)   Compute the maximum risk of the MLE.

(ii)   Compute the average risk of the MLE with respect to the uniform distribution on $[0, 1]$ (which is the parameter space).

Maximum risk optimality and average risk optimality are intricately connected. For example, for any prior λ,

$$\mathcal{R}_\lambda(S) \le \mathcal{R}_{\max}(S), \quad \text{for all } S,$$

and this immediately implies that

$$\mathcal{R}_\lambda^* \le \mathcal{R}_{\max}^*.$$

Problem 13.13 Show that an estimator S is minimax when there is a sequence of priors (λ_k) such that $\liminf_{k \to \infty} \mathcal{R}_{\lambda_k}^* \ge \mathcal{R}_{\max}(S)$.

Problem 13.14 Use the previous problem to show that a Bayes estimator with constant risk function is necessarily minimax.

Problem 13.15 In a binomial experiment, derive a minimax estimator. To do so, find a prior in the beta family such that the resulting Bayes estimator has constant risk function. Using R, produce a graph comparing the risk functions of this estimator and that of the MLE for various values of n.

13.3.2 Admissibility

We say that an estimator S is *inadmissible* if there is an estimator T such that

$$\mathcal{R}_\theta(T) \leq \mathcal{R}_\theta(S), \quad \text{for all } \theta \in \Theta,$$

and the inequality is strict for at least one $\theta \in \Theta$. Otherwise, if no such estimator T exists, we say that the estimator S is *admissible*.

At least in theory, if an estimator is inadmissible, it can be replaced by another estimator that is uniformly better in terms of risk. However, even then, there might be other reasons for using an inadmissible estimator, such as simplicity or ease of computation.

Admissibility is interrelated with maximum risk and average risk optimality.

Problem 13.16 Show that an estimator that is unique Bayes for some prior is necessarily admissible.

Problem 13.17 Show that an estimator that is unique minimax is necessarily admissible.

13.3.3 Risk Unbiasedness

We say here that an estimator S is *risk unbiased* if

$$\mathbb{E}_\theta[\mathcal{L}(S,\theta)] \leq \mathbb{E}_\theta[\mathcal{L}(S,\theta')].$$

In words, this means that S is on average as close (as measured by the loss function) to the true value of the parameter as any other value.

Mean Unbiased Estimators

Suppose that $\mathcal{L}$ is the squared error loss, meaning $\mathcal{L}(s,\theta) = (s - \varphi(\theta))^2$.

Problem 13.18 Show that S is risk unbiased if and only if

$$\mathbb{E}_\theta(S) = \varphi(\theta), \quad \text{for all } \theta \in \Theta.$$

Such estimators are said to be *mean-unbiased*, or more commonly, simply *unbiased*.

From the bias-variance decomposition (12.4), an estimator with small mean squared error (MSE) has necessarily small bias (and also small variance). Therefore, a small bias is desirable. However, strict unbiasedness is not necessarily desirable, for a small MSE does not imply unbiasedness. In fact, unbiased estimators may not even exist.

Problem 13.19 Consider a binomial experiment as in Example 12.9. Show that there is an unbiased estimator of $\varphi(\theta)$ if and only if φ is a polynomial of degree at most n.

Median Unbiased Estimators

Suppose that $\mathcal{L}$ is the absolute loss, meaning $\mathcal{L}(s, \theta) = |s - \varphi(\theta)|$.

Problem 13.20 Show that S is risk unbiased if and only if, for all $\theta \in \Theta$, $\varphi(\theta)$ is a median of S under $\mathbb{P}_\theta$.

Such estimators are said to be *median-unbiased*.

Problem 13.21 Show that if S is median-unbiased for $\varphi(\theta)$, then for any strictly monotone function $g: \mathbb{R} \to \mathbb{R}$, $g(S)$ is median-unbiased for $g(\varphi(\theta))$. Show by exhibiting a counter-example that this is no longer true if 'median' is replaced with 'mean'.

Problem 13.22 Consider a binomial experiment as before and the estimation of θ. Is the MLE median-unbiased? Explore this question both analytically and via numerical simulations on a computer.

13.4 Notions of Optimality for Tests

Our discussion of optimality for estimators has parallels for tests. Indeed, once the level is under control, the larger the power (i.e., the more the test rejects) the better. However, this is quantified by a power function and not a simple number.

We consider a statistical model as in (12.1) and consider testing $\mathcal{H}_0$: $\theta \in \Theta_0$. Unless otherwise specified, Θ_1 is the complement of Θ_0 in Θ, as in (12.8).

Remark 13.23 (Level requirement) We assume that all tests that appear below have level α. It is crucial that the tests being compared satisfy this requirement, for otherwise the comparison may not be accurate or even fair.

13.4.1 Minimum Power and Average Power

There are various ways of reducing the power function to a single number.

Minimum Power

A first avenue is to consider the *minimum power*,

$$\inf_{\theta \in \Theta_1} \mathbb{P}_\theta(\phi = 1).$$

In a number of classical models, Θ is a domain of a Euclidean space and the power function of any test is continuous over Θ. In such a setting, if there is no 'separation' between Θ_0 and Θ_1, then any test has minimum power bounded from above by its size, which is not very interesting. The consideration of minimum power is thus only relevant when there is a 'separation' between the null and alternative sets.(More formally, we say that there is no separation between two sets of a Euclidean space if their closures intersect. For example, there is no separation between the intervals $[0, 1)$ and $(1, 2]$.)

Average Power

A second avenue is to consider the *average power*. Let λ be a density on Θ_1. We can then average the power with respect to λ,

$$\int_{\Theta_1} \mathbb{P}_\theta(\phi = 1)\lambda(\theta)\mathrm{d}\theta.$$

Problem 13.24 (Binomial experiment) Consider a binomial experiment as in Example 12.4.2 and the testing of $\Theta_0 := [0, \theta_0]$ versus $\Theta_1 := (\theta_1, 1]$, where $\theta_0 \le \theta_1$ are given. For the level-α test based on rejecting for large values of Y:

(i) Compute the minimum power when $\theta_1 > \theta_0$ (so there is a separation).
(ii) Compute the average power for the uniform distribution on Θ_1.

Answer these questions analytically and also numerically, using R.

13.4.2 Admissibility

We say that a test ϕ is *inadmissible* if there is a test ψ such that

$$\mathbb{P}_\theta(\phi = 1) \le \mathbb{P}_\theta(\psi = 1), \quad \text{for all } \theta \in \Theta_1,$$

and the inequality is strict for at least one $\theta \in \Theta_1$.

At least in theory, if a test is inadmissible, it can be replaced by another test that is uniformly better in terms of power. However, even then, there might be other reasons for using an inadmissible test, such as simplicity or ease of computation.

13.4.3 Uniformly Most Powerful Tests

Consider testing $\theta \in \Theta_0$. A test ϕ is said to be *uniformly most powerful* (UMP) among level-α tests if ϕ itself has level α and is at least as powerful as any other level-α test, meaning that for any other test ψ with level α,

$$\mathbb{P}_\theta(\phi = 1) \geq \mathbb{P}_\theta(\psi = 1), \quad \text{for all } \theta \in \Theta_1.$$

This is clearly the best one can hope for. However, a UMP test seldom exists.

Simple vs Simple

A very particular case where a UMP test exists is when both the null set and alternative set are singletons. We say that a hypothesis is *simple* if the corresponding parameter subset is a singleton; it is said to be *composite* otherwise. Suppose, therefore, that $\Theta_0 = \{\theta_0\}$ and $\Theta_1 = \{\theta_1\}$, and let $\mathbb{Q}_j$ be short for $\mathbb{P}_{\theta_j}$.

The following is a consequence of the *Neyman–Pearson Lemma*[55] – one of the most celebrated results in the theory of tests. Recall that a likelihood ratio test is any test that rejects for large values of the likelihood ratio.

Theorem 13.25. *In the present context, any LRT is UMP at level α equal to its size.*

To understand where the result comes from, suppose we want to test at a prescribed level α in a situation where the sample space is discrete. In that case, we want to solve the following optimization problem

$$\begin{aligned} \text{maximize} \quad & \sum_{\omega \in \mathcal{R}} \mathbb{Q}_1(\omega) \\ \text{subject to} \quad & \sum_{\omega \in \mathcal{R}} \mathbb{Q}_0(\omega) \leq \alpha. \end{aligned}$$

The optimization is over subsets $\mathcal{R} \subset \Omega$, which represent candidate rejection regions. In that case, it makes sense to rank $\omega \in \Omega$ according to its likelihood ratio value $L(\omega) := \mathbb{Q}_1(\omega)/\mathbb{Q}_0(\omega)$. If we denote by $\omega_1, \omega_2, \ldots$ the elements

---

[55] Named after Jerzy Neyman (1894–1981) and Egon Pearson (1895–1980).

of Ω in decreasing order, meaning that $L(\omega_1) \geq L(\omega_2) \geq \cdots$, and define the regions $\mathcal{R}_k = \{\omega_1, \ldots, \omega_k\}$, then it makes intuitive sense to choose

$$k_\alpha := \max\{k : \mathbb{Q}_0(\omega_1) + \cdots + \mathbb{Q}_0(\omega_k) \leq \alpha\}.$$

This is correct when the level can be exactly achieved in this fashion. Otherwise, the optimization is more complex, leading to a linear program.

In a different setting where the sample space is a subset of some Euclidean space, suppose that $\mathbb{Q}_1$ has density f_1 and that $\mathbb{Q}_0$ has density f_0. In that case, the likelihood ratio is defined as $L := f_1/f_0$. If L has continuous distribution under $\mathbb{Q}_0$, by choosing the critical value c appropriately, any prescribed level α can be attained. As a consequence, if c is chosen such that $\mathbb{Q}_0(L > c) = \alpha$, then the test with rejection region $\{L > c\}$ is UMP at level α. However, the same cannot always be done if L does not have a continuous distribution under the null hypothesis.

Monotone Likelihood Ratio Property

As in Section 12.2.2, assume a discrete model for concreteness, although what follows applies more broadly. We consider a situation where Θ is an interval of $\mathbb{R}$.

The family of distributions $\{\mathbb{P}_\theta : \theta \in \Theta\}$ is said to have the *monotone likelihood ratio* (MLR) property in T if the model is identifiable and, for any $\theta < \theta'$, $\mathbb{P}_{\theta'}/\mathbb{P}_\theta$ is monotone increasing in T, meaning there is a non-decreasing function $g_{\theta,\theta'}: \mathbb{R} \to \mathbb{R}$ satisfying

$$\frac{\mathbb{P}_{\theta'}(\omega)}{\mathbb{P}_\theta(\omega)} = g_{\theta,\theta'}(T(\omega)), \quad \text{for all } \omega \in \Omega.$$

Theorem 13.26. *Assume that the family $\{\mathbb{P}_\theta : \theta \in \Theta\}$ has the MLR property in T and that the null hypothesis is of the form $\Theta_0 = \{\theta \in \Theta : \theta \leq \theta_0\}$. Then any test of the form $\{T > t\}$ has size $\mathbb{P}_{\theta_0}(T > t)$, and is UMP at level α equal to its size.*

Problem 13.27 Show that in the context of Theorem 13.26, the LR is a non-decreasing function of T and that, therefore, an LRT is UMP at level α equal to its size.

Problem 13.28 (Binomial experiment) Show that the MLR property holds in a binomial experiment, and that in particular, for testing $\mathcal{H}_0 : \theta_* \leq \theta_0$, a LRT is UMP at its size. Show that the same is true of a test based on the maximum likelihood estimator.

13.4.4 Unbiased Tests

As we said above, a UMP test rarely exists. In particular, it does not exist in the most of the important two-sided situations, including if the MLR property holds.

Consider a setting where Θ is an interval of the real line and the null hypothesis to be tested is $\mathcal{H}_0 : \theta_* = \theta_0$ for some given θ_0 in the interior of Θ. (If θ_0 is one of the boundary points, the situation is one-sided.) In such a situation, a UMP test would have to be at least as good as a UMP test for the one-sided null $\mathcal{H}_0^{\leq} : \theta_* \leq \theta_0$ and at least as good as a UMP test for the one-sided null $\mathcal{H}_0^{\geq} : \theta_* \geq \theta_0$. In most cases, this proves to be impossible.

In some sense this competition is unfair, because a test for the one-sided null such as $\mathcal{H}_0^{\leq}$ is ill-suited for the two-sided null $\mathcal{H}_0$. Indeed, if a test for $\mathcal{H}_0^{\leq}$ has level α, then the probability that it rejects when $\theta_* \leq \theta_0$ is bounded from above by α.

To prevent one-sided tests from competing in two-sided testing problems, one may restrict attention to so-called unbiased tests. A test is said to be *unbiased* at level α if it has level α, and the probability of rejecting at any $\theta \in \Theta_1$ is bounded from below by α.

While the number of situations where a UMP test exists is rather limited, there are many more situations where there exists a test that is UMP among unbiased tests. Such a test is said to be *uniformly most powerful unbiased* (UMPU).

An important class of situations where this occurs includes the case of *general exponential families*, where we work with a family of densities $\{f_\theta : \theta \in \Theta\}$ of the form

$$f_\theta(\omega) = A(\theta) \exp(a(\theta)T(\omega)) h(\omega),$$

where, in addition, $a : \Theta \to \mathbb{R}$ is strictly increasing. Suppose in this setting that the null set Θ_0 is a closed subinterval of Θ, possibly a singleton.

Theorem 13.29. *In the present setting, any test with rejection region of the form $\{T < t_1\} \cup \{T > t_2\}$, with $t_1 < t_2$, is UMPU at its size.*

Problem 13.30 Show that a binomial experiment leads to a general exponential family.

Remark 13.31 In a binomial experiment, an equal-tailed two-sided LRT is approximately UMPU for $\theta_* = \theta_0$ as long as θ_0 is not too close to 0 or 1. (The larger the number of trials n, the closer θ_0 can be to these extremes.)

14

One Proportion

Estimating a proportion is one of the most basic problems in statistics. Although basic, it arises in a number of important real-life situations:

- *Election polls* are conducted to estimate the proportion of people that will vote for a particular candidate.
- In *quality control*, the proportion of defective items manufactured at a particular plant or assembly line needs to be monitored, and one may resort to statistical inference to avoid having to check every single item.
- *Clinical trials* are conducted in part to estimate the proportion of people that would benefit (or suffer serious side effects) from receiving a particular treatment.

The situation is commonly modeled as sampling from an urn. The resulting distribution, as we know, depends on the contents of the urn and on how the sampling is done. The subsequent statistical analysis flows from the sampling model.

14.1 Binomial Experiments

We start with the binomial experiment of Example 12.2, which has served as our running example in the last two chapters. Remember that this is an experiment where a θ-coin is tossed a predetermined number of times n, and the goal is to infer the value of θ based on the outcome of the experiment (the data).

To summarize what we have learned about this model so far, in Chapter 12, we derived the maximum likelihood estimator (MLE), which is the sample proportion of heads, that is, $S = Y/n$. We then focused on testing null hypotheses of the form $\theta_* \leq \theta_0$ for a given θ_0. We considered tests which reject for large values of the MLE, meaning with rejection regions of the form $\{S > c\}$. Using the correspondence between tests and confidence intervals, we derived the (one-sided) confidence interval (12.15).

Problem 14.1 Adapt the arguments to testing null hypotheses of the form $\theta_* \geq \theta_0$ for a given θ_0 and derive the corresponding confidence interval at level $1 - \alpha$. The interval will be of the form

$$I = [0, \bar{S}). \tag{14.1}$$

14.1.1 Two-Sided Tests and Confidence Intervals

We now consider the two-sided situation. In brief, we study the problem of testing null hypotheses of the form $\mathcal{H}_0 : \theta_* = \theta_0$ for a given $\theta_0 \in (0, 1)$ and then derive a confidence interval as in Section 12.4.7. (If $\theta_0 = 0$ or $= 1$, the problem is one-sided.) We still use the MLE as test statistic.

Tests

For the null $\theta_* = \theta_0$, both large and small values of Y (the number of heads) are evidence against the null hypothesis. This leads us to consider tests of the form

$$\phi = \{Y < a \text{ or } Y > b\}. \tag{14.2}$$

The critical values, a and b, are chosen to control the level at some prescribed α, meaning

$$\mathbb{P}_{\theta_0}(\phi = 1) \leq \alpha.$$

Equivalently,

$$\mathbb{P}_{\theta_0}(Y < a) + \mathbb{P}_{\theta_0}(Y > b) \leq \alpha. \tag{14.3}$$

Note that a number of choices are possible. Two natural ones are

- *Equal tail*   This choice corresponds to choosing

$$a = \max\{a' : \mathbb{P}_{\theta_0}(Y < a') \leq \alpha/2\},$$
$$b = \min\{b' : \mathbb{P}_{\theta_0}(Y > b') \leq \alpha/2\},$$

  where the maximization and minimization are over integers.
- *Minimum length*   This choice corresponds to minimizing $b - a$ subject to the constraint (14.3).

Problem 14.2 Derive the likelihood ratio test (LRT) in the present context. You will find it is of the form (14.2) for particular critical values a and b (to be derived explicitly). How does the LRT compare with the equal-tail and minimum-length tests above?

Confidence Intervals

Let ϕ_θ be a test for $\theta_* = \theta$ as constructed above, meaning with rejection region of the form $\{Y < a_\theta\} \cup \{Y > b_\theta\}$, where

$$\mathbb{P}_\theta(Y < a_\theta) + \mathbb{P}_\theta(Y > b_\theta) \le \alpha.$$

As in Section 12.4.7, based on this family of tests we can obtain a $(1 - \alpha)$-confidence interval of the form

$$I = (\underline{S}, \bar{S}). \tag{14.4}$$

For building a confidence interval, the minimum-length choice for a_θ and b_θ is particularly appealing.

Problem 14.3 Derive this confidence interval.

The one-sided intervals (12.15) and (14.1), and the two-sided interval (14.4) with the equal-tail construction, are due to Clopper and Pearson [29]. The construction yields an exact interval in the sense that the desired confidence level is achieved.

R corner The Clopper–Pearson interval (one-sided or two-sided), and the related test, can be computed in R using the function binom.test.

We describe below other traditional ways of constructing confidence intervals. Compared to the Clopper–Pearson construction, they are less labor intensive although they are not as precise. They were particularly useful in the pre-computer age, and some of them are still in use.

Remark 14.4 We focus here on confidence intervals, from which we know tests can be derived as in Section 12.4.7.

14.1.2 Confidence Interval based on Chebyshev's Inequality

As we saw in Example 12.10, Chebyshev's inequality provides a simple, although crude way to derive a confidence interval. Indeed, to recapitulate, Chebyshev's inequality tells us that

$$\frac{|S - \theta|}{\sqrt{\theta(1 - \theta)/n}} < z, \tag{14.5}$$

with probability at least $1 - 1/z^2$ under $\mathbb{P}_\theta$.

Problem 14.5 Prove that (14.5) is equivalent to

$$\theta \in I_z := \left(S_z \pm z \frac{\sigma_z}{\sqrt{n}} \right), \tag{14.6}$$

where

$$S_z := \frac{S + z^2/2n}{1 + z^2/n}, \quad \sigma_z^2 := \frac{S(1 - S) + z^2/4}{(1 + z^2/n)^2}.$$

In particular, I_z defined in (14.6) is a $(1 - 1/z^2)$-confidence interval for θ.

14.1.3 Confidence Intervals based on the Normal Approximation

The Chebyshev's confidence interval is commonly believed to be too conservative, and practitioners have instead relied on the normal approximation to the binomial distribution instead of Chebyshev's inequality, which is deemed too crude. Let Φ denote the distribution function of the standard normal distribution (which we saw in (5.1)). Then, using the Central Limit Theorem, for any $z > 0$,

$$\mathbb{P}_\theta \left(\frac{|S - \theta|}{\sqrt{\theta(1 - \theta)/n}} < z \right) \approx 2\Phi(z) - 1, \tag{14.7}$$

as long as the sample size n is large enough. (More formally, the left-hand side converges to the right-hand side as $n \to \infty$.)

Wilson's Normal Interval

The construction of this interval, proposed by Wilson [205], is based on the large-sample approximation (14.7) and the derivations of Problem 14.5, which together imply that the interval defined in (14.6) has approximate confidence level $2\Phi(z) - 1$.

R corner Wilson's interval (one-sided or two-sided) and the related test, can be computed in R using the function prop.test.

The simpler variants that follow are also based on the approximation (14.7). However, they offer no advantage compared to Wilson's interval, except for simplicity if calculations must be done by hand. These constructions start by noticing that (14.7) implies

$$\mathbb{P}_\theta(\theta \in J_\theta) \approx 1 - \alpha, \tag{14.8}$$

where

$$J_\theta := \left(S \pm z_{1-\alpha/2} \frac{\sigma_\theta}{\sqrt{n}} \right), \tag{14.9}$$

where $z_u := \Phi^{-1}(u)$ and $\sigma_\theta^2 := \theta(1 - \theta)$. At this point, J_θ is *not* a confidence interval as its computation depends on θ, which is unknown.

Conservative Normal Interval

In our derivation of the interval (12.7), we used the fact that $\theta \in [0, 1] \mapsto \theta(1 - \theta)$ is maximized at $\theta = 1/2$. Therefore,

$$J_\theta \subset J_{1/2} = \left(S \pm \frac{z_{1-\alpha/2}}{2\sqrt{n}}\right).$$

$J_{1/2}$ can be computed without knowledge of θ and, because of (14.8), it achieves a confidence level of at least $1 - \alpha$ in the large-sample limit. Unless the true value of θ happens to be equal to $1/2$, this interval will be conservative in large samples.

Plug-in Normal Interval

It is very tempting to replace θ in (14.9) with S. After all, S is a consistent estimator of θ. The resulting interval is

$$J_S = \left(S \pm z_{1-\alpha/2} \frac{\sigma_S}{\sqrt{n}}\right).$$

J_S is a bona fide confidence interval. Moreover, (14.8), coupled with Slutsky's theorem (Problem 8.40), implies that J_S achieves a confidence level of $1 - \alpha$ in the large-sample limit. Note that this construction relies on two approximations.

Problem 14.6 Verify the claims made here.

This is a popular interval, and its half-width (traditionally at $\alpha = 0.95$) is often reported in polls and referred to as the *margin of error*.

14.2 Hypergeometric Experiments

Consider an experiment where balls are repeatedly drawn from an urn containing r red balls and b blue balls a predetermined number of times n. The total number of balls in the urn, $v := r + b$, is assumed known. The goal is to infer the proportion of red balls in the urn, namely, r/v. If the draws are with replacement, this is a binomial experiment with probability parameter $\theta = r/v$, a case that was treated in Section 14.1. We assume here that the draws are without replacement.

Let Y denote the number of red balls that are drawn. Assume that $n < v$, for otherwise the situation is descriptive as the experiment reveals the contents of the urn and there is no inference left to do. We call the resulting experiment a *hypergeometric experiment* because Y is hypergeometric and sufficient for this experiment.

Problem 14.7 Prove that Y is indeed sufficient in this model.

14.2.1 Maximum Likelihood Estimator (MLE)

Let y denote the realization of Y, meaning $y = Y(\omega)$ with ω denoting the observed outcome of the experiment. Recalling the definition of falling factorials (2.6), the likelihood is given by (see (2.9))

$$\text{lik}(r) = \frac{(r)_y (v - r)_{n-y}}{(v)_n},$$

where we used the fact that $b = v - r$. (As we saw before, although the likelihood is a function of y also, this dependency is left implicit to focus on the parameter r.)

Although it may look intimidating, this is a tame function. It suffices to consider r in the range $y \le r \le v - n + y$, for otherwise the likelihood is zero. (This is congruent with the fact that, having drawn y red balls and $n - y$ blue balls, we know that there were that many red and blue balls in the urn to start with.) For $r < v - n + y$, we have

$$\frac{\text{lik}(r + 1)}{\text{lik}(r)} = \frac{(r + 1)(v - r - n + y)}{(r - y + 1)(v - r)},$$

so that

$$\text{lik}(r + 1) \le \text{lik}(r) \iff r \ge \frac{yv - n + y}{n}.$$

Similarly, for $r > y$, we have

$$\text{lik}(r - 1) \le \text{lik}(r) \iff r \le \frac{yv + y}{n}.$$

We conclude that any r that maximizes the likelihood satisfies

$$-\frac{n - y}{nv} \le \frac{r}{v} - \frac{y}{n} \le \frac{y}{nv}. \tag{14.10}$$

More than necessary, this is also sufficient for r to maximize the likelihood.

Problem 14.8 Verify that r (integer) satisfies this condition if and only if r/v is closest to y/n. Show that there is only one such r, except when $y/n = k/(v + 1)$ for some integer k, in which case there are two such r. Conclude that, in any case, any r satisfying (14.10) maximizes the likelihood.

Note that r/v is the proportion of red balls in the urn, and thus the MLE is in essence the same as in a binomial experiment with $\theta = r/v$ as parameter, except that the proportion of reds is here an integer multiple of $1/v$.

14.2.2 Confidence Intervals

The various constructions of a confidence interval that we presented in the context of a binomial experiment apply almost verbatim in the context of a hypergeometric experiment. This is because the same normal approximation that applies to the binomial distribution with parameters (n, θ) also applies to the hypergeometric distribution with parameters $(n, r, v - r)$, with r/v in place of θ, if

$$v \to \infty, \quad r \to \infty, \quad n \to \infty,$$
$$\text{with } r/v \to \theta \in (0, 1), \quad n/r \to 0.$$

Problem 14.9 Prove this under the more stringent condition that $n/\sqrt{r} \to 0$. [Use Problem 2.36 and the normal approximation to the binomial.]

Problem 14.10 Derive the exact (Clopper–Pearson) one-sided and then two-sided confidence intervals for a hypergeometric experiment. Then implement this as a function in R.

14.2.3 Comparison with a Binomial Experiment

We already argued that a hypergeometric experiment with parameters $(n, r, v - r)$ is very similar to a binomial experiment with parameters (n, θ) with $\theta = r/v$. In finer detail, however, it would seem that a hypergeometric experiment, where sampling is without replacement, allows for more precise inference compared to the corresponding binomial experiment, where sampling is with replacement and therefore seemingly wasteful to a certain degree. Indeed, if the balls are numbered (which we can always assume, at least as a thought experiment) and we have already drawn ball number i, then drawing it again does not provide any additional information on the contents of the urn.

Sampling without replacement is indeed preferable, because the resulting confidence intervals are narrower.

Problem 14.11 Verify numerically that the Clopper–Pearson two-sided interval is narrower in a hypergeometric experiment compared to the corresponding binomial experiment.

14.3 Negative Binomial/Hypergeometric Experiments

Negative Binomial Experiments

Consider an experiment that consists in tossing a θ-coin until a predetermined number of heads, m, has been observed. Thus, the number of trials is not set in advance, in contrast with a binomial experiment. The goal, as before, is to infer the value of θ based on the result of such an experiment. In practice, such a design might be appropriate in situations where θ is believed to be small.

Let N denote the number of tails until m heads are observed. We call the resulting experiment a *negative binomial experiment* because N is negative binomial with parameters (m, θ) and sufficient for this experiment.

Problem 14.12 Prove that N is indeed sufficient.

Problem 14.13 Prove that the MLE for θ is $m/(m + N)$. Note that this is still the observed proportion of heads in the sequence, just as in a binomial experiment.

Problem 14.14 Derive the exact (Clopper–Pearson) one-sided and then two-sided confidence intervals for a negative binomial experiment. (These intervals have a simple closed form expression when $m = 1$, that is, in a *geometric experiment*.) Then implement this as a function in R.

Negative Hypergeometric Experiments

When the experiment consists in repeatedly sampling without replacement from an urn, with r red and b blue balls, until m red balls are collected, we talk of a *negative hypergeometric experiment*, in particular because the key distribution in this case is the negative hypergeometric distribution.

Problem 14.15 Consider and solve the previous three problems in the present context.

14.4 Sequential Experiments

We present here another classical experimental design where the number of trials is not set in advance of conducting the experiment, that may be appropriate in surveillance applications (e.g., epidemiological monitoring, quality control, etc.). The setting is again that of a θ-coin being repeatedly tossed. As before, we let Y_n and $S_n = Y_n/n$ denote the number of heads and the proportion of heads in the first n tosses, respectively.

14.4.1 Sequential Probability Ratio Test

Suppose we want to decide between two hypotheses

$$\mathcal{H}_0^{\leq} : \theta_* \leq \theta_0, \quad \text{versus} \quad \mathcal{H}_1^{\geq} : \theta_* \geq \theta_1, \tag{14.11}$$

where $0 \leq \theta_0 < \theta_1 \leq 1$ are given.

Example 14.16 (Multistage testing) Such designs are used in mastery tests where a human subject's knowledge and command of some material or topic is tested on a computer. In such a context, each question is a trial and it is answered correctly ('heads') or incorrectly ('tails'), and θ_1 and θ_0 are the thresholds for Pass/Fail, respectively. (Note that, in such a real-life situation, it is not at all obvious that the trials are iid.)

The *sequential probability ratio test* (SPRT) (aka *sequential likelihood ratio test*) was proposed by Abraham Wald (1902–1950) for this situation [196], except that he originally considered testing

$$\mathcal{H}_0^{=} : \theta_* = \theta_0, \quad \text{versus} \quad \mathcal{H}_1^{=} : \theta_* = \theta_1,$$

However, the same test can be applied verbatim to (14.11), which is more general. The procedure is based on the sequence of likelihood ratio (LR) test statistics (as the number of trials increases). The method is general and is here specialized to the case of Bernoulli trials.

The likelihood ratio statistic for $\mathcal{H}_0^{=}$ versus $\mathcal{H}_1^{=}$ is

$$L_n := \left(\frac{\theta_1}{\theta_0}\right)^{Y_n} \left(\frac{1-\theta_1}{1-\theta_0}\right)^{n-Y_n}.$$

The test makes a decision in favor of $\mathcal{H}_0^{=}$ if $L_n < c_{n,0}$, and makes a decision in favor of $\mathcal{H}_1^{=}$ if $L_n > c_{n,1}$, where the thresholds $c_{n,0} < c_{n,1}$ are predetermined based on the desired level and power. This testing procedure amounts to stopping the trials when there is enough evidence against either $\mathcal{H}_0^{=}$ or $\mathcal{H}_1^{=}$.

More specifically, given $\alpha_0, \alpha_1 \in (0, 1)$, $c_{n,0}$ and $c_{n,1}$ are chosen so that

$$\begin{aligned} \mathbb{P}_{\theta_0}(L_n > c_{n,1}) &\leq \alpha_0, \\ \mathbb{P}_{\theta_1}(L_n < c_{n,0}) &\leq \alpha_1. \end{aligned} \tag{14.12}$$

These thresholds can be determined numerically in an efficient manner using a bisection search.

Problem 14.17 In R, write a function taking as input $(n, \theta_0, \theta_1, \alpha_0, \alpha_1)$ and returning (approximate) values for $c_{n,0}$ and $c_{n,1}$. Try your function on simulated data.

Proposition 14.18. *With $c_{n,0} = \alpha_1/(1 - \alpha_0)$ and $c_{n,1} = (1 - \alpha_1)/\alpha_0$, it holds that the probability of an error is controlled by $\alpha_0 + \alpha_1$, meaning,*

$$\mathbb{P}_{\theta_0}(L_n > c_{n,1}) + \mathbb{P}_{\theta_1}(L_n < c_{n,0}) \leq \alpha_0 + \alpha_1.$$

Problem 14.19 Show that, although the procedure is designed for testing $\mathcal{H}_0^=$ versus $\mathcal{H}_1^=$, it applies to testing $\mathcal{H}_0^\leq$ versus $\mathcal{H}_1^\geq$ above, meaning that if (14.12) holds, then it also holds that

$$\mathbb{P}_\theta(L_n > c_{n,1}) \leq \alpha_0, \quad \text{for all } \theta \leq \theta_0,$$
$$\mathbb{P}_\theta(L_n < c_{n,0}) \leq \alpha_1, \quad \text{for all } \theta \geq \theta_1.$$

14.4.2 Experiments with Optional Stopping

In [148], Randi debunks a number of experiments claimed to exhibit paranormal effects. He recounts experiments where a self-proclaimed psychic tries to influence the outcome of a computer-generated sequence of coin tosses. Randi questions the validity of these experiments because the subject had the option of stopping or continuing the experiment at will.

Much more disturbing is the fact that such strategies are commonly employed by scientists, most notably, in clinical trials [8, 132]. And indeed, if optional stopping is allowed in the experiment but not taken into account in the analysis, the resulting conclusions can be grossly incorrect.

Problem 14.20 Consider an experiment where a fair coin is flipped in front of a self-proclaimed psychic who claims to be able to influence the coin so it comes up heads more often than tails. If you were that individual, propose a stopping rule that would (artificially) bend the outcome of the experiment in your favor if, not knowing any better, the experimenter analyzes the outcome as if the sample size had been predetermined. Perform some numerical experiments to evaluate your strategy.

14.5 Additional Problems

Problem 14.21 (Proportion test) In the context of the binomial experiment of Example 12.2, a *proportion test* can be defined based on the fact that the sample proportion is approximately normal (Theorem 5.4). Based on this, obtain a p-value for testing $\theta_* \leq \theta_0$. Note that the p-value will only be valid in the large-sample limit. (In R, this test is implemented in the prop.test function.)

Problem 14.22 (A comparison of confidence intervals) A numerical comparison of various confidence intervals for a proportion is presented in [137]. Perform simulations to reproduce Table I, rows 1, 3, and 5, in the article. [Of course, due to randomness, the numbers resulting from your numerical simulations will be a little different.]

Problem 14.23 Detail the calculations (implicitly) done in the "Feedback Experiments" section of the paper [43].

15

Multiple Proportions

When a die with m faces is rolled, the result of each trial can take one of m possible values. The same is true in the context of an urn experiment, when the balls in the urn are of m different colors. Such models are broadly applicable. Indeed, even 'yes/no' polls almost always include at least one other option like 'not sure' or 'no opinion'. See Table 15.1 for an example. These data can be plotted, for instance, as a *bar chart* or a *pie chart*, as shown in Figure 15.1.

Table 15.1 *Washington Post – ABC News poll of 1003 adults in the US (March 7–10, 2012). "Do you think a political leader should or should not rely on his or her religious beliefs in making policy decisions?"*

Should	Should not	Depends	No opinion
31%	63%	3%	3%

Another situation where discrete variables arise is when two or more coins are compared in terms of their chances of landing heads, or more generally, when two or more (otherwise identical) dice are compared in terms of their chances of landing on a particular face. In terms of urn experiments, the analogue is a situation where balls are drawn from multiple urns. This sort of experiment can be used to model clinical trials where several treatments are compared and the outcome is dichotomous. See Table 15.2 for an example. The corresponding data are plotted as a *side-by-side bar chart* in Figure 15.2.

When the coins are tossed together, or when the dice are rolled together, we might want to test for independence. And there are various ways to do that depending on the study's design.

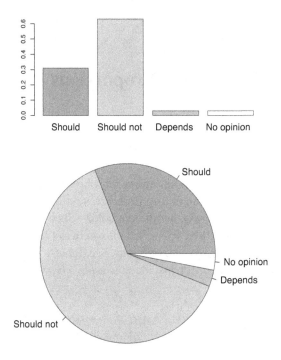

Figure 15.1 A bar chart and a pie chart of the data appearing in Table 15.1.

Table 15.2 *The study [147] examined the impact of supplementing newborn infants with vitamin A on early infant mortality. This was a randomized, double-blind trial, performed in two rural districts of Tamil Nadu, India, where newborn infants (11619 in total) received either vitamin A or a placebo. The primary response was mortality at six months.*

	Death	No death
Placebo	188	5645
Vitamin A	146	5640

15.1 One-Sample Goodness-of-Fit Testing

We consider a die with m faces with distinct labels, say $1, \ldots, m$, and for $s \in \{1, \ldots, m\}$, we let θ_s denote the probability that in a given trial the die lands on s. The outcome of the experiment is of the form $\omega = (\omega_1, \ldots, \omega_n)$,

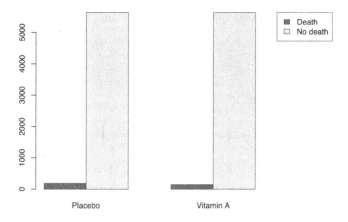

Figure 15.2 A side-by-side bar chart of the data appearing in Table 15.2.

where $\omega_i = s$ if the ith roll resulted in face s. We assume that the rolls are independent. Let $Y_1, \ldots, Y_m$ denote the counts defined in (6.4). In Section 6.2.2, we saw that $(Y_1, \ldots, Y_m)$ has the multinomial distribution with parameters $(n, \theta_1, \ldots, \theta_m)$. In what follows, we will let $y_s = Y_s(\omega)$, and $\hat{\theta}_s := y_s/n$, which are the observed counts and observed proportions, respectively.

Problem 15.1 Show that $(Y_1, \ldots, Y_m)$ is sufficient for $(\theta_1, \ldots, \theta_m)$. Note that, when focusing the counts rather than the trials themselves, all that is lost is the order of the trials, which at least intuitively is not informative since the trials are assumed to be iid.

Problem 15.2 Show that $(Y_1/n, \ldots, Y_m/n)$ is the maximum likelihood estimator (MLE) for $(\theta_1, \ldots, \theta_m)$.

Various questions may arise regarding the parameter vector $\boldsymbol{\theta}$. These can be recast in the context of multiple testing (Chapter 20). We present here a more classical treatment focusing on *goodness-of-fit testing*, where the central question is whether the underlying distribution is a given distribution or, said differently, how well a given distribution fits the data. In detail, given $\boldsymbol{\theta}_0 = (\theta_{0,1}, \ldots, \theta_{0,m})$, we are interested in testing

$$\mathcal{H}_0 : \boldsymbol{\theta}^* = \boldsymbol{\theta}_0,$$

where, as before, $\boldsymbol{\theta}^* = (\theta_1^*, \ldots, \theta_m^*)$ denotes the true value of the parameter.

We first try a likelihood approach. In the variant (12.10), the likelihood ratio is here given by

$$\frac{\hat{\theta}_1^{y_1} \cdots \hat{\theta}_m^{y_m}}{\theta_{0,1}^{y_1} \cdots \theta_{0,m}^{y_m}},$$

and taking the logarithm, this becomes

$$\ell = L(\omega) := \sum_{s=1}^{m} y_s \log\left(\frac{y_s}{n\theta_{0,s}}\right). \tag{15.1}$$

The likelihood ratio test (LRT) rejects for large values of L.

Remark 15.3 (Observed and expected counts) While $y_1, \ldots, y_m$ are the *observed counts*, $n\theta_{0,1}, \ldots, n\theta_{0,m}$ are often referred to as the *expected counts (under the null hypothesis)*. This is because $\mathbb{E}_{\theta_0}(Y_s) = n\theta_{0,s}$ for all s. In view of (15.1), the likelihood ratio compares observed with expected counts.

15.1.1 Normal-Approximation P-Value

Recall the chi-squared family of distributions introduced in Section 5.9.3. We assume, without much loss of generality, that $\theta_{0,s} > 0$ for all s.

Theorem 15.4. *Under the null hypothesis, in the large-sample limit were $n \to \infty$, $2L$ converges weakly to the chi-squared distribution with $m - 1$ degrees of freedom.*

In view of this theorem, the p-value can be approximated as follows

$$\mathbb{P}_{\theta_0}(L \geq \ell) \approx 1 - H_{m-1}(2\ell),$$

where H_{m-1} is the distribution function of the chi-squared distribution with $m - 1$ degrees of freedom. Modulo some slight correction that improves the accuracy of this numerical approximation, this is how the p-value was computed in the pre-computer age.

15.1.2 Monte Carlo P-Value

In the computer age, this p-value can be estimated by Monte Carlo simulation, as we first saw in Section 10.1. The general procedure is detailed in Algorithm 15.1. (A variant of the algorithm consists in directly generating values of the test statistic under its null distribution.)

Algorithm 15.1   Monte Carlo p-value

Input: data ω, test statistic T, null distribution $\mathbb{P}_0$, number of Monte Carlo samples B
Output: an estimate of the p-value

Compute $t = T(\omega)$
For $b = 1, \ldots, B$
   generate ω_b from $\mathbb{P}_0$
   compute $t_b = T(\omega_b)$
Return

$$\widehat{\mathrm{pv}}_{\mathrm{MC}} := \frac{\#\{b : t_b \geq t\} + 1}{B + 1}. \tag{15.2}$$

Proposition 15.5. *The Monte Carlo p-value* (15.2) *is itself a valid p-value in the sense of* (12.12).

Problem 15.6 Prove Proposition 15.5. [The conclusions of Problem 8.53 will prove useful.]

Problem 15.7 In R, write a function that takes as input the vector of observed counts, the null parameter vector, and a number of Monte Carlo replicates, and returns an estimate of the p-value for the LRT. [Use the function rmultinom to generate Monte Carlo counts under the null.]

15.1.3 Bootstrap P-Value

In some situations, the null distribution may not be perfectly specified. For example, in genetics, we may want to know if a particular population is at equilibrium as in Problem 13.7. Or we may want to know if, in an experiment yielding counts, a Poisson distribution provides a good fit to the data as in Problem 12.28 and Problem 12.29. Such problems can be formulated as testing the null hypothesis that the underlying distribution is in some given family, which we refer to as the *null family of distributions*. We only consider the situation where the null distributions have finite support. (The situation where the support is infinite, as in the case with the Poisson family, can be dealt with by truncation or merging.)

We still assume the support to be $\{1, \ldots, m\}$, and denote the null family of distribution by Θ_0, which is therefore a set of probability vectors on

$\{1, \ldots, m\}$. Based on a sample of size n, we are interested in testing

$$\mathcal{H}_0 : \boldsymbol{\theta}^* \in \Theta_0.$$

We assume that an MLE exists for $\boldsymbol{\theta}$ under the null family Θ_0, and denote it by $\hat{\boldsymbol{\theta}}_0 = (\hat{\theta}_{0,1}, \ldots, \hat{\theta}_{0,m})$. In that case, in the variant (12.10), the log likelihood ratio is given by

$$\ell = L(\omega) := \sum_{s=1}^{m} y_s \log\left(\frac{y_s}{n\hat{\theta}_{0,s}}\right),$$

and the LRT rejects for large values of L.

It is known that, when the null family of distributions is parametrized by d free parameters and is 'smooth' in some way, under the null hypothesis, $2L$ converges weakly to the chi-squared distribution with $m - 1 - d$ degrees of freedom. (This assumes that $d < m - 1$, and the reader is invited to check that this is condition is not restrictive.) This is essentially how the p-value was approximated in the old days.

Alternatively, it is also possible to approximate the p-value by simulation on a computer. The idea is to estimate the p-value by Monte Carlo simulation, as before, but now using the estimated null distribution, $\hat{\boldsymbol{\theta}}_0$ above, to generate the samples. This process, of performing Monte Carlo simulations based on an estimated distribution, is generally called a *bootstrap*. We note that the resulting *bootstrap p-value* is not exactly valid in finite samples – even though, in the present situation, it is exactly valid in the large-sample limit.

Problem 15.8 We apply the methodology just described to assess whether a Poisson distribution provides a good fit to the yeast data of Table 4.1. Start by formulating the question as a hypothesis testing problem. Then compute a bootstrap p-value of the LRT. Repeat with the data of Table 4.2.

15.2 Multi-Sample Goodness-of-Fit Testing

In Section 15.1, we assumed that we were provided with a null distribution and were tasked with determining how well that distribution fitted the data. However, in applications, often no such reference distribution is available. This is the case in clinical trials where the efficacy of a treatment is compared to an existing treatment or a placebo, such as in the example of Table 15.2. In other situations, more than two groups are to be compared. A question, then, is whether these groups of observations were generated by the same distribution. The difference here with the basic premise of Section 15.1 is that this hypothesized common distribution is not given.

An abstract model for this setting is that of an experiment involving g dice, with the jth die rolled n_j times, all rolls being independent. As before, each die has m faces labeled $1, \ldots, m$. Let θ_j denote the probability vector of the jth die. Our goal is to test the following null hypothesis:

$$\mathcal{H}_0 : \theta_1 = \cdots = \theta_g .$$

This is sometimes referred to as *testing for homogeneity*.

Let $\omega_{ij} = s$ if the ith roll of the jth die results in s, so that $\omega = (\omega_{ij})$ are the data, and define the (per group) counts

$$Y_{sj}(\omega) := \#\{i : \omega_{ij} = s\}.$$

Problem 15.9 Show that these counts are jointly sufficient.

Problem 15.10 Show that (Y_{sj}/n_j) is the MLE for (θ_{sj}).

We define the total sample size as $n := \sum_{j=1}^{g} n_j$. We also define the following counts

$$y_{sj} := Y_{sj}(\omega), \quad Y_s := \sum_{j=1}^{g} Y_{sj}, \quad y_s := Y_s(\omega) = \sum_j y_{sj},$$

leading to the following proportions

$$\hat{\theta}_{sj} := y_{sj}/n_j, \quad \hat{\theta}_s := y_s/n, \quad \hat{\theta}_j := (\hat{\theta}_{1j}, \ldots, \hat{\theta}_{mj}), \quad \hat{\theta} := (\hat{\theta}_1, \ldots, \hat{\theta}_m).$$

15.2.1 Likelihood Ratio

Because of independence, the likelihood of all the observations combined is just the product of the likelihoods, one for each die (see (6.5)).

Problem 15.11

(i)   Prove that, without any constraints on the parameters, the likelihood is maximized at $(\hat{\theta}_1, \ldots, \hat{\theta}_g)$.

(ii)   Prove that, under the constraint that $\theta_1 = \cdots = \theta_g$, this is maximized at $(\hat{\theta}, \ldots, \hat{\theta})$.

(iii) Deduce that the likelihood ratio is given by

$$\frac{\prod_{j=1}^{g} \prod_{s=1}^{m} \hat{\theta}_{sj}^{y_{sj}}}{\prod_{s=1}^{m} \hat{\theta}_s^{y_s}} = \prod_{j=1}^{g} \prod_{s=1}^{m} \left(\frac{\hat{\theta}_{sj}}{\hat{\theta}_s} \right)^{y_{sj}} .$$

Taking the logarithm, this becomes

$$\ell = L(\omega) := \sum_{j=1}^{g} \sum_{s=1}^{m} y_{sj} \log \left(\frac{y_{sj}}{n_j y_s / n} \right). \tag{15.3}$$

The LRT rejects for large values of L.

Remark 15.12 (Estimated expected counts) The (y_{sj}) are the observed counts. The expected counts are not available since the common null distribution is not given. Nonetheless, they can be estimated. Indeed, under the null hypothesis where θ denotes the common distribution, the expected counts are

$$\mathbb{E}_\theta(Y_{sj}) = n_j \theta_s,$$

and we can estimate this by plugging in $\hat\theta_s$ in place of θ_s, leading to the following *estimated expected counts (under the null hypothesis)*

$$\mathbb{E}_\theta(Y_{sj}) \approx n_j \hat\theta_s = n_j y_s / n.$$

In view of (15.3), the likelihood ratio statistic compares the observed counts with the expected counts.

15.2.2 Normal-Approximation P-Value

We assume, without loss of generality, that all values are possible under the null distribution, meaning that $\theta_s > 0$ for all s.

Theorem 15.13. *Under the null hypothesis, in the large-sample limit were $n_j \to \infty$ for all j, $2L$ converges weakly to the chi-squared distribution with $(g-1)(m-1)$ degrees of freedom.*

As before, the theorem allows us to approximate the p-value as follows

$$\mathbb{P}_{(\theta,\dots,\theta)}(L \geq \ell) \approx 1 - H_{(g-1)(m-1)}(2\ell).$$

15.2.3 Bootstrap P-Value

As in Section 15.1.3, a bootstrap approach is available to approximate the p-value by simulation on a computer. The idea is the same and consists in estimating the p-value by Monte Carlo simulation using the estimated null distribution, $\hat\theta$ above, to generate the samples across all groups. We again note that the resulting bootstrap p-value is not exactly valid in finite samples, but is exactly valid in the large-sample limit.

Problem 15.14 In R, write a function that takes as input the list of observed counts as a g-by-m matrix of counts and the number of bootstrap samples to be drawn, and returns an estimate of the p-value. Apply your function to the dataset of Table 15.2.

15.3 Completely Randomized Experiments

The data presented in Table 15.2 can be analyzed using the methodology for comparing groups presented in Section 15.2, as done in Problem 15.14. We present a different perspective which leads to a different analysis. It is reassuring to know that the two analyses will yield similar results as long as the group sizes are not too small.

The typical null hypothesis in such a situation is that the treatment and the placebo are equally effective. We describe a model where inference can only be done for the group of subjects – while a generalization to a larger population would be contingent on this sample of individuals being representative of the said population.

Suppose that the group sizes are n_1 and n_2, for a total sample size of $n = n_1 + n_2$. The result of the experiment can be summarized in a table of counts, Table 15.3, often called a *contingency table*.

If $z = z_1 + z_2$ denotes the total number of successes, in the present model it is assumed to be deterministic since we are drawing inferences on the group of subjects in the study. What is random is the group labeling, and if treatment and placebo truly have the same effect on these individuals, then the group labeling is completely arbitrary. Thus, this is the null hypothesis to be tested.

There is no model for the alternative, so we cannot derive the likelihood ratio, for example. However, it is fairly clear what kind of test statistic we should be using. In fact, a good option is the same statistic (15.3), which in the context of Table 15.3 takes the form

$$z_1 \log\left(\frac{nz_1}{n_1 z}\right) + (n_1 - z_1) \log\left(\frac{n(n_1 - z_1)}{n_1(n-z)}\right)$$
$$+ z_2 \log\left(\frac{nz_2}{n_2 z}\right) + (n_2 - z_2) \log\left(\frac{n(n_2 - z_2)}{n_2(n-z)}\right).$$

Another option is the odds ratio, which after applying a log transformation takes the form

$$\log\left(\frac{z_1}{n_1 - z_1}\right) - \log\left(\frac{z_2}{n_2 - z_2}\right).$$

More generally, consider a randomized experiment where the subjects are assigned to one of g treatment groups and the response can take m possible values. With the notation of Section 15.2, the jth group is of size n_j, for a total sample size of $n := n_1 + \cdots + n_g$. In the null model, the total counts are here taken to be deterministic (while there are random in Section 15.2),

Table 15.3 *A prototypical contingency table summarizing the result of a completely randomized experiment with two groups and two possible outcomes.*

	Success	Failure	Total
Group 1	z_1	$n_1 - z_1$	n_1
Group 2	z_2	$n_2 - z_2$	n_2
total	z	$n - z$	n

while the group labeling is random. The test statistic of choice remains (15.3).

Problem 15.15 Write down the contingency table (in general form) using the notation of Section 15.2.

15.3.1 Permutation P-Value

Regardless of the test statistic that is chosen, the corresponding p-value is obtained under the null model. Since the group sizes are set, the null model amounts to permuting the labels. This procedure is an example of re-randomization testing, developed further in Section 22.1.1.

Let Π denote the set of all permutations of the labels. There are

$$|\Pi| = \frac{n!}{n_1! \cdots n_g!}$$

such permutations in the present setting. Importantly, we permute the labels placed on the rolls, which then yield new counts. (We do *not* permute the counts.) Let T be a test statistic whose large values provide evidence against the null hypothesis. We let t denote the observed value of T and, for a permutation $\pi \in \Pi$, we let t_π denote the value of T applied to the corresponding permuted data. Then the *permutation p-value* is defined as

$$\mathsf{pv}_{\text{perm}} := \frac{\#\{\pi : t_\pi \geq t\}}{|\Pi|}. \tag{15.4}$$

Proposition 15.16. *The permutation p-value* (15.4) *is a valid p-value in the sense of* (12.12).

Problem 15.17 Prove Proposition 15.16. [The conclusions of Problem 8.53 will prove useful.]

Monte Carlo Estimation

Unless the group sizes are very small, $|\Pi|$ is impractically large, and this leads one to estimate the p-value by sampling permutations uniformly at random from Π. This may be called a *Monte Carlo permutation p-value.* The general procedure is detailed in Algorithm 15.2.

Algorithm 15.2   Monte Carlo Permutation p-value

Input: data ω, test statistic T, group Π of permutations that leave the null invariant, number of Monte Carlo samples B
Output: an estimate of the p-value

Compute $t = T(\omega)$
For $b = 1, \ldots, B$
    draw π_b uniformly at random from Π
    permute ω according to π_b to get ω_b
    compute $t_b = T(\omega_b)$
Return

$$\widehat{\mathrm{pv}}_{\mathrm{perm}} := \frac{\#\{b : t_b \geq t\} + 1}{B + 1}. \tag{15.5}$$

Proposition 15.18. *The Monte Carlo permutation p-value* (15.5) *is a valid p-value in the sense of* (12.12).

Problem 15.19 Prove this proposition. [The conclusions of Problem 8.53 will prove useful.]

Problem 15.20 In R, write a function that implements Algorithm 15.2. Apply your function to the data in Table 15.2.

Remark 15.21 (Conditional inference) This proposition holds true also in the setting of Section 15.2. The use of a permutation p-value there is an example of conditional inference, which is discussed in Section 22.1.

15.4 Matched-Pairs Experiments

Consider a randomized matched-pairs design where two treatments are compared (Section 11.2.5). The outcome is binary ('success' or 'failure'). The sample is of the form $(\omega_{11}, \omega_{21}), \ldots, (\omega_{n1}, \omega_{n2})$, where ω_{ij} is the response from the subject in pair i that received Treatment j, with $\omega_{ij} = 1$ indicating success. If no other information on the subjects is taken into account in the analysis, the data can be summarized in a 2-by-2 contingency table (Table 15.4) displaying the counts

$$y_{st} := Y_{st}(\omega) := \#\{i : (\omega_{i1}, \omega_{i2}) = (s, t)\}.$$

IID Assumption

At this point we cannot claim that the counts are jointly sufficient. This is the case, however, in a situation where the pairs can be assumed to be sampled uniformly at random from a population. In that case, the pairs can be taken to be iid and we may define θ_{st} as the probability of observing the pair (s, t). The treatment (one-sided) effect is defined as $\theta_{10} - \theta_{01}$. The null hypothesis of no treatment effect is $\theta_{10} = \theta_{01}$.

It is rather natural to ignore the pairs where the two subjects responded in the same way and base the inference on the pairs where the subjects responded differently, which leads to rejecting for large values of $Y_{10} - Y_{01}$ while conditioning on (Y_{11}, Y_{00}). After all, $(Y_{10} - Y_{01})/n$ is unbiased for $\theta_{10} - \theta_{01}$. This is the *McNemar test* [128] and it is known to be uniformly most powerful among unbiased tests (UMPU) in this situation.

R corner McNemar's test is implemented in the function mcnemar.test.

Problem 15.22 Argue that rejecting for large values of $Y_{10} - Y_{01}$ given (Y_{11}, Y_{00}) is equivalent to rejecting for large values of Y_{10} given (Y_{11}, Y_{00}). Further, show that given $(Y_{11}, Y_{00}) = (y_{11}, y_{00})$, Y_{10} has the binomial distribution with parameters $k := n - y_{11} - y_{00}$ and $p := \theta_{10}/(\theta_{10} + \theta_{01})$. Conclude that the McNemar test reduces to testing $p = 1/2$ in a binomial experiment with parameters (k, p).

Problem 15.23 Is the McNemar test the LRT in the present context?

Remark 15.24 (Observational studies) Although we worked in the context of a randomized experiment, the test may be applied in the context of an observational study with the caveat that the conclusion is conservatively

Table 15.4 *A prototypical contingency table summarizing the result of a matched-pairs experiment with two possible outcomes.*

| | | Treatment B | |
		Success	Failure
Treatment A	Success	y_{11}	y_{10}
	Failure	y_{01}	y_{00}

understood as being in terms of association instead of causality. (Domain-specific expertise is often needed to determine whether the iid assumption is reasonable in a particular observational study.)

Beyond the IID Assumption

In some situations it may not be realistic to assume that the pairs constitute a representative sample from a population. Even then, a permutation approach remains valid due to the initial randomization. The key observation is that, if there is no treatment effect, then ω_{i1} and ω_{i2} are exchangeable by design. The idea then is to condition on the observed ω_{ij} and permute within each pair. Then, under the null hypothesis, any such permutation is (conditionally) equally likely, and this is exploited in the derivation of a p-value. This procedure is again an example of re-randomization testing (Section 22.1.1).

In more detail, a permutation in the present context transforms the (observed) data,

$$(\omega_{11}, \omega_{12}), \dots, (\omega_{n1}, \omega_{n2}),$$

into

$$(\omega_{1\pi_1(1)}, \omega_{1\pi_1(2)}), \dots, (\omega_{n\pi_n(1)}, \omega_{n\pi_n(2)}),$$

with $(\pi_i(1), \pi_i(2)) = (1, 2)$ or $= (2, 1)$ for each i. Let Π denote the class of $\pi = (\pi_1, \dots, \pi_n)$ where each π_i is a permutation of $\{1, 2\}$. Note that $|\Pi| = 2^n$.

Suppose we still reject for large values of $T := Y_{10} - Y_{01}$, which after all remains appropriate. The observed value of this statistic is $t := y_{10} - y_{01}$. Let t_π denote the value of this statistic computed on the data permuted by applying $\pi \in \Pi$. Then the *permutation p-value* is defined as in (15.4).

Proposition 15.25. *This is a valid p-value in the sense of* (12.12).

Problem 15.26 Prove this proposition. [The conclusions of Problem 8.53 will prove useful.]

Computing this p-value may be challenging, as there are many possible permutations and one would in principle have to consider every single one of them. Luckily, this is not necessary.

Problem 15.27 Find an efficient way of computing the p-value.

Problem 15.28 Show that, in fact, this permutation test is equivalent to the McNemar test.

15.5 Fisher's Exact Test

Fisher [13] describes in [63] a now famous experiment meant to illustrate his concept of null hypothesis. The setting is that of a lady who claims to be able to distinguish whether milk or tea was added to the cup first. To test her professed ability, she is given 8 cups of tea, in four of which milk was added first. With full information on how the experiment is being conducted, she is asked to choose 4 cups where she believes milk was poured first. The resulting counts, reproduced from Fisher's original account, are displayed in Table 15.5.

Table 15.5 *The columns are labeled by the liquid that was poured first (milk or tea) and the rows are labeled by the lady's guesses.*

	Truth	
Lady's guess	Milk	Tea
Milk	3	1
Tea	1	3

The null hypothesis is that of no association between the true order of pouring and the woman's guess, while the alternative is that of a positive association.[56] In particular, under the null hypothesis, the lady is purely guessing and so her guesses are independent of the truth. This gives the null distribution, which leads to a permutation p-value.

Remark 15.29 Thus, a p-value is obtained by permutation, exactly as in Section 15.3, because here too the total counts are all fixed. However, the situation is not exactly the same. The difference is subtle: here, all the totals are fixed by design; there, the totals are fixed because of the focus on the subjects in the study.

In such a 2×2 setting, it is possible to test for a positive association. Indeed, first note that, since the margin totals are fixed (all equal to 4), the number in the top-left corner ('milk', 'milk') determines all the others, so it suffices to consider this number (which is obviously sufficient). Now, clearly, a large value of that number indicates a positive association. This leads us (and Fisher before us) to rejecting for large values of this statistic.

---

[56] If an alternative of no association is preferred, a two-sided test results.

Consider a general 2×2 setting, where there are n cups, with k of them receiving milk first. The lady is to select k cups that she believes received the milk first. Then the contingency table would look something like this

	Truth		
Lady's guess	Milk	Tea	Total
Milk	y_{11}	y_{12}	k
Tea	y_{21}	y_{22}	$n-k$
Total	k	$n-k$	n

The test statistic is Y_{11}, whose large values weigh against the null hypothesis, and the resulting test is often referred to as *Fisher's exact test*.

Problem 15.30 Prove that, under the null hypothesis, Y_{11} is hypergeometric with parameters $(k, k, n-k)$.

Thus, computing the p-value can be done exactly, without having to enumerate all possible permutations.

Problem 15.31 Derive the p-value for the original experiment in closed form, and confirm your answer numerically using R.

R corner Fisher's exact test is implemented in the function fisher.test.

Remark 15.32 Fisher's test can be applied in the context of Section 15.3, and doing so is an example of conditional inference (Section 22.1).

15.6 Association in Observational Studies

Consider the poll summarized in Table 15.6 and depicted as a *segmented bar chart* in Figure 15.3. The description of the polling procedure that appears on the website, and additional practical considerations, lead one to believe that the interviewees were selected without a priori knowledge of their political party affiliation. We will assume this is the case.

It is safe to assume that one of the main reasons for collecting these data is to determine whether there is an *association* between party affiliation and views on climate change, and the steps that the US Government should take to address this issue, if any.[57]

---

[57] The entire poll questionnaire, not shown here, includes other questions related to climate change. We refer the reader to the original source: https://www.cbsnews.com/news/global-warming-and-the-paris-climate-change-conference/

Table 15.6 *New York Times – CBS News poll of 1030 adults in the US (November 18–22, 2015). "Do you think the US should or should not join an international treaty requiring America to reduce emissions in an effort to fight global warming?"*

	Should	Should not	Other
Republicans	42%	52%	6%
Democrats	86%	9%	5%
Independents	66%	25%	9%

Figure 15.3 A segmented bar chart of the data appearing in Table 15.6.

More formally, a lack of association in such a setting is modeled as *independence*. The variables here are A for 'party membership', equal to either 'Republican', 'Democrat', or 'Independent'; and B for 'opinion', equal to either 'should', 'should not', or 'other'.

Remark 15.33 (Factors) In statistics, a categorical variable is often called a *factor* and the values it takes are called *levels*. Thus, A is a factor with levels {'Republican', 'Democrat', 'Independent'}.

The raw data here is of the form $\{(a_i, b_i) : i = 1, \ldots, n\}$, where $n = 1030$ is the sample size. Table 15.6 provides the percentages. We are told that there were 254 republicans, 304 democrats, and 472 independents. With this information, we can recover the table of counts up to rounding error. See Table 15.7.

An abstract model for the present setting is that of an experiment where two dice are rolled *together n* times. Die A has faces labeled $1, \ldots, m_a$, while Die B has faces labeled $1, \ldots, m_b$. (As before, the labeling by integers is for convenience, as the faces could be labeled by any other symbols.) The outcome of the experiment is $\omega = ((a_1, b_1), \ldots, (a_n, b_n))$.

Table 15.7 *The table of counts corresponding to the poll summarized in Table 15.6.*

	Should	Should not	Other
Republicans	107	132	15
Democrats	261	27	15
Independents	312	118	42

Define the cross-counts as

$$y_{st} := Y_{st}(\omega) := \#\big\{i : (a_i, b_i) = (s, t)\big\}.$$

Assuming the rolls are iid, which we do, these counts are jointly sufficient, and have the multinomial distribution with parameters n and (θ_{st}), where θ_{st} is the probability that a roll results in (s, t). These counts are often displayed in a contingency table.

If (A, B) denotes the result of a roll, the task is *testing for the independence* of A and B. Under (θ_{st}), A has (marginally) the multinomial distribution with parameters n and (θ_s^a), while B has (marginally) the multinomial distribution with parameters n and (θ_t^b), where

$$\theta_s^a := \sum_{t=1}^{m_b} \theta_{st}, \qquad \theta_t^b := \sum_{s=1}^{m_a} \theta_{st}.$$

The null hypothesis of independence can be formulated as

$$\mathcal{H}_0 : \theta_{st} = \theta_s^a \theta_t^b, \quad \text{for all } (s, t) \in \{1, \ldots, m_a\} \times \{1, \ldots, m_b\}.$$

Remark 15.34 Contrast the present situation with that of Section 15.3, where only the null distribution is modeled.

Remark 15.35 ($k \geq 3$ variables) We focus on two discrete variables, but what follows extends without conceptual difficulty to any number of discrete random variables. When there are k factors with $m_1, \ldots, m_k$ levels, respectively, the counts are simply organized in a $m_1 \times \cdots \times m_k$ array. The methodology developed below applies with only obvious changes to testing whether the variables are mutually independent. But when there are $k \geq 3$ variables, other questions can be considered, for example, whether the two variables are independent conditional on the remaining variables.

15.6.1 Likelihood Ratio

Recall that (y_{st}) denotes the observed counts. Based on these, define the proportions $\hat{\theta}_{st} = y_{st}/n$, the marginal counts

$$y_s^a = \sum_{t=1}^{m_b} y_{st}, \quad y_t^b = \sum_{s=1}^{m_a} y_{st},$$

and the corresponding marginal proportions $\hat{\theta}_s^a = y_s^a/n$ and $\hat{\theta}_t^b = y_t^b/n$.

Problem 15.36 Show that the likelihood ratio in the variant (12.10) is equal to

$$\prod_{s=1}^{m_a} \prod_{t=1}^{m_b} \left(\frac{\hat{\theta}_{st}}{\hat{\theta}_s^a \hat{\theta}_t^b} \right)^{y_{st}},$$

Taking the logarithm, this becomes

$$\sum_{s=1}^{m_a} \sum_{t=1}^{m_b} y_{st} \log \left(\frac{y_{st}}{y_s^a y_t^b / n} \right), \tag{15.6}$$

and dividing by n, this becomes

$$\sum_{s=1}^{m_a} \sum_{t=1}^{m_b} \hat{\theta}_{st} \log \left(\frac{\hat{\theta}_{st}}{\hat{\theta}_s^a \hat{\theta}_t^b} \right).$$

The LRT rejects for large values of this statistic.

Problem 15.37 (Estimated expected counts) Justify calling $y_s^a y_t^b / n$ the *estimated expected count* for (s, t).

Remark 15.38 (Deterministic or random group sizes) Compare (15.3) and (15.6). They are *identical* as functions of the counts. This is rather surprising, perhaps shocking, given that the statistic is rather peculiar and the counts are, at first glance, quite different. Although in both cases the counts are organized in a table, in Section 15.2 they are indexed by (value, group), while in the present section they are indexed by (A value, B value). This can be explained by viewing 'group' in the setting of Section 15.2 as a variable. Indeed, from this perspective the only difference between the two settings is that, in Section 15.2, the group sizes are predetermined, while here they are random. The fact that the likelihood ratios coincide can be explained by the fact that the group sizes are not informative. However, despite the fact that they are the same, the corresponding p-value is derived in (slightly) different ways. See below. Even then, the use of the p-value of Section 15.2 in the present context could be justified based on conditional inference (Section 22.1).

15.6.2 Bootstrap P-Value

In Section 15.2.3, we presented a form of bootstrap tailored to the situation there. The situation is a little different here since the group sizes are not predetermined and another form of bootstrap is more appropriate. In the end, however, these two methods will yield similar p-values as long as the sample sizes are not too small.

The motivation for using a bootstrap approach is the same. Indeed, if we were given the marginal distributions, (θ_s^a) and (θ_t^b), we would simply sample their product, as this is the null distribution in the present situation. However, these distributions are not available to us, but we have estimates, $(\hat{\theta}_s^a)$ and $(\hat{\theta}_t^b)$, and the bootstrap method consists in estimating the p-value by Monte Carlo simulation by repeatedly sampling from the corresponding product distribution, $\hat{\theta}_{st} := \hat{\theta}_s^a \hat{\theta}_t^b$, which is our estimate for the distribution under the null hypothesis.

Problem 15.39 In R, write a function which takes as input the matrix of observed counts and the number of bootstrap samples to be drawn, and returns an estimate of the p-value. Apply your function to the dataset of Table 15.7.

15.6.3 Simpson's Paradox

In Problem 2.42, we saw that inequalities involving probabilities could be reversed when conditioning. This has consequences in real life.

Example 15.40 (Berkeley admissions) In 1973, the Graduate Division at the University of California, Berkeley, received a number of applications. Ignoring incomplete applications, there were 8442 male applicants of whom 3738 were admitted (44%), compared with 4321 female applicants of whom 1494 were admitted (35%). The difference is not only statistically highly significant but also substantial with a difference of almost 10% in the admission rate when comparing men and women. This appeared to be damning evidence of gender bias. However, a breakdown of admission rates by department revealed a much more nuanced picture, where in fact few departments showed any significant difference in admission rates. In the end, the authors of [15] concluded that there was little evidence for gender bias, and that the numbers could be transparently explained by the fact that women tended to apply to departments with lower admission rates.[58]

---

[58] The authors did not publish the entire dataset for reasons of privacy and university policy, but the published data are available in R as UCBAdmissions.

Example 15.41 (Race and death-penalty) Consider the following table, taken from [146], which examined the issue of race in death penalty sentences in murder cases in the state of Florida from 1976 through 1987:

defendant	death (yes/no)
Caucasian	53/430
African-American	15/176

Thus, in the aggregate, Caucasians were more likely to be sentenced to the death penalty. However, after stratifying by the race of the victim, the table becomes:

Victim	Defendant	Death (yes/no)
Caucasian	Caucasian	53/414
Caucasian	African-American	11/37
African-American	Caucasian	0/16
African-American	African-American	4/139

Thus, in the disaggregate, African-Americans were more likely to be sentenced to the death penalty, particularly in cases where the victim was Caucasian.

This type of (Simpson) reversal is not uncommon in practice. Hence, the analyst needs to use extreme caution when deriving causal inferences based on observational data, or any other situation where randomization was not properly applied, as this can easily lead to confounding. (In Example 15.40, a confounder is the department, while in Example 15.41, a confounder is the victim's race.)

15.7 Tests of Randomness

Consider an experiment where the outcome is a sequence of symbols of length n, denoted $\omega = (\omega_1, \ldots, \omega_n)$, with $\omega_i \in \Omega_0$.

In Section 15.1, we focused on the situation where this sequence is the result of drawing repeatedly from an unknown distribution, and that distribution was then the object of our interest. We now turn to the question of whether the sequence was generated iid from some distribution. When this is the case, the sequence is said to be 'random', and procedures addressing this question are called *tests of randomness*.[59]

---

[59] This terminology is rather confusing. Indeed, as we saw in Section 9.1 with Markov chains, a sequence of random variables can be random (i.e., not deterministic) and not be iid. We only use this terminology because it is well-established.

The question of interest is whether the distribution that generated ω, denoted $\mathbb{P}$, is the product of its marginals, and whether these marginals are all the same. Let $\mathcal{P}_0$ denote the class of iid distributions on $\Omega := \Omega_o^n$, so that any $\mathbb{P} \in \mathcal{P}_0$ must be of the form $\mathbb{P}_o^{\otimes n}$ for some distribution $\mathbb{P}_o$ on Ω_o. We want to test the null hypothesis that $\mathbb{P} \in \mathcal{P}_0$.

Example 15.42 (Binary setting) In a binary setting where Ω_o has cardinality 2 and is taken to be $\{0, 1\}$ without loss of generality, $\mathcal{P}_0$ is the family of Bernoulli trials of length n.

IID-ness vs Independence

We are testing iid-ness and not independence. In fact, testing for independence is ill-posed without further restricting the model. To see this, suppose the outcome is $\omega = (\omega_1, \ldots, \omega_n)$. This could have been the result of sampling from the point mass at ω, for which independence trivially holds, and the available data, ω, are clearly not enough to discard this possibility.

IID-ness vs Exchangeability

$\mathbb{P}$ is exchangeable if it is invariant with respect to permutation, meaning that, for any permutation $\pi = (\pi_1, \ldots, \pi_n)$ of $(1, \ldots, n)$, $\mathbb{P}(\omega) = \mathbb{P}(\omega_\pi)$, where $\omega_\pi := (\omega_{\pi_1}, \ldots, \omega_{\pi_n})$. As we know, this property is more general than iid-ness, but it turns out that the available data, ω, are not sufficient to tell the two apart. For illustration, we place ourself in the binary setting of Example 15.42. Within that setting, consider the distribution $\mathbb{P}$ where, with probability $1/2$, we generate an iid sequence of length n from $\text{Ber}(1/4)$, while with probability $1/2$, we generate an iid sequence of length n from $\text{Ber}(3/4)$. Thus, $\mathbb{P}$ here is not an iid distribution. However, the outcome will be a realization of an iid distribution – either $\text{Ber}(1/4)^{\otimes n}$ or $\text{Ber}(3/4)^{\otimes n}$.

Thus, what we can hope to test is exchangeability. (That said, to adhere to tradition, we will use 'randomness' in place of 'exchangeability'.) The tests that follow all take a conditional inference approach by conditioning on the values of the ω_i without regard to their order. This leads to obtaining their p-values by permutation.

We present, below, a few of randomness that are tailored to the discrete setting. Such tests are important for evaluating the accuracy of a generator of pseudo-random numbers. Examples include the *DieHard* and *DieHarder* suites of George Marsaglia and Robert Brown, and some tests developed by the US National Institute of Standards and Technology (NIST) [158].

R corner In R, some of these tests are available in the RDieHarder package.

15.7.1 Tests based on Runs

Some tests of randomness are based on runs, where a *run* is a sequence of identical symbols. Take the binary setting and consider the following outcome sequence (of length $n = 20$ here) where there are 9 runs in total:

$$\underbrace{1}\ \underbrace{000}\ \underbrace{1}\ \underbrace{0000}\,\underbrace{1111}\ \underbrace{0}\ \underbrace{1}\ \underbrace{00}\ \underbrace{111}$$

Number of Runs Test

This test rejects for small values of the total number of runs. Intuitively, a small number of runs indicates less 'mixing'. The test dates back to Wald and Wolfowitz [197], who proposed the test for the purpose of two-sample goodness-of-fit testing (Section 17.3.4).

Problem 15.43 The conditional null distribution of this statistic is known in closed form in the binary setting. To derive this distribution, first consider the number of 0-runs. (There are 4 such runs in the sequence displayed above.) Derive its null distribution. Then use that to derive the null distribution of the total number of runs.

Problem 15.44 What kinds of alternatives do you expect this test to be powerful against? What would be a two-sided version of this test?

Longest Run Test

This test rejects for large values of the length of the longest run (equal to 4 in the sequence displayed above). Intuitively, the presence of a long run indicates less 'mixing'. The test is due to Mosteller [134].

Remark 15.45 (Erdős–Rényi Law) The conditional null distribution of this statistic, denoted L_n, is not known in an amiable form, although it can be estimated by Monte Carlo permutation in practice. However, the asymptotic behavior of L_n under the unconditional null is well-understood, at least in the binary setting. The first-order behavior was derived by Erdös and Rényi [59], which in the context of Bernoulli trials with parameter θ is given by

$$\frac{L_n}{\log n} \ \xrightarrow{\text{P}} \ \frac{1}{\log(1/\theta)}, \quad \text{as } n \to \infty.$$

(Although L_n does not have a limiting distribution, it has a family of limiting distributions [6].)

Problem 15.46 What kinds of alternatives do you expect this test to be powerful against? What would be a two-sided version of this test?

15.7.2 Tests based on Transitions

The following class of tests are designed with Markov chain alternatives in mind. The simplest such test is based on counting transitions $a \to b$, meaning instances where $(\omega_i, \omega_{i+1}) = (a, b)$, where $a, b \in \Omega_0$. The test consists in computing a test statistic for independence applied to the pairs

$$(\omega_1, \omega_2), (\omega_2, \omega_3), \ldots, (\omega_{n-1}, \omega_n),$$

and then obtaining a p-value by permutation (which is typically estimated by Monte Carlo, as usual).

Problem 15.47 In R, write a function that takes in the observed sequence and a number of Monte Carlo replicates, and outputs the p-value just described. Compare this procedure with the number-of-runs test in the binary setting.

Problem 15.48 In the binary setting, perform some numerical simulations to evaluate the power of this procedure against a distribution corresponding to starting at 0 or 1 with probability $1/2$ each (which is the stationary distribution) and then running the Markov chain with the following transition matrix $n - 1$ times

$$\begin{pmatrix} q & 1-q \\ 1-q & q \end{pmatrix}$$

(i) Show that the resulting distribution is exchangeable if and only if $q = 1/2$. (In fact, in that case the distribution is iid.)

(ii) Evaluate the power by applying the procedure of the previous problem to various settings: try $n \in \{10, 10^2, 10^3\}$ and for each n choose a grid of values for q in $[1/2, 1]$ so that a transition from powerless to powerful as q decreases towards 1 is clearly visible. Repeat each setting 10^3 times. Draw a power curve for each n. (Note that, as q approaches 0, the sequence is more and more mixed.)

Problem 15.49 The test procedures described here are based on first-order transitions (of the form $a \to b$). How would test procedures based on second-order transitions look like? Implement such a procedure in R, and apply it to the setting of the previous problem.

15.8 Further Topics

15.8.1 Pearson's Approximation

Before the advent of computers, computing logarithms was not trivial. Karl Pearson[60] (1857–1936) suggested an approximation to the log LRs appearing in (15.1) and (15.3) that could be computed using simpler calculations [141].

Take (15.1) for simplicity. Pearson's approximation is based on two facts:

(i) Under the null hypothesis, $\hat{\theta}_s \to_P \theta_{0,s}$ as the sample size increases (due to the Law of Large Numbers), and this is true for all s.

(ii) Based on a Taylor development of the logarithm,

$$x\log(x/x_0) = x - x_0 + \frac{(x - x_0)^2}{2x_0} + O(x - x_0)^3,$$

when $x \to x_0 > 0$.

Problem 15.50 Use these facts to show that, under the null hypothesis,

$$\sum_{s=1}^{m} Y_s \log\left(\frac{Y_s}{n\theta_{0,s}}\right) = (1/2 + R_n) \sum_{s=1}^{m} \frac{(Y_s - n\theta_{0,s})^2}{n\theta_{0,s}},$$

where R_n is an unspecified term that converges to 0 in probability as $n \to \infty$. The sum of the right-hand side is Pearson's statistic.

15.9 Additional Problems

Problem 15.51 Another reasonable way to obtain a test statistic in the context of Section 15.1 is to come up with an estimator $\hat{\theta}$ for θ (for example the MLE) and use as test statistic $\mathscr{L}(\hat{\theta}, \theta_0)$, where $\mathscr{L}$ is some predetermined loss function, e.g., $\mathscr{L}(\theta, \theta_0) = \|\theta - \theta_0\|^2$, where $\|\cdot\|$ denotes here the Euclidean norm. In R, perform some simulations to compare the resulting test with the Pearson test of Section 15.8.1.

Problem 15.52 (Daily 3 lottery) The *Daily 3* is a lottery game run by the State of California. Each day of the year, three digits are drawn independently and uniformly at random from $\{0, \ldots, 9\}$. Note that the order matters. According the website: "The draws are conducted using an Automated Draw Machine, which is a state-of-the-art computer used to draw winning numbers."

---

[60] This is the father of Egon Pearson [55].

- *Independent digits*   Suppose we are willing to assume that the digits are generated independently. The question that remains, then, is whether they are generated uniformly at random.
- *Independent daily draws*   Suppose we are not willing to assume a priori that the digits are generated independently, but we are willing to assume that the daily draws (each consisting of 3 digits) are generated independently.

In both situations, test for uniformity using the function of Problem 15.7 based on all past winning numbers.[61]

Problem 15.53 (continued) Although the second test makes fewer assumptions, it is not necessarily better than the first test in terms of power. Indeed, while by construction the first test is insensitive to any dependency between the digits in a draw, it is more powerful than the second test if there is no such dependency. Perform some numerical simulations to probe this claim.

Problem 15.54 Consider a goodness-of-fit situation with m possible values for each trial and assume that n trials are performed. Suppose we want to test

$$\mathcal{H}_0 : \text{the distribution is uniform,}$$

versus

$$\mathcal{H}_1 : \text{the distribution has support of size } \lfloor m/2 \rfloor.$$

These hypotheses are seemingly quite 'far apart', but in fact this really depends on how large m is compared to n.

(i)   Show that, if $m, n \to \infty$ with $n \ll \sqrt{m}$, then no test has any power in the limit, meaning that any level α test has limiting power at most α.

(ii)   Confirm this with numerical experiments. In R, perform some simulations to evaluate the power of the LRT. Set the level at $\alpha = 0.01$. Try $m \in \{10^2, 10^3, 10^4\}$ and $n = \lfloor \sqrt{m} \rfloor$.

Problem 15.55 (Group sizes and power)   Consider a simple setting where we want to compare two coins in terms of their chances of landing heads. Coin j is a θ_j-coin and is tossed n_j times, for $j \in \{1, 2\}$. Fix the total sample size at $n = n_1 + n_2 = 100$. Also, fix θ_1 at $1/2$. For $n_2 \in \{10, 20, 30, 40, 50\}$, evaluate the power of the LRT as a function of θ_2 carefully chosen on a grid that changes with n_2. A possible way to present the results is to set the level at $\alpha = 0.10$ and draw the (estimated) power curve as a function of θ_2 for each n_2.

---

[61] https://calottery.com/play/draw-games/daily-3

Problem 15.56 (The number π) The digits defining π in base 10 behave very much like a sequence of iid random variables from the uniform distribution on $\{0,\ldots,9\}$. Consider the first $n = 20000$ digits.[62]

(i)   Ignoring the order, could these numbers be construed as a sample from the uniform distribution?

(ii) Taking the order into account, do these numbers 'look like' they were sampled iid from some distribution?

Problem 15.57 (Gauss–Kuzmin distribution) Let X be uniform in $[0,1]$ and let (K_m) denote the coefficients in the continued fraction expansion of X, meaning that

$$X = \cfrac{1}{K_1 + \cfrac{1}{K_2 + \dots}}.$$

Then (K_m) converges weakly to the *Gauss-Kuzmin distribution*, defined by its mass function

$$f(k) := -\log_2\left(1 - 1/(k+1)^2\right), \quad \text{for } k \ge 1.$$

($\log_2$ denotes the logarithm in base 2.) Perform some simulations in R to numerically corroborate this statement.

Problem 15.58 (Racial discrimination in the labor market) The article [13] describes an experiment where resumés are sent in response to real job ads in Boston and Chicago with randomly assigned African-American- or White- sounding names. Look at the data summarized in Table 1 of that paper. Identify and then apply the most relevant testing procedure introduced in the present chapter.

Problem 15.59 (Proportions test) The authors of [13] used a procedure not introduced in the present chapter called the *proportions test*, which is based on the fact that the difference of the sample proportions is approximately normal. In general, consider an experiment as in Table 15.3. Define $\bar{Z}_j = Z_j/n_j$ and $\bar{Z} = (Z_1 + Z_2)/(n_1 + n_2)$. Show that, under the null hypothesis,

$$\frac{\bar{Z}_1 - \bar{Z}_2}{\sqrt{\bar{Z}(1-\bar{Z})(1/n_1 + 1/n_2)}} \xrightarrow{\mathcal{L}} \mathcal{N}(0,1),$$

in the large-sample limit where n_1 and n_2 diverge to infinity. Based on this, obtain a p-value. (Note that the p-value will only be valid in the large-sample limit.) Apply the resulting testing procedure to the data of Table 1 in [13]. [In R, this test is implemented in the prop.test function.]

---

[62] Available at https://oeis.org/A000796/b000796.txt

Problem 15.60 Consider drawing n times without replacement from an urn with v_j balls marked with the number j, where $j \in \{1,\dots,m\}$. Let Y_j denote the number of times that a ball numbered j was drawn. Show that $(Y_1,\dots,Y_m)$ is sufficient for $(v_1,\dots,v_m)$. Assume that the total number of balls in the urn, $v := v_1 + \cdots + v_m$, is known. Can you derive the MLE for $(v_1,\dots,v_m)$?

Problem 15.61 (Sequential permutation test) Computing a permutation p-value is typically a very heavy burden, and is often out of reach. In that case, one resorts to Monte Carlo sampling to estimate the permutation p-value, but in situations where the rejection level is quite small (as is typically the case in multiple testing, see Chapter 20), this requires sampling very many permutations, which can also be costly. Can you see a way of operating sequentially, as in Section 14.4.1, so that if there is early evidence that the p-value is small/large the test rejects/accepts before sampling the a priori required number of permutations? For background, see [117].

16

One Numerical Sample

We consider, in this chapter, an experiment that yields real-valued data. We model such data as a sample of independent and identically distributed (real-valued) random variables, $X_1, \ldots, X_n$, with common distribution denoted P, having distribution function F, and density f when absolutely continuous. We will let $x_1, \ldots, x_n$ denote a realization of $X_1, \ldots, X_n$, and denote $\boldsymbol{X} = \boldsymbol{X}_n = (X_1, \ldots, X_n)$ and $\boldsymbol{x} = \boldsymbol{x}_n = (x_1, \ldots, x_n)$, the latter gathering the observed data. Throughout, we will let X denote a generic random variable with distribution P. The distribution P is assumed to belong to some class of distributions on the real line, which will be taken to be all such distributions whenever that class is not specified. The goal, as usual, is to infer this distribution, or some of its features, from the observed data.

Example 16.1 (Exoplanets) The *Extrasolar Planets Encyclopaedia* (accessible online at https://exoplanet.eu/) offers a catalog of confirmed exoplanets together with some characteristics of these planets (e.g., mass).

Remark 16.2 There is a statistical model in the background, $(\Omega, \Sigma, \mathcal{P})$, which will be left implicit, except that $\mathbb{P} \in \mathcal{P}$ will be used on occasion to denote the (true) underlying distribution.

16.1 Order Statistics

Order $X_1, \ldots, X_n$ to get the so-called *order statistics*, typically denoted

$$X_{(1)} \le \cdots \le X_{(n)}.$$

In particular, $X_{(1)} = \min(X_1, \ldots, X_n)$ and $X_{(n)} = \max(X_1, \ldots, X_n)$. Each $X_{(k)}$ is indeed a bona fide statistic since it is a function of $\boldsymbol{X}$.

Problem 16.3 Derive the distribution of the random vector $(X_1, \ldots, X_n)$ given that $(X_{(1)}, \ldots, X_{(n)}) = (y_1, \ldots, y_n)$, where $y_1 \le \cdots \le y_n$. [Note that this distribution only depends on $(y_1, \ldots, y_n)$.]

230

This proves that the order statistics are jointly sufficient, regardless of the assumed statistical model. That the order statistics are sufficient is intuitively clear. Indeed, when reducing the sample to the order statistics, all that is lost is the order, and that order does not carry any information because the sample is assumed to be iid.

Problem 16.4 Show that, if the underlying distribution P is continuous, then $X_1, \ldots, X_n$ are all distinct with probability 1, implying in particular that, with probability 1,

$$X_{(1)} < \cdots < X_{(n)}.$$

16.2 Empirical Distribution

The *empirical distribution* is defined as the uniform supported on the sample itself, $\{x_1, \ldots, x_n\}$, meaning

$$\widehat{\mathsf{P}}_x(\mathcal{A}) := \frac{\#\{i : x_i \in \mathcal{A}\}}{n}, \quad \text{for } \mathcal{A} \subset \mathbb{R}. \tag{16.1}$$

We remark that a sample of size k can be generated from $\widehat{\mathsf{P}}_x$ by sampling uniformly at random *with replacement* k times from $\{x_1, \ldots, x_n\}$.

R corner In R, the function sample can be used to sample (with or without replacement) from a finite set, where a finite set is represented by a vector.

As a function of $X_1, \ldots, X_n$, $\widehat{\mathsf{P}}_X$ is a random distribution on the real line. (We sometimes drop the subscript in what follows.)

Problem 16.5 (Consistency of the empirical distribution) Using the Law of Large Numbers, show that, for any Borel set $\mathcal{A} \in \mathcal{B}$,

$$\widehat{\mathsf{P}}_{X_n}(\mathcal{A}) \overset{\mathrm{P}}{\longrightarrow} \mathsf{P}(\mathcal{A}), \quad \text{as } n \to \infty. \tag{16.2}$$

16.2.1 Empirical Distribution Function

The *empirical distribution function* is the distribution function of the empirical distribution defined in (16.1), and is given by

$$\widehat{\mathsf{F}}_x(x) := \frac{1}{n} \sum_{i=1}^{n} \{x_i \le x\}.$$

See Figure 16.1 for an illustration.

R corner The function ecdf takes a sample (in the form of a numerical vector) and returns the empirical distribution function.

Problem 16.6 When the observations are all distinct, show that $\widehat{F}_x$ is a step function jumping an amount of $1/n$ at each x_i, and

$$\widehat{F}_x(x_{(i)}) = \frac{i}{n}, \quad \text{for all } i = 1, \ldots, n.$$

In general, if $x_{(i-1)} < x_{(i)} = \cdots = x_{(i+k-1)} < x_{(i+k)}$, then $\widehat{F}_x$ jumps an amount of k/n at $x_{(i)}$

Seen as a function of $X_1, \ldots, X_n$, $\widehat{F}_X$ is a random distribution function. (We sometimes drop the subscript in what follows.)

Problem 16.7 (Consistency of the empirical distribution function) Using the Law of Large Numbers, show that, for any $x \in \mathbb{R}$,

$$\widehat{F}_{X_n}(x) \xrightarrow{\text{P}} F(x), \quad \text{as } n \to \infty. \tag{16.3}$$

Thus, the empirical distribution function is a pointwise consistent estimator of the distribution function. In fact, the convergence is uniform over the whole real line.

Theorem 16.8 (Glivenko–Cantelli Theorem[63]). *In the present context of an empirical distribution function based on an iid sample of size n with distribution function* F, *denoted X_n,*

$$\sup_{x \in \mathbb{R}} \left| \widehat{F}_{X_n}(x) - F(x) \right| \xrightarrow{\text{P}} 0, \quad \text{as } n \to \infty.$$

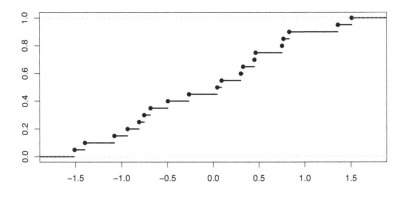

Figure 16.1 A plot of the empirical distribution function of a sample of size $n = 20$ drawn from the standard normal distribution.

[63] Named after Valery Glivenko (1897–1940) and Cantelli[33].

In fact, the convergence happens at the $\sqrt{n}$ rate.

Theorem 16.9 (Dvoretzky–Kiefer–Wolfowitz [55]). *In the context of Theorem 16.8, assuming that* F *is continuous, for all* $t \geq 0$,

$$P\left(\sup_{x \in \mathbb{R}} \left| \widehat{F}_{X_n}(x) - F(x) \right| \geq t/\sqrt{n} \right) \leq 2\exp(-2t).$$

Remark 16.10 (Continuous interpolation) Even if the underlying distribution function is continuous, its empirical counterpart is a step function. For this reason, it is sometimes preferred to use a continuous variant of the empirical distribution function. A popular one is the function that linearly interpolates the points

$$(x_{(1)}, 1/n), (x_{(2)}, 2/n), \ldots, (x_{(n)}, 1).$$

(This assumes the observations are distinct.) The function can be defined to take the value 1 at $x > x_{(n)}$, but it is not clear how to define this function at $x < x_{(1)}$. An option is to define it as 0 there, but in case the resulting function is discontinuous at $x_{(1)}$. If the underlying distribution is known to be supported on the positive real line, for example, then the function can be made to linearly interpolate $(0,0)$ and $(x_{(1)}, 1/n)$, and take the value 0 at $x < 0$.

16.2.2 Empirical Quantile Function

The *empirical quantile function* is simply the quantile function of the empirical distribution, or equivalently, the pseudo-inverse defined in (3.9) of the empirical distribution function. (If one prefers the variant of the empirical distribution function defined in Remark 16.10, then its pseudo-inverse should be preferred also.)

Problem 16.11 The function quantile in R computes quantiles in a number of ways. In fact, the method offers no fewer than 9 ways of doing so.

(i)   What type of quantile corresponds to our definition?

(ii)   What type of quantile corresponds to the pseudo-inverse (3.12) of the empirical distribution function?

(iii)   What type of quantile corresponds to the pseudo-inverse of the piecewise linear variant of the empirical distribution function defined in Remark 16.10?

Remark 16.12 When the observations are distinct, according to the definition given in (3.11), $x_{(i)}$ is a u-quantile of the empirical distribution for any $(i-1)/n \leq u \leq i/n$.

Problem 16.13 (Consistency of the empirical quantile function) Show that, at any point u where F is continuous and strictly increasing,

$$\widehat{\mathsf{F}}^{-}_{X_n}(u) \xrightarrow{\mathrm{P}} \mathsf{F}^{-}(u), \quad \text{as } n \to \infty,$$

[This is based on the consistency of $\widehat{\mathsf{F}}$, see (16.3), and arguments similar to those underlying Problem 8.49.]

16.2.3 Histogram

Assume that P has a density, denoted f. Is there an empirical equivalent to f? Obviously, the empirical distribution, being discrete, does not have a density but rather a mass function.

The idea behind the construction of a *histogram* is to consider averages of the mass function over short intervals so that it provides a piecewise constant approximation to the underlying density. See Figure 16.2 for an illustration.

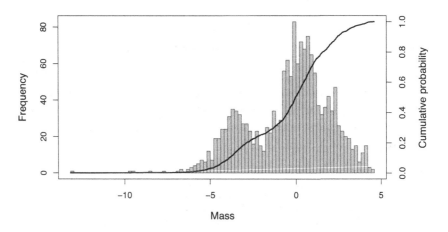

Figure 16.2 A histogram and a plot of the distribution function of the data described in Example 16.1, meaning of the mass of 1582 exoplanets discovered in or before 2017. The mass is measured in Jupiter mass (M_{Jup}), presented here in logarithmic scale for clarity.

This approximation turns out to be pointwise consistent under some conditions. See Figure 16.3 for an illustration.

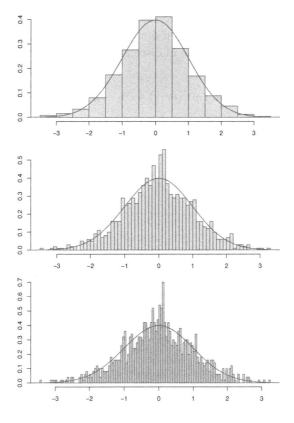

Figure 16.3 Histograms of a sample of size $n = 10^3$ drawn from the standard normal distribution with different number of bins. (The bins themselves where automatically chosen by the function hist.)

Consider a strictly increasing sequence $(a_k : k \in \mathbb{Z})$, which defines a partition of the real line into the intervals $\{(a_{k-1}, a_k] : k \in \mathbb{Z}\}$, which often called *bins* in the present context. Define the corresponding *counts*, also called *frequencies*, as

$$y_k := \#\{i : x_i \in (a_{k-1}, a_k]\}, \quad k \in \mathbb{Z},$$

with the corresponding random variables being denoted $(Y_k : k \in \mathbb{Z})$. We have that $Y_k \sim \mathrm{Bin}(n, p_k)$ with

$$p_k := \mathsf{F}(a_k) - \mathsf{F}(a_{k-1}),$$

which can therefore be estimated by $\hat{p}_k := y_k/n$.

We consider the question of consistency in the context where f is continuous. (The discussion that follows extends without difficulty to the case where f has a finite number of discontinuities.) In that case, if $a_k - a_{k-1}$ is small, then

$$p_k = F(a_k) - F(a_{k-1}) \approx (a_k - a_{k-1})f(a_k). \qquad (16.4)$$

The approximation is in fact exact to first order.

The histogram with bins defined by (a_k) is the piecewise constant function

$$\hat{f}_{X_n}(x) := \frac{y_k}{n(a_k - a_{k-1})}, \quad \text{when } x \in (a_{k-1}, a_k].$$

For $x \in (a_{k-1}, a_k]$,

$$\hat{f}_{X_n}(x) \xrightarrow{\text{P}} \frac{p_k}{(a_k - a_{k-1})} \approx f(a_k), \quad \text{as } n \to \infty,$$

where the approximation is valid when $a_k - a_{k-1}$ is small, as seen in (16.4).

For $\hat{f}_{X_n}$ to be consistent for f, it is thus necessary that the bins become smaller and smaller as the sample size increases. Below we let the bins depend on n and denote $(a_{k,n})$ the sequence defining the bins.

Problem 16.14 Suppose that

$$\max_k (a_{k,n} - a_{k-1,n}) \to 0, \quad \text{with } \min_k (a_{k,n} - a_{k-1,n}) \gg 1/n. \qquad (16.5)$$

Show that, at any point $x \in \mathbb{R}$,

$$\hat{f}_{X_n}(x) \xrightarrow{\text{P}} f(x), \quad \text{as } n \to \infty.$$

To better appreciate the crucial role that (16.5) plays, consider the regular grid $a_{k,n} = k/n$, so that all bins have size $1/n$. Show that, when $f(x) > 0$,

$$\mathbb{P}(\hat{f}_{X_n}(x) = 0) \to 1/e, \quad \text{as } n \to \infty.$$

Remark 16.15 (Choice of bins) Choosing the bins automatically is in general a complex task. Often, a regular partition is chosen, for example $a_k = kh$, and even then, the choice of $h > 0$ is nontrivial. It is known that, if the function has a bounded first derivative, a bin size of order $h \propto n^{-1/3}$ is best. Although this can provide some guidance, the best choice depends on the underlying density, resulting in a chicken-and-egg problem.[64]

---

[64] It is possible to choose the bin size h by cross-validation, as Rudemo proposes in [157]. We provide some details in the closely related context of kernel density estimation in Section 16.10.5.

R corner The function hist computes and (by default) plots a histogram based on the data. The function offers the possibility of manually choosing the bins as well as three methods for choosing the bins automatically.

16.3 Inference about the Median

Recall the definition of a u-quantile of P or, equivalently, F, given in (3.11). We called *median* any $1/2$-quantile. In particular, by definition, x is a median of F if (recall (3.10))

$$F(x) \geq 1/2 \text{ and } \widetilde{F}(x) \geq 1/2,$$

or, equivalently in terms of $X \sim P$,

$$\mathbb{P}(X \leq x) \geq 1/2 \text{ and } \mathbb{P}(X \geq x) \geq 1/2.$$

Problem 16.16 Show that these inequalities are in fact equalities when F is continuous at any of its median points.

Problem 16.17 Show that the set of medians is the interval $[a, b)$ where $a := \inf\{x : F(x) \geq 1/2\}$ and $b := \sup\{x : F(x) = F(a)\}$. (Use the convention $[a, a) = \{a\}$.) Conclude that there is a unique median if and only if the distribution function is strictly increasing at any of its median points.

In what follows, to ease the exposition, we consider the case where there is a unique median (denoted μ). The reader is invited to examine how what follows generalizes to the case where the median is not unique.

16.3.1 Sample Median

We already have an estimator of the median, namely, $\widehat{F}_X^-(1/2)$, which is consistent if F is strictly increasing and continuous at μ (Problem 16.13). Any such estimator for the median can be called a *sample median*. Any reasonable definition leads to a consistent estimator.

R corner In R, the function median computes the median of a sample based on a different definition of pseudo-inverse, specifically $\widehat{F}_x^\ominus$ as defined in (3.12). Note that, if all the data points are distinct,

$$\widehat{F}_x^\ominus(1/2) = \begin{cases} x_{(n+1)/2} & \text{if } n \text{ is odd,} \\ \frac{1}{2}\left(x_{(n/2)} + x_{(n/2+1)}\right) & \text{if } n \text{ is even.} \end{cases}$$

16.3.2 Confidence Interval

A (good) confidence interval can be built for the median without really any assumption on the underlying distribution. The interval is of the form $[X_{(k)}, X_{(l)}]$ for some $k \le l$ chosen as functions of the desired level of confidence.

We start with

$$\mathbb{P}(X_{(k)} \le \mu \le X_{(l)}) = \mathbb{P}(X_{(k)} \le \mu) - \mathbb{P}(X_{(l)} < \mu).$$

Let

$$q_k := \text{Prob}\{\text{Bin}(n, 1/2) \ge k\}.$$

We have

$$\mathbb{P}(X_{(k)} \le \mu) = \mathbb{P}\big(\#\{i : X_i \le \mu\} \ge k\big) \ge q_k,$$

because $\#\{i : X_i \le \mu\}$ is binomial with success probability $\mathbb{P}(X \le \mu) \ge 1/2$, and similarly,

$$\mathbb{P}(X_{(l)} < \mu) = \mathbb{P}\big(\#\{i : X_i < \mu\} \ge l\big) \le q_l,$$

because $\mathbb{P}(X < \mu) \le 1/2$. Thus,

$$\mathbb{P}(X_{(k)} \le \mu \le X_{(l)}) \ge q_k - q_l.$$

Hence, $[X_{(k)}, X_{(l)}]$ is a confidence interval for μ at level $q_k - q_l$. Choosing k as the largest integer such that $q_k \ge 1 - \alpha/2$ and l the smallest integer such that $q_l \le \alpha/2$, we obtain a $(1 - \alpha)$-confidence interval for μ.

Problem 16.18 (One-sided interval) Derive a one-sided $(1 - \alpha)$-confidence interval for the median following the same reasoning.

Problem 16.19 In R, write a function that takes as input the data points, the desired confidence level, and the type of interval, and returns the corresponding confidence interval for the median. Try your function on a sample of size $n \in \{10, 20, 30, \ldots, 100\}$ from the exponential distribution with rate 1. Repeat each setting $N = 200$ times and plot the average length of the confidence interval as a function of n.

16.3.3 Sign Test

Suppose we want to test $\mathcal{H}_0 : \mu = \mu_0$. In Section 12.4.7, we saw how to derive a p-value from a procedure for constructing confidence intervals.

Problem 16.20 In R, write a function that takes as input the data points and μ_0, and returns the p-value based on the above procedure for building a confidence interval for the median.

Depending on what variant of the median is used, and on whether there are observed values at the median, the resulting test procedure coincides with, or is very close to, the following test known as the *sign test*. Let

$$Y_+ = \#\{i : X_i > \mu_0\}, \quad Y_- = \#\{i : X_i < \mu_0\}, \quad Y_0 = \#\{i : X_i = \mu_0\}.$$

The test rejects for large values of $S := \max(Y_+, Y_-)$, with the p-value computed conditional on Y_0.

Remark 16.21 The name of the test comes from looking at the sign of $X_i - \mu_0$ and counting how many are positive, negative, or zero.

Problem 16.22 Show that, if the underlying distribution has median μ_0,

$$\mathbb{P}(Y_+ \geq k \mid Y_0 = y_0) \leq \text{Prob}\{\text{Bin}(n - y_0, 1/2) \geq k\},$$
$$\mathbb{P}(Y_- \geq k \mid Y_0 = y_0) \leq \text{Prob}\{\text{Bin}(n - y_0, 1/2) \geq k\},$$

and deduce that

$$\mathbb{P}(S \geq k \mid Y_0 = y_0) \leq 2\,\text{Prob}\{\text{Bin}(n - y_0, 1/2) \geq k\}.$$

[This upper bound can be used to obtain a (conservative) p-value.]

Problem 16.23 Derive the sign test and its (conservative) p-value for testing $\mathcal{H}_0 : \mu \leq \mu_0$.

Problem 16.24 In R, write a function that takes in the data points and μ_0, and the type of alternative, and returns the (conservative) p-value for the corresponding sign test.

16.3.4 Inference about a Quantile

Whatever was said thus far about estimating or testing about the median can be extended to any u-quantile with $0 < u < 1$.

Problem 16.25 Repeat for the 1st quartile what was done for the median.

Estimating the 0-quantile or the 1-quantile amounts to estimating the boundary points of the support of the distribution.

16.4 Possible Difficulties

We consider some emblematic situations where estimating the median is difficult. We do the same for the mean, as a prelude to studying its inference. In the process, we provide some insights into why inference about the mean is much more complicated than inference for the median.

16.4.1 Difficulties with the Median

A difficult situation for inference about the median is when the underlying distribution is flat at the median. For $\theta \in [0, 1]$, consider the following density

$$f_\theta(x) = (1 - \theta) \{x \in [0, 1]\} + \theta \{x \in [2, 3]\},$$

and let F_θ denote the corresponding distribution function.

Problem 16.26 Show that sampling from f_θ amounts to drawing $\xi \sim \mathrm{Ber}(\theta)$ and then drawing $X \sim \mathrm{Unif}(0, 1)$ if $\xi = 0$, and $X \sim \mathrm{Unif}(2, 3)$ if $\xi = 1$.

Problem 16.27 Show that F_θ is flat at its median if and only if $\theta = 1/2$. When this is the case, show that any point in $[1, 2]$ is a median. When this is not the case, meaning when $\theta \neq 1/2$, show that the median is unique and derive it as a function of θ.

Assume we have an iid sample from f_θ of size n and that our goal is to draw some inference about 'the' median.

Problem 16.28 Show that, when $\theta = 1/2$ and n is odd, the sample median belongs to $[0, 1]$ with probability $1/2$, while it belongs to $[2, 3]$ with probability $1/2$.

The difficulty is only in appearance, however. Indeed, the sample median will converge, as the sample size increases, to a median of the underlying distribution and, more importantly, the confidence interval of Section 16.3.2 has the desired confidence no matter what, although it can be quite wide.

Problem 16.29 In R, generate a sample from f_θ of size $n = 101$ (so it is odd) and produce a 95% confidence interval for the median using the function of Problem 16.19. Do that for

$$\theta \in \{0.2, 0.4, 0.45, 0.49, 0.5, 0.51, 0.55, 0.6, 0.8\}.$$

Repeat each setting a few times to get a feel for the randomness.

Remark 16.30 The situation is qualitatively the same when

$$f_\theta(x) = (1 - \theta) f_0(x) + \theta f_1(x),$$

with f_0 and f_1 being densities with disjoint supports.

16.4.2 Difficulties with the Mean

While estimating the median does not pose particular difficulties despite some cases where it is 'unstable', estimating the mean poses very real difficulties, to the point that the problem is almost ill-posed.

For a prototypical example, consider the family of densities

$$f_\theta(x) = (1 - \theta) \{x \in [0, 1]\} + \theta \{[h(\theta), h(\theta) + 1]\},$$

parameterized by $\theta \geq 0$, where $h \colon \mathbb{R}_+ \to \mathbb{R}_+$ is some function.

Problem 16.31 Show that f_θ has mean $1/2 + \theta h(\theta)$.

As before, sampling from f_θ amounts to generating $\xi \sim \mathrm{Ber}(\theta)$, and then drawing $X \sim \mathrm{Unif}([0, 1])$ if $\xi = 0$, and $X \sim \mathrm{Unif}([h(\theta), h(\theta) + 1])$ if $\xi = 1$.

Problem 16.32 In a sample of size n from f_θ, show that the number of points generated from $\mathrm{Unif}([0, 1])$ is binomial with parameters $(n, 1 - \theta)$.

Consider the situation where h is such that $\theta h(\theta) \to \infty$ as $\theta \to 0$, and choose $\theta = \theta_n$ such that $n\theta_n \to 0$. In that case, f_{θ_n} has mean $1/2 + \theta_n h(\theta_n) \to \infty$ as $n \to \infty$, while with probability tending to 1, the entire sample is drawn from $\mathrm{Unif}([0, 1])$, which has mean $1/2$.

Remark 16.33 A very similar difficulty is at the core of the Saint Petersburg Paradox discussed in Section 7.10.3. We saw there that using the median instead of the mean offers an attractive way out of the apparent paradox.

16.5 Bootstrap World

Inference about the mean may be performed using a bootstrap. The reasoning is as follows. If we could sample from P at will, we would be able to estimate any feature of P (including mean, median, quantiles, etc.) by Monte Carlo simulation, and the accuracy of our inference would only be limited by the amount of computational resources at our disposal. Doing this is not possible since P is unknown, but we can estimate it by the empirical distribution. This is justified by the fact that the empirical distribution is a consistent estimator, as seen in (16.2).

The *bootstrap world* is the fictitious, parallel world we construct for the purpose of performing inference. In that world, the empirical distribution is the underlying distribution. The accuracy of the resulting inference depends on how strong of a parallel we can draw between the bootstrap world and the real world.

The beauty of such a construction is that, in principle, we know everything in the bootstrap world, since we have access to the empirical distribution, and, in particular, we can use Monte Carlo simulations to compute any feature of interest of that distribution – which then serves to draw some inference on the corresponding feature of the distribution in the real world.

An asterisk $\ast$ next of a symbol representing some quantity is often used to denote the corresponding quantity in the bootstrap world. For example, if μ denotes the median, then μ^* will denote the median of the empirical distribution, which is none other than the empirical median. In particular, we will use P^* in place of $\widehat{\mathsf{P}}_x$ in what follows to denote the empirical distribution.

16.5.1 Bootstrap Sample

Sampling from P^* is relatively straightforward, since it is the uniform distribution on $\{x_1, \ldots, x_n\}$. A sample (of same size n) drawn from the empirical distribution is called a *bootstrap sample* and denoted $X_1^*, \ldots, X_n^*$. As we saw earlier, a bootstrap sample is generated by sampling *with replacement* n times from $\{x_1, \ldots, x_n\}$.

Remark 16.34 (Ties in the bootstrap sample) Even when all the observations are distinct, a bootstrap sample may include some ties.

Problem 16.35 Compute the probability that there are no ties in a bootstrap sample when $x_1, \ldots, x_n$ are all distinct.

16.5.2 Bootstrap Distribution

Let T be a statistic of interest, and let P_T denote its distribution (meaning, the distribution of $T(X_1, \ldots, X_n)$). Having observed $x_1, \ldots, x_n$, resulting in the empirical distribution P^*, the *bootstrap distribution* of T is the distribution of $T(X_1^*, \ldots, X_n^*)$. We denote this distribution by P_T^*. It is used to estimate the distribution of T.

In practice, only rarely can we obtain P_T^* in closed form. Instead, P_T^* is typically estimated by Monte Carlo simulation, which is available to us since we may sample from P^* at will.

Problem 16.36 In R, generate a sample of size $n \in \{5, 10, 20, 50\}$ from the exponential distribution with rate 1. Let T be the sample mean and estimate its bootstrap distribution by Monte Carlo using $B = 10^4$ replicates. For each n, draw a histogram of this estimate, overlay the density given by the normal approximation, and overlay the density of the gamma distribution with shape parameter n and rate n, which is the (real) distribution of T.

16.5.3 Bootstrap Estimate for the Bias

Suppose a particular statistic T is meant to estimate a feature of the underlying distribution, denoted $\varphi(\mathsf{P})$. Its bias is thus (recall (12.3))

$$b := \mathbb{E}(T) - \varphi,$$

where $\mathbb{E}(T)$ is shorthand for $\mathbb{E}(T(X_1, \ldots, X_n))$ where $X_1, \ldots, X_n$ are iid from P, and φ is shorthand for $\varphi(\mathsf{P})$.

It turns out that b can be estimated by bootstrap. Indeed, in the bootstrap world the corresponding quantity is

$$b^* := \mathbb{E}^*(T) - \varphi^*,$$

where $\mathbb{E}^*(T)$ is shorthand for $\mathbb{E}(T(X_1^*, \ldots, X_n^*))$ where $X_1^*, \ldots, X_n^*$ are iid from P^*, and φ^* is shorthand for $\varphi(\mathsf{P}^*)$. (We assume here that φ applies to discrete distributions.)

In the bootstrap world, we know P^*, and therefore we know b^*, at least in principle. In practice, though, b^* is typically estimated by Monte Carlo simulation by repeatedly sampling from P^*. This estimate for b^* serves as an estimate for b.

16.5.4 Bootstrap Estimate for the Variance

Suppose that we are interested in estimating the variance of a given statistic T. The bootstrap estimate is simply its variance in the bootstrap world, namely, its variance under P^*, or in formula

$$\text{Var}^*(T) = \mathbb{E}^*(T^2) - (\mathbb{E}^*(T))^2.$$

As before, this is estimated by Monte Carlo simulation based on repeatedly sampling from P^*, and that estimate itself serves as an estimate for the variance of T in the real world.

16.6 Inference about the Mean

While the inference about the median can be performed sensibly without really any assumption on the underlying distribution, the same cannot be said of the mean. The reason is that the mean is not as well-behaved as the median, as we saw in Section 16.4. Thus, it might be preferable to focus on the median rather than the mean whenever possible. However, some situations call for inference about the mean. We assume in this section that P has a mean, which we denote by μ.

16.6.1 Sample Mean

A statistic of choice here is the *sample mean* defined as the average of $x_1, \ldots, x_n$, namely $\bar{x} := \frac{1}{n} \sum_{i=1}^{n} x_i$. This is also the mean of the empirical distribution. We will denote the corresponding random variable $\bar{X}$, or $\bar{X}_n$ if we want to emphasize that it was computed from a sample of size n.

The Law of Large Numbers implies that the sample mean is consistent for the mean, that is

$$\bar{X}_n \xrightarrow{\text{P}} \mu, \quad \text{as } n \to \infty.$$

16.6.2 Normal Confidence Interval

Assume that P has variance σ^2. The Central Limit Theorem implies that

$$\frac{\bar{X}_n - \mu}{\sigma / \sqrt{n}} \xrightarrow{\mathcal{L}} \mathcal{N}(0, 1), \quad \text{as } n \to \infty.$$

This in turn implies that, if z_u denotes the u-quantile of the standard normal distribution, we have

$$\mu \in \left[\bar{X}_n - z_{1-\alpha/2} \frac{\sigma}{\sqrt{n}}, \bar{X}_n - z_{\alpha/2} \frac{\sigma}{\sqrt{n}} \right] \tag{16.6}$$

with probability converging to $1 - \alpha$ as $n \to \infty$.

If σ^2 is known, then the interval in (16.6) is a bona fide confidence interval and its level is $1 - \alpha$ in the large-sample limit, although the confidence level at a given sample size n will depend on the underlying distribution.

Problem 16.37 Bound the confidence level from below using Chebyshev's inequality.

If σ^2 is unknown, we estimate it using the *sample variance*, which may be defined as

$$S^2 := \frac{1}{n} \sum_{i=1}^{n} (X_i - \bar{X})^2. \tag{16.7}$$

This is the variance of the empirical distribution. By Slutsky's theorem, in conjunction with the Central Limit Theorem, it holds that

$$\frac{\bar{X}_n - \mu}{S_n/\sqrt{n}} \xrightarrow{\mathcal{L}} \mathcal{N}(0,1), \quad \text{as } n \to \infty,$$

which in turn implies that

$$\mu \in \left[\bar{X}_n - z_{1-\alpha/2} \frac{S_n}{\sqrt{n}}, \bar{X}_n - z_{\alpha/2} \frac{S_n}{\sqrt{n}} \right] \tag{16.8}$$

with probability converging to $1 - \alpha$ as $n \to \infty$.

Remark 16.38 (Unbiased sample variance) The following variant is sometimes used in place of (16.7)

$$\frac{1}{n-1} \sum_{i=1}^{n} (X_i - \bar{X})^2. \tag{16.9}$$

(This is what the R function var computes.) This variant happens to be unbiased (Problem 16.89). In practice, the two variants are, of course, very close to each other, unless the sample size is quite small.

Student Confidence Interval

When $X_1, \ldots, X_n$ are iid from a normal distribution,

$$\frac{\bar{X} - \mu}{S/\sqrt{n-1}}$$

has the Student distribution with parameter $n - 1$. In particular, if t_u^n denotes the u-quantile of this distribution, then

$$\mu \in \left[\bar{X}_n - t_{1-\alpha/2}^{n-1} \frac{S_n}{\sqrt{n-1}}, \bar{X}_n \; t_{\alpha/2}^{n-1} \frac{S_n}{\sqrt{n-1}} \right] \tag{16.10}$$

with probability $1 - \alpha$ when the underlying distribution is normal. In general, this is only true in the large-sample limit.

Problem 16.39 Show that the Student distribution with n degrees of freedom converges to the standard normal distribution as n increases. Deduce that, for any $u \in (0, 1)$, $t_u^n \to z_u$ as $n \to \infty$.

Remark 16.40 The Student confidence interval (16.10) appears to be more popular than the normal confidence interval (16.8).

R corner The family of Student distributions is available via the functions dt (density), pt (distribution function), qt (quantile function), and rt (pseudo-random number generator). The Student confidence intervals and the corresponding tests can be computed using the function t.test.

16.6.3 Bootstrap Confidence Interval

When σ is known, the confidence interval in (16.6) relies on the fact that the distribution of $\bar{X} - \mu$ is approximately normal with mean 0 and variance σ^2/n (that is, if n is large enough).

Remark 16.41 The random variable $\bar{X} - \mu$ is often called a *pivot*. It is *not* a statistic, as it cannot be computed from the data alone (since μ is unknown).

Let us carefully examine the process of deriving this confidence interval. Let Q denote the distribution of $\bar{X} - \mu$. Let q_u denote a u-quantile of Q. In view of Problem 3.14 and Problem 3.15,

$$\mathbb{P}(\bar{X} - \mu \le q_{1-\alpha/2}) \ge 1 - \alpha/2, \quad \mathbb{P}(\bar{X} - \mu < q_{\alpha/2}) \le \alpha/2,$$

so that

$$\mathbb{P}(q_{\alpha/2} \le \bar{X} - \mu \le q_{1-\alpha/2}) \ge 1 - \alpha,$$

or, equivalently,

$$\mu \in \left[\bar{X} - q_{1-\alpha/2}, \bar{X} - q_{\alpha/2} \right], \tag{16.11}$$

with probability at least $1 - \alpha$. The issue here, of course, is that this construction relies on Q, which is unknown, so that this interval is *not* a bona fide confidence interval. In (16.6), Q is approximated by a normal distribution. Here we estimate Q by bootstrap instead.

The bootstrap estimation of Q is done, as usual, by going to the bootstrap world, where the parallel is rather natural in the present context:

Real world	Bootstrap world
$\bar{X}$ = mean of n-sample from P	$\bar{X}^*$ = mean of n-sample from P*
μ = mean of P	μ^* = mean P*

As we know, $\mu^* = \bar{x}$, which makes the correspondence above particularly simple. Let Q^* denote the distribution function of $\bar{X}^* - \mu^*$, which is the bootstrap-world equivalent of Q.

Remark 16.42 In practice, Q^* is itself estimated by Monte Carlo simulation from P^*, and that estimate is our estimate for Q. Below, we reason as if we knew Q^*, or equivalently, as if we had infinite computational power and we had the luxury of drawing an infinite number of Monte Carlo samples. Doing so allows us to separate statistical issues from computational issues.

Having computed Q^*, a confidence interval is built as before. Let q_u^* denote the u-quantile of Q^*. The bootstrap $(1 - \alpha)$-confidence interval for μ is obtained by plugging q^* in place of q in (16.11), resulting in

$$\left[\bar{X} - q_{1-\alpha/2}^*, \bar{X} - q_{\alpha/2}^* \right]. \tag{16.12}$$

(Note that it is $\bar{X}$ and *not* $\bar{X}^*$. The latter does not really have a meaning since it denotes the average of a generic bootstrap sample and the procedure is based on drawing many such samples.)

Problem 16.43 In R, write a function that takes in the data, the desired confidence level, and a number of Monte Carlo replicates, and returns the bootstrap confidence interval (16.12).

16.6.4 Bootstrap Studentized Confidence Interval

Instead of using $\bar{X} - \mu$ as pivot as we did in Section 16.6.3, we now use $(\bar{X} - \mu)/S$. The process of deriving a bootstrap confidence interval is then completely parallel. (We only repeat it for the reader's convenience, however the reader is invited to anticipate what follows.)

Redefine Q as the distribution of $(\bar{X} - \mu)/S$, and q_u as the u-quantile of this Q. Then

$$\mu \in \left[\bar{X} - S q_{1-\alpha/2}, \bar{X} - S q_{\alpha/2} \right], \tag{16.13}$$

with probability at least $1 - \alpha$. In (16.8), Q is approximated by a normal distribution. Here we estimate Q by bootstrap instead.

Suppose we have observed $x_1, \ldots, x_n$. In the bootstrap world, the equivalent of $(\bar{X} - \mu)/S$ is $(\bar{X}^* - \mu^*)/S^*$, where S^* is the sample standard deviation of a bootstrap sample of size n. Let Q^* denote the distribution function of $(\bar{X}^* - \mu^*)/S^*$ and let q_u^* denote its u-quantile. The bootstrap Studentized $(1 - \alpha)$-confidence interval for μ is obtained by plugging q^* in place of q in (16.13), resulting in

$$\left[\bar{X} - S q^*_{1-\alpha/2}, \bar{X} - S q^*_{\alpha/2}\right]. \tag{16.14}$$

(Note that it is $\bar{X}$ and S, and *not* $\bar{X}^*$ and S^*.)

Problem 16.44 In R, write a function that takes in the data, the desired confidence level, and number of Monte Carlo replicates, and returns the bootstrap confidence interval (16.14).

Remark 16.45 (Comparison) The Studentized version is typically more accurate. You are asked to numerically probe this in Problem 16.94.

16.6.5 Bootstrap Tests

Suppose we want to test some null hypothesis about the mean such as $\mathcal{H}_0 : \mu = \mu_0$. Here are two natural bootstrap approaches:

- *Via a confidence interval*   In Section 12.4.7, we saw how to derive a p-value from a confidence interval procedure, which in the present context could be (16.12) or the Studentized version (16.14).
- *Direct*   Suppose we want to reject for large values of $\bar{X}$. The idea is to construct a *null version of the bootstrap world* and estimate the distribution of $\bar{X}$ under the null hypothesis by its distribution in that bootstrap world. A natural way to build such a bootstrap world is to translate P^* by $\mu_0 - \mu^*$, obtaining P_0^*, which has mean μ_0 by construction. It turns out that this approach is equivalent to testing based on the confidence interval (16.12). The same procedure applied, instead, to the test statistic $(\bar{X} - \mu_0)/S$ is, analogously, equivalent to the testing based on the confidence interval (16.14).

16.6.6 Inference about Moments and other Parameters

The bootstrap approach to drawing inference about the mean generalizes to other moments, and more generally, to any expectation such as $\mu_\psi :=$ $\mathbb{E}(\psi(X))$, where ψ is a given function. This is because, if we define $Y_i = \psi(X_i)$, then $Y_1, \ldots, Y_n$ are iid with mean μ_ψ.

Problem 16.46 Define a bootstrap confidence interval for the 2nd moment, first without and then with Studentization.

16.7 Inference about the Variance and other Parameters

A confidence interval for the variance σ^2 can be obtained by computing a confidence interval for the mean and a confidence interval for the 2nd moment, and combining these to get a confidence interval for the variance, since

$$\text{Var}(X) = \mathbb{E}(X^2) - (\mathbb{E}(X))^2.$$

However, there is a more direct route, which is typically preferred.

In general, consider a feature of interest $\varphi(\mathsf{P})$. We assume that φ applies to discrete distributions. In that case, a natural estimator is the plug-in estimator $\varphi(\mathsf{P}^*)$. A bootstrap confidence interval can then be derived as we did for the mean in Section 16.6.3, using $\varphi(\mathsf{P}^*) - \varphi(\mathsf{P})$ as pivot.

Indeed, let v_u denote the u-quantile of this pivot. Then

$$\varphi(\mathsf{P}) \in \left[\varphi(\mathsf{P}^*) - v_{\alpha/2}, \varphi(\mathsf{P}^*) - v_{1-\alpha/2} \right]$$

with probability at least $1 - \alpha$. Since v_u is not available, we go to the bootstrap world to fetch a proxy. There, the analogue to the pivot is $\varphi(\mathsf{P}^{**}) - \varphi(\mathsf{P}^*)$, where P^{**} is the empirical distribution function of a sample of size n from P^*. We then estimate v_u by v_u^*, the u-quantile of the bootstrap distribution of $\varphi(\mathsf{P}^{**}) - \varphi(\mathsf{P}^*)$, resulting in

$$\varphi(\mathsf{P}) \in \left[\varphi(\mathsf{P}^*) - v_{\alpha/2}^*, \varphi(\mathsf{P}^*) - v_{1-\alpha/2}^* \right]$$

with approximate probability $1 - \alpha$ under suitable circumstances.

Problem 16.47 We consider this general procedure to derive a confidence interval for the variance. First, describe the procedure on paper. Then, implement the procedure in R. Perform some simple numerical experiments to test its accuracy.

Remark 16.48 A bootstrap Studentized confidence interval can also be constructed based on $(\varphi(\mathsf{P}^*) - \varphi(\mathsf{P}))/D$ as pivot, where D is an estimate for the standard deviation of $\varphi(\mathsf{P}^*)$. Unless a simpler estimator is available, we can always use the bootstrap estimate for the standard deviation. If this is our D, then the construction of the Studentized confidence interval requires the computation of a bootstrap estimate for the variance within the bootstrap world. A direct implementation of this requires a loop within a loop, which may be computationally burdensome.

16.8 Goodness-of-Fit Testing and Confidence Bands

Suppose we want to test whether the underlying distribution is a given distribution, often called the *null distribution* and denoted P_0 henceforth, with distribution function F_0 and (when applicable) density f_0. Recall that we considered this problem in the discrete setting in Section 15.1.

If this happens in the context of a family of distributions that admits a simple parameterization, the hypothesis testing problem will likely be approachable via a standard test, in particular, the likelihood ratio test (LRT). This is the case, for example, when we assume that the underlying distribution is of the form $\mathrm{Beta}(\theta, 1)$ for some $\theta > 0$, and we are testing whether the distribution is the uniform distribution on $[0, 1]$, which is equivalent in this context to testing whether $\theta = 1$.

We place ourselves in a setting where no simple parameterization is available. In that case, a plug-in approach points to comparing the empirical distribution with the null distribution – since the empirical distribution is always available as an estimate of the underlying distribution. We discuss two approaches for doing so: one based on comparing distribution functions and another one based on comparing densities.

Remark 16.49 (Tests for uniformity) Under the null distribution, the transformed data $U_1, \ldots, U_n$, with $U_i := F_0(X_i)$, are iid uniform in $[0, 1]$. In principle, therefore, testing for a particular null distribution can be reduced to *testing for uniformity*, that is, the special case where the null distribution is $P_0 = \mathrm{Unif}(0, 1)$. For pedagogical reasons, we chose not to work with this reduction in what follows. (See Section 16.10.1 for additional details.)

16.8.1 Tests based on the Distribution Function

A goodness-of-fit test based on comparing distribution functions is generally based on a statistic of the form

$$\Delta(\widehat{F}_X, F_0),$$

where Δ is a measure of dissimilarity between distribution functions. There are many such dissimilarities and we present a few classical examples. In each case, large values of the statistic weigh against the null hypothesis.

Kolmogorov–Smirnov Test

This test[65] uses the supremum norm as a measure of dissimilarity, namely

$$\Delta(\mathsf{F},\mathsf{G}) := \sup_{x\in\mathbb{R}}|\mathsf{F}(x) - \mathsf{G}(x)|. \qquad (16.15)$$

Problem 16.50 Show that

$$\Delta(\widehat{\mathsf{F}}_X,\mathsf{F}_0) = \max_{i=1,\dots,n} \max\left\{\frac{i}{n} - \mathsf{F}_0(X_{(i)}), \mathsf{F}_0(X_{(i)}) - \frac{i-1}{n}\right\}.$$

Calibration is (obviously) by Monte Carlo under F_0. This calibration by Monte Carlo necessitates the use of a computer, yet the method was in use before the advent of computers. What made this possible is the following.

Proposition 16.51. *The distribution of $\Delta(\widehat{\mathsf{F}}_X,\mathsf{F}_0)$ under F_0 does not depend on F_0 as long as F_0 is continuous.*

Proof We prove the result in the special case where F_0 is strictly increasing on the real line. The key point is that, under F_0, $\mathsf{F}_0(X) \sim \mathrm{Unif}(0,1)$. Let $U_i = \mathsf{F}_0(X_i)$ and let $\widehat{\mathsf{G}}$ denote the empirical distribution function of $U_1,\dots,U_n$. Also, let $\widehat{\mathsf{F}}$ be shorthand for $\widehat{\mathsf{F}}_X$. For $x \in \mathbb{R}$, let $u = \mathsf{F}_0(x)$, and derive

$$\widehat{\mathsf{F}}(x) - \mathsf{F}_0(x) = \widehat{\mathsf{F}}(\mathsf{F}_0^{-1}(u)) - \mathsf{F}_0(\mathsf{F}_0^{-1}(u))$$
$$= \widehat{\mathsf{G}}(u) - u.$$

Thus, because $\mathsf{F}_0 \colon \mathbb{R} \to (0,1)$ is one-to-one, we have

$$\sup_{x\in\mathbb{R}}|\widehat{\mathsf{F}}(x) - \mathsf{F}_0(x)| = \sup_{u\in(0,1)}|\widehat{\mathsf{G}}(u) - u|.$$

Although the computation of $\widehat{\mathsf{G}}$ surely depends on F_0 (since F_0 is used to define the U_i), clearly its distribution under F_0 does not. Indeed, it is simply the empirical distribution function of an iid sample of size n from $\mathrm{Unif}(0,1)$. Note that the function $u \mapsto u$ coincides with the distribution function of $\mathrm{Unif}(0,1)$ on the interval $(0,1)$. □

Remark 16.52 In the pre-computer days, the distribution of $\Delta(\widehat{\mathsf{F}}_X,\mathsf{F}_0)$ under F_0 was obtained in the special case where F_0 is the distribution function of $\mathrm{Unif}(0,1)$. There are recursion formulas for the exact computation of the p-value. The large-sample limiting distribution, known as the *Kolmogorov distribution*, was derived by Kolmogorov [108] and tabulated by Smirnov [170], and was used for larger sample sizes. Further details are provided in Problem 16.92.

---

[65] Named after Kolmogorov [1] and Nikolai Smirnov (1900–1966).

R corner In R, the test is implemented in ks.test, which returns a warning if there are ties in the data. The Kolmogorov distribution function is available in the package kolmim.

Remark 16.53 The recursion formulas just mentioned are only valid when there are no ties in the data, a condition which is satisfied (with probability one) in the context of Proposition 16.51, as F_0 is assumed there to be continuous. Although some strategies are available for handling ties, a calibration by Monte Carlo simulation is always available and accurate.

Problem 16.54 In R, write a function that mimics ks.test but instead returns a Monte Carlo p-value based on a specified number of replicates.

Cramér–von Mises Test

This test[66] uses the following dissimilarity measure

$$\Delta(F, G)^2 := \mathbb{E}\left[(F(X) - G(X))^2\right], \quad X \sim G. \tag{16.16}$$

Problem 16.55 Show that

$$\Delta(\widehat{F}_X, F_0)^2 = \frac{1}{12n^2} + \frac{1}{n} \sum_{i=1}^{n} \left[\frac{2i-1}{2n} - F_0(X_{(i)})\right]^2.$$

As before, calibration is done by Monte Carlo simulation under F_0.

Problem 16.56 Show that Proposition 16.51 applies.

Anderson–Darling Tests

These tests[67] use dissimilarities of the form

$$\Delta(F, G) := \sup_{x \in \mathbb{R}} w(x) |F(x) - G(x)|, \tag{16.17}$$

or of the form

$$\Delta(F, G)^2 := \mathbb{E}\left[w(x)^2 (F(X) - G(X))^2\right], \quad X \sim G,$$

where w is a (non-negative) weight function. In both cases, a common choice is

$$w(x) := [G(x)(1 - G(x))]^{-1/2}.$$

This is motivated by the fact that, under the null hypothesis, for any $x \in \mathbb{R}$, $\widehat{F}_X(x) - F_0(x)$ has mean 0 and variance $F_0(x)(1 - F_0(x))$.

Problem 16.57 Prove this assertion.

---

[66] Named after Harald Cramér (1893–1985) and Richard von Mises (1883–1953).
[67] Named after Theodore Anderson (1918–2016) and Donald Darling (1915–2014).

16.8.2 Confidence Bands

Confidence bands are the equivalent of confidence intervals when the object to be estimated is a function rather than a real number. A confidence band can be obtained by inverting a test based on a measure of dissimilarity Δ, for example (16.15), following the process described in Section 12.4.7. In what follows, we assume that the underlying distribution, F, is continuous and that Δ is such that Proposition 16.51 applies.

Let δ_u denote the u-quantile of $\Delta(\widehat{F}_X, F)$, which does not depend on F by Proposition 16.51. Within the space of continuous distribution functions, define the following (random) subset

$$\mathcal{B} := \left\{ G : \Delta(\widehat{F}_X, G) \le \delta_{1-\alpha} \right\}.$$

This is the acceptance region for the level-α test defined by Δ. This is thus the analogue of (12.14), and following the arguments provided in Section 12.4.7, we get that, $F \in \mathcal{B}$ with probability $1 - \alpha$ under P.

Remark 16.58 The region $\mathcal{B}$ is called a 'confidence band' because, if all the distribution functions in $\mathcal{B}$ are plotted, it yields a band as a subset of the plane (at least this is the case for the most popular measures of dissimilarity).

Problem 16.59 In the particular case of the supremum norm (16.15), the band is particularly easy to compute or draw, because it can be defined pointwise. Indeed, show that in this case the band (as a subset of $\mathbb{R}^2$) is defined as

$$\left\{ (x, p) : |p - \widehat{F}(x)| \le \delta_{1-\alpha} \right\},$$

where δ_u is the u-quantile of the Kolmogorov distribution. In R, write a function that takes in the data and the desired confidence level, and plots the empirical distribution function as a solid black line and the corresponding band in grey. [The function polygon can be used to draw the band.] Try your function on simulated data from the standard normal distribution, with sample sizes $n \in \{10, 10^2, 10^3\}$. Each time, overlay the real distribution function plotted as a red line. See Figure 16.4 for an illustration.

16.8.3 Tests based on the Density

A goodness-of-fit test based on comparing densities rejects for large values of a test statistic of the form

$$\Delta(\hat{f}_X, f_0),$$

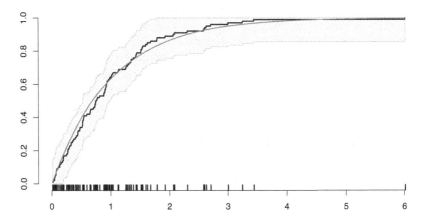

Figure 16.4 A 95% confidence band for the distribution function based on a sample of size $n = 10^3$ from the exponential distribution with rate $\lambda = 1$. The staircase dark line in the middle of the band is the empirical distribution function, the smoother line is the underlying distribution function, and the shaded grey region the confidence band. The marks on the horizontal axis identify the sample points.

where $\hat{f}_X$ is an estimator for a density of the underlying distribution, f_0 is a density of F_0, and Δ is a measure of dissimilarity between densities such as the *total variation distance*,

$$\Delta(f, g) := \int_{\mathbb{R}} |f(x) - g(x)|\mathrm{d}x,$$

or the *Kullback–Leibler divergence* (recall Problem 7.101),

$$\Delta(f, g) := \int_{\mathbb{R}} f(x) \log(f(x)/g(x))\mathrm{d}x.$$

Remark 16.60 If a histogram is used as a density estimator, the procedure is similar to binning the data and then using a goodness-of-fit test for discrete data (Section 15.1). (Note that this approach is possible even if the null distribution does not have a density.)

Problem 16.61 In R, write a function that implements the procedure of Remark 16.60. Use the function hist to bin the data and let it choose the bins automatically. The function returns a Monte Carlo p-value based on a specified number of replicates.

16.8.4 Testing against a Family of Distributions

So far in this section, we dealt with a situation where there is a single null distribution to test against. In Remark 16.49, we even saw that, as long as the null distribution is continuous, we can assume it to be the uniform distribution on $[0, 1]$, which is why the problem is sometimes referred to as 'testing for uniformity'.

We now consider a situation where there is a null family of distributions to test against. For example, we may want to know whether the sample comes from a normal distribution, a problem often referred to as *testing for normality*.

Suppose, therefore, that we have access to a sample, $X = (X_1, \ldots, X_n)$, assumed iid from a continuous distribution, and our task is to test whether the underlying distribution is in some given (null) family of distributions, $\{F_\theta : \theta \in \Theta\}$, against the alternative that is *not* in that family.

Assuming we have an estimator for θ, denoted $S(X)$, and working with distribution functions as in Section 16.8.1, a plug-in approach leads to using a test statistic of the form

$$\Delta(\widehat{F}_X, F_{S(X)}). \tag{16.18}$$

We still reject for large values of this statistic, although the main difference here is that a p-value is obtained by bootstrap, since the null distribution needs to be estimated.

Lilliefors Test

This is a test for normality that is derived from using the supremum norm as a measure of dissimilarity (16.15). The parameters (mean and variance) are obtained by maximum likelihood or the method of moments.

Although the p-value is in principle obtained by bootstrap as explained above, here the test statistic has the same distribution under the null hypothesis regardless of the sample values. This is because the normal family is a location-scale family, and the supremum norm is invariant with respect to affine transformations.

Problem 16.62 Prove that the statistic (16.18), with Δ denoting the supremum norm (16.15), has the same distribution under any normal distribution.

R corner In R, the test is implemented in the function lillie.test in the package nortest.

16.9 Censored Observations

In some settings, the observations are censored. An emblematic example is that of clinical trials where patient survival is the primary outcome. In such a setting, patients might be lost in the course of the study for other reasons, such as the patient moving too far away from any participating center, or simply by the termination of the study. Other fields where censoring is common include, for example, quality control where the reliability of some manufactured item is examined. The study of such settings is called *Survival Analysis*.

Example 16.63 (Stop smoking) The article [178] reports on a randomized clinical trial involving adults who want to stop smoking. According to the description provided in the article: "Participants were allocated by blocked randomization to receive either the nicotine patch alone for a standard 10-week, tapering course [64 subjects, 13 lost] or the combination of nicotine patch, nicotine oral inhaler, and bupropion ad libitum [63 subjects, 18 lost]. [...] Participants lost to follow-up were classified as smoking from the last point of contact." The time from entrance into the study until relapse was one of the measured outcomes. In the context of a survival analysis, 'relapse' is the event here, and the (right) censoring comes from the fact that subjects were only followed up for six months. (In clinical trials it is very common to compare multiple groups. In the context of this chapter, each group would considered separately.)

We consider a model of independent right-censoring. In the context of a clinical trial, let T_i denote the time to event (say death) for Subject i, with $T_1, \ldots, T_n$ assumed iid from some distribution H. These are not observed directly as they may be subject to censoring. Let $C_1, \ldots, C_n$ denote the censoring times, assumed to be iid from some distribution G. The actual observations are $(X_1, \delta_1), \ldots, (X_n, \delta_n)$, where

$$X_i = \min(T_i, C_i), \quad \delta_i = \{X_i = T_i\},$$

so that $\delta_i = 1$ indicates that the ith case was observed uncensored. The goal is to infer the underlying distribution H, or some features of H such as its median.

Problem 16.64 Recall the definition of the survival function given in (3.8). Show that $X_1, \ldots, X_n$ are iid with survival function $\bar{F}(x) := \bar{H}(x)\bar{G}(x)$.

16.9.1 Kaplan–Meier Estimator

We consider the case where the variables are discrete. We assume they are supported on the positive integers without loss of generality. In a survival context, this is the case, for example, when survival time is the number days to death since the subject entered the study. In that special case, it is possible to derive the MLE for H. Denote the corresponding survival and mass functions by $S(t) = 1 - H(t)$ and $h(t) = H(t) - H(t-1)$, respectively. Let $I = \{i : \delta_i = 1\}$, which indexes the uncensored survival times.

Problem 16.65 Show that the likelihood, as a function of H, takes the following form

$$\prod_{i \in I}(H(X_i) - H(X_i - 1)) \times \prod_{i \notin I}(1 - H(X_i)).$$

Express the likelihood as a function of S. And then as a function of h.

As usual, we need to maximize this with respect to H. For a positive integer t, define the *hazard rate*

$$\lambda(t) = \mathbb{P}(T = t \mid T > t - 1)$$
$$= 1 - \mathbb{P}(T > t \mid T > t - 1)$$
$$= h(t)/S(t - 1).$$

By the Law of Multiplication (1.16),

$$S(t) = \mathbb{P}(T > t) = \prod_{k=1}^{t} \mathbb{P}(T > k \mid T > k - 1)$$
$$= \prod_{k=1}^{t}(1 - \lambda(k)), \qquad (16.19)$$

so that

$$h(t) = S(t - 1) - S(t)$$
$$= \lambda(t) \prod_{k=1}^{t-1}(1 - \lambda(k)).$$

In particular, λ determines h, and therefore H.

Problem 16.66 Deduce that the likelihood, as a function of λ, is given by

$$\prod_{t \geq 1} \lambda(t)^{D_t}(1 - \lambda(t))^{Y_t - D_t},$$

where D_t is the number of events at time t, meaning

$$D_t := \#\{i : X_i = t, \delta_i = 1\},$$

and Y_t is the number of subjects still alive (said to be *at risk*) at time t, meaning

$$Y_t := \#\{i : X_i \geq t\}.$$

Then show that the maximizer is $\hat{\lambda}$ given by

$$\hat{\lambda}(t) := D_t/Y_t.$$

The corresponding estimate for S is obtained by plugging $\hat{\lambda}$ in (16.19), resulting in the MLE for S being given by

$$\widehat{S}(t) := \prod_{k=1}^{t} (1 - D_k/Y_k). \tag{16.20}$$

This estimator is known as the *Kaplan–Meier estimator*, proposed by Edward Kaplan (1920–2006) and Paul Meier (1924–2011) in the late 1950s. See Figure 16.5 for an illustration based on Example 16.63.

Remark 16.67 Bootstrap confidence bands for this estimator are discussed in [3]. The situation is a bit complex and we omit details.

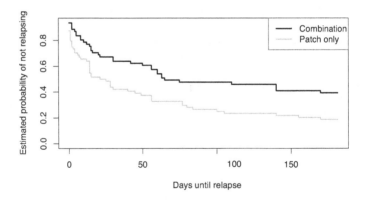

Figure 16.5 The Kaplan–Meier estimators of the survival functions for the two groups described in Example 16.63.

16.10 Further Topics

16.10.1 Reduction to a Uniform Sample

Assume that $X_1, \ldots, X_n$ are iid from a continuous distribution F. If we define $U_i := \mathsf{F}(X_i)$, $U_1, \ldots, U_n$ are iid Unif$(0, 1)$. Thus, the study of the order statistics $(X_{(1)}, \ldots, X_{(n)})$ reduces to the study of the uniform order statistics $(U_{(1)}, \ldots, U_{(n)})$, which is the object of *Empirical Process Theory*.

It is known, for example, that $U_{(i)}$ has the beta distribution with parameters $(i, n + 1 - i)$, which in particular implies that

$$\mathbb{E}(U_{(i)}) = \frac{i}{n+1}, \quad \mathrm{Var}(U_{(i)}) = \frac{i(n+1-i)}{(n+1)^2(n+2)}.$$

Problem 16.68 Prove that $U_{(i)}$ is concentrated near its mean by showing that, for all $t > 0$,

$$\mathbb{P}\big(|U_{(i)} - i/(n+1)| \geq t/\sqrt{n}\big) \leq 1/4t^2.$$

(See also Problem 16.95.)

It is also known that $U_{(i)}$ is approximately normal when n is large and i/n is not too close to 0 or 1. In particular, the following is true for the median.

Proposition 16.69 (Asymptotic normality of the sample median). *Assume that $(X_i : i \geq 1)$ are iid from a distribution with median θ and a distribution function having a strictly positive derivative at θ, denoted $f(\theta)$. Then x*

$$\sqrt{n}\big(Med(X_1, \ldots, X_n) - \theta\big) \xrightarrow{\mathcal{L}} \mathcal{N}\Big(0, \frac{1}{4f(\theta)^2}\Big), \quad as \ n \to \infty.$$

Problem 16.70 Show that the median is indeed uniquely defined in the context of this proposition.

As mentioned in Remark 16.49, when testing for a particular null distribution F_0, we can work with the transformed sample $U_i := \mathsf{F}_0(X_i)$ and test for uniformity. The test that Berk and Jones [10] proposed directly exploits the fact that the distribution of each order statistic is known the null hypothesis. Specifically, the test rejects for *small* values of $\min(S^-, S+)$, where $S^+ := \min_i \mathsf{G}_i(U_{(i)})$ and $S^- := \min_i(1 - \mathsf{G}_i(U_{(i)}))$, where G_i is the distribution function of Beta$(i, n + 1 - i)$. The asymptotic distribution of this test statistic under the null hypothesis is derived in [133].

16.10.2 Other Forms of Bootstrap

There are other forms of bootstrap besides the one that relies on the empirical distribution (often called the *empirical bootstrap*). We briefly describe a few of them.

Smooth Bootstrap

The empirical distribution is always available as an estimate of the underlying distribution, but being discrete, it is not 'smooth'. A smooth bootstrap is based on a smoother estimate of the underlying distribution, for example, the piecewise linear empirical distribution function (Remark 16.10) or a kernel density estimate (Section 16.10.5).

A common way to implement the latter is as follows. Let $x_1, \ldots, x_n$ denote the observations. Given a distribution with density K, generate a (smooth) bootstrap sample by first generating an iid sample from K, denoted $W_1, \ldots, W_n$, and then adding that to the observations, resulting in the bootstrap sample $X_1^*, \ldots, X_n^*$ with $X_i^* := x_i + W_i$. By (6.7) and Section 16.10.5, conditional on the observations, $X_1^*, \ldots, X_n^*$ are iid with density

$$f^*(x) := \frac{1}{n} \sum_{i=1}^{n} \mathsf{K}(x_i - x),$$

which is a kernel density estimate for f.

Remark 16.71 When the task is to estimate a parameter of interest, $\varphi(\mathsf{F})$, with φ not defined on discrete distributions, the empirical bootstrap is not applicable, but a smooth bootstrap might.

Parametric Bootstrap

If we know (or rather, if we are willing to assume) that F is in some parametric family of distributions, say $\{\mathsf{F}_\theta : \theta \in \Theta\}$, then a possible approach is to estimate F with $\mathsf{F}_{\hat{\theta}}$, where $\hat{\theta}$ is an estimator for θ, for example, the MLE. (This is, in fact, what we did in Section 15.1.3 and Section 15.2.3 in the context of discrete distributions.)

16.10.3 Method of Moments

Suppose an experiment results in an iid sample, denoted $X_1, \ldots, X_n$, having distribution P_θ on $\mathbb{R}$, where $\theta \in \Theta$ is unknown and needs to be estimated. This is, for example, the case of the binomial experiment of Example 12.2 if we define $X_i = 1$ when the ith trial results in heads and $X_i = 0$ otherwise. Let $\widehat{\mathsf{P}}_X$ denote the empirical distribution.

We already saw maximum likelihood estimation. A competing approach is the *method of moments*. Having chosen some distributional features of interest (e.g., various moments), the idea is to find a value of θ such that, in terms these features, P_θ is close to the empirical distribution.

A feature here is a real-valued function on distributions on $\mathbb{R}$. Based on k such features, denoted by $\Lambda_1, \ldots, \Lambda_k$, define the *method of moments estimator* (MME) to be

$$S := \arg\min_{\theta \in \Theta} \sum_{j=1}^{k} \left[\Lambda_j(P_\theta) - \Lambda_j(\widehat{P}_X) \right]^2.$$

(This assumes the minimization problem has a unique solution.)

Remark 16.72 (Classical method of moments) The name of this approach comes from the fact that the features are traditionally chosen to be moments, meaning, $\Lambda_j(P) = \mathbb{E}_P(X^j)$, where $\mathbb{E}_P$ denotes the expectation under $X \sim P$.

In some classical settings, it is possible to find $\theta \in \Theta$ such that $\Lambda_j(P_\theta) = \Lambda_j(\widehat{P}_X)$ for all $j = 1, \ldots, k$.

Problem 16.73 (Binomial experiment) Consider the binomial experiment of Example 12.2. Define the MME based solely on the 1st moment, meaning that $k = 1$ and $\Lambda_1(P) = \mathbb{E}_P(X) = \text{mean}(P)$, and show that the resulting estimator coincides with the MLE.

Problem 16.74 More generally, assume that $\Theta \subset \mathbb{R}$ and that θ is the mean of P_θ, which is how the Poisson family is typically parameterized. Show that the classical MME for θ, with $k = 1$, is the sample mean. Extend this to the case where $\Theta = \mathbb{R} \times \mathbb{R}_+$ and $\theta = (\mu, \sigma^2)$, with μ being the mean and σ^2 the variance of P_θ, which is how the normal family is typically parameterized.

16.10.4 Prediction Intervals

In Section 16.3.2 and Section 16.6, our goal was to construct a confidence interval for the location parameter of interest, respectively, the median and the mean. Consider instead the problem of constructing an interval for a new observation sampled from the same underlying distribution. Such an interval is called a *prediction interval*.

In what follows, we let $X_1, \ldots, X_n, X_{n+1}$ be iid from a distribution P on the real line, where $X_n := (X_1, \ldots, X_n)$ plays the role of the available sample, while X_{n+1} plays the role of a new datum to become available later in the future.

Problem 16.75 Suppose that P is the normal distribution with unknown mean μ and unknown variance σ^2. Let $\bar{X}_n$ and S_n denote the sample mean and standard deviation based on X_n. Show that

$$X_{n+1} \in \left[\bar{X}_n - t_{1-\alpha/2}^{n-1} S_n \sqrt{\tfrac{n-1}{n+1}}, \bar{X}_n - t_{\alpha/2}^{n-1} S_n \sqrt{\tfrac{n-1}{n+1}} \right]$$

with probability $1 - \alpha$.

Compared to the confidence interval for the mean given in (16.10), the prediction interval above is much wider. In fact, its half-width converges (in probability) to $\sigma z_{1-\alpha/2}$, and therefore does not converge to zero as $n \to \infty$. (As before, z_u denotes the u-quantile of the standard normal distribution.) This is to be expected. Indeed, even when μ and σ are known, we cannot do better asymptotically.

Problem 16.76 Show that $[\mu - z_{1-\alpha/2}\sigma, \mu - z_{\alpha/2}\sigma]$ is the shortest interval I such that $\mathbb{P}(X \in I) \geq 1 - \alpha$, when X is sampled from $\mathcal{N}(\mu, \sigma^2)$.

When no parametric family is assumed, one can rely on the empirical distribution to obtain a prediction interval. One way to do so is via the appropriate sample quantiles.

Problem 16.77 Let $\widehat{F}_n^-$ denote the empirical quantile function based on X_n. Prove that

$$X_{n+1} \in \left[\widehat{F}_n^-(\alpha/2), \widehat{F}_n^-(1 - \alpha/2) \right] \tag{16.21}$$

with probability tending to $1 - \alpha$ as $n \to \infty$. Propose another variant such that, asymptotically, the interval is shortest (at the same confidence level).

The prediction interval (16.21) has asymptotically the prescribed confidence level. If the level must be guaranteed in finite samples, one can use the Dvoretzky–Kiefer–Wolfowitz bound.

Problem 16.78 Derive a prediction interval based on Theorem 16.9 that satisfies the prescribed level of confidence in finite samples. [The shorter the better.] How does the interval behave in the large-sample limit?

The following may provide a more satisfying option.

Proposition 16.79. *With $X_{(1)} \leq \cdots \leq X_{(n)}$ denoting the ordered sample, it holds that*

$$X_{n+1} \in \left[X_{(\lfloor n\alpha/2 \rfloor)}, X_{(\lceil n(1-\alpha/2) \rceil)} \right] \tag{16.22}$$

with probability at least $1 - \alpha$.

Proof For a sample $x_1, \ldots, x_n \in \mathbb{R}$, let

$$I(x_1, \ldots, x_n) = \left[x_{(\lfloor n\alpha/2 \rfloor)}, x_{(\lceil n(1-\alpha/2) \rceil)} \right],$$

where $x_{(1)} \leq \cdots \leq x_{(n)}$ denote the ordered sample. For $i \in \{1, \ldots, n+1\}$, let

$$Y_i = \begin{cases} 1 & \text{if } X_i \in I(X_1, \ldots, X_{i-1}, X_{i+1}, \ldots, X_{n+1}); \\ 0 & \text{otherwise.} \end{cases}$$

Because the event (16.22) can be equivalently stated as $Y_{n+1} = 1$, it suffices to prove that $\mathbb{P}(Y_{n+1} = 1) \geq 1 - \alpha$. Since Y_{n+1} takes values in $\{0, 1\}$, $\mathbb{P}(Y_{n+1} = 1) = \mathbb{E}(Y_{n+1})$. Although $Y_1, \ldots, Y_{n+1}$ are *not* independent, they are exchangeable, and, in particular, they have the same expectation, so that

$$\mathbb{E}(Y_{n+1}) = \mathbb{E}\left(\frac{Y_1 + \cdots + Y_{n+1}}{n+1} \right).$$

We then conclude with the following problem. $\qquad\qquad\qquad\qquad\qquad\square$

Problem 16.80 For $x_1, \ldots, x_{n+1} \in \mathbb{R}$, define

$$y_i = \begin{cases} 1, & \text{if } x_i \in I(x_1, \ldots, x_{i-1}, x_{i+1}, \ldots, x_n), \\ 0, & \text{otherwise.} \end{cases}$$

Show that

$$\frac{y_1 + \cdots + y_{n+1}}{n+1} \geq 1 - \alpha.$$

Remark 16.81 (Conformal prediction) The approach underlying the construction of the prediction interval (16.22) can be seen as an example of *conformal prediction* [194, 165], which may be seen as a general approach based on inverting a permutation test for goodness-of-fit comparing samples $\{X_1, \ldots, X_n\}$ and $\{X_{n+1}\}$ using an approach similar to that of Section 12.4.7 for inverting a test to obtain a confidence interval.

16.10.5 Kernel Density Estimation

Consider a sample, $X_1, \ldots, X_n$, drawn iid from a density f that we want to estimate. Let K be a function on $\mathbb{R}$, which here plays the role of *kernel function*, and for $a > 0$ define

$$\mathsf{K}_a(x) = a^{-1} \mathsf{K}(x/a).$$

The key is the following result.

Proposition 16.82. *Suppose that f is continuous at x and bounded on $\mathbb{R}$, and choose K such that $\int_{\mathbb{R}} K(x)dx = 1$. Then*

$$\int_{\mathbb{R}} K_a(z-x)f(z)dz \to f(x), \quad \text{as } a \to 0.$$

Since,

$$\int_{\mathbb{R}} K_a(z-x)f(z)dz = \mathbb{E}_f(K_a(X-x)), \qquad (16.23)$$

where $\mathbb{E}_f$ denotes the expectation with respect to $X \sim f$, this suggests estimating f with

$$\hat{f}_a(x) := \frac{1}{n}\sum_{i=1}^{n} K_a(x_i - x). \qquad (16.24)$$

(Note that, if K is even, as most kernel functions used in practice are, then (16.23) is the convolution of K_a and f.)

The method has one parameter $a > 0$, called the *bandwidth*, that plays the exact same role as the bin size in the construction of a histogram with bins of equal size. In fact, when K is the so-called rectangular (aka flat, aka box) kernel, namely, $K(x) = \{x \in [-1/2, 1/2]\}$, the estimate is quite similar to a histogram with bin size h.

R corner The function density in R offers a number of choices for the kernel K. The default is the Gaussian kernel $K(x) := \frac{1}{\sqrt{2\pi}}\exp(-x^2/2)$.

A kernel density estimate is at least as smooth the kernel used to define it. In particular, a kernel density estimate with the Gaussian kernel is infinitely differentiable.

Choice of Bandwidth

The choice of bandwidth a has generated a lot of proposals. It is directly related to the choice of bin size in the construction of a histogram, and the same chicken-and-egg situation arises.

For example, here too, the optimal choice for a is of order $n^{-1/3}$ when f has bounded slope. To see why, suppose that f is bounded by c_0 and has slope bounded by c_1 everywhere, meaning

$$f(x) \le c_0, \quad \text{for all } x \in \mathbb{R},$$
$$|f(x) - f(z)| \le c_1 |x - z|, \quad \text{for all } x, z \in \mathbb{R}.$$

Assume that the kernel is non-negative, has support in $[-1/2, 1/2]$, and is bounded by some $c_2 > 0$ in absolute value.

The mean squared error (MSE) at $x \in \mathbb{R}$ is

$$\mathsf{mse}_a(x) := \mathbb{E}\big[(\hat{f}_a(x) - f(x))^2\big],$$

where the expectation is with respect to the sample defining $\hat{f}_a$.

Problem 16.83 Derive the following *bias-variance decomposition*

$$\mathsf{mse}_a(x) = \big(\mathbb{E}[\hat{f}_a(x)] - f(x)\big)^2 + \mathrm{Var}[\hat{f}_a(x)].$$

For the bias, by (16.24),

$$\begin{aligned}
\mathbb{E}[\hat{f}_a(x)] - f(x) &= \mathbb{E}[\mathsf{K}_a(X - x)] - f(x) \\
&= \int_{\mathbb{R}} \mathsf{K}_a(z - x) f(z) \, \mathrm{d}z - f(x) \\
&= \int_{\mathbb{R}} \mathsf{K}_a(z - x)\big(f(z) - f(x)\big) \, \mathrm{d}z,
\end{aligned}$$

using the fact that K_a integrates to 1. Hence, using Jensen's inequality,

$$\begin{aligned}
\big|\mathbb{E}[\hat{f}_a(x)] - f(x)\big| &\le \int_{\mathbb{R}} \mathsf{K}_a(z - x)\big|f(z) - f(x)\big| \, \mathrm{d}z \\
&\le \int_{\mathbb{R}} \mathsf{K}_a(z - x) c_1 |z - x| \, \mathrm{d}z \\
&\le c_1 a \int_{[-a/2, a/2]} \mathsf{K}_a(z) \, \mathrm{d}z \\
&= c_1 a,
\end{aligned}$$

using the bound on the slope of f, and then the fact that the kernel is non-negative, supported on $[-1/2, 1/2]$, and integrates to 1.

For the variance, by (16.24) and independence,

$$\mathrm{Var}[\hat{f}_a(x)] = \frac{1}{n} \mathrm{Var}[\mathsf{K}_a(X - x)],$$

with

$$\begin{aligned}
\mathrm{Var}[\mathsf{K}_a(X - x)] &\le \mathbb{E}[\mathsf{K}_a(X - x)^2] \\
&= \int_{\mathbb{R}} \mathsf{K}_a(z - x)^2 f(z) \, \mathrm{d}z \\
&\le \frac{c_2}{a} c_0 \int_{\mathbb{R}} \mathsf{K}_a(z - x) \, \mathrm{d}z \\
&= c_2 c_0 / a,
\end{aligned}$$

using the bound on K and the bound on f, and the fact that the kernel integrates to 1.

Thus,

$$\mathsf{mse}_a(x) \le (c_1 a)^2 + \frac{c_2 c_0}{na},$$

and the right-hand side is minimized at $c_3 n^{-1/3}$ where $c_3 := (c_0 c_2 / 2c_1^2)^{1/3}$.

These calculations are crude and refinements are definitely possible. However, the order of magnitude of the optimal bandwidth is known to be $\propto n^{-1/3}$, with a multiplicative constant that depends on the (unknown) density. That constant is important in practice and makes it necessary to choose the bandwidth based on the data.

Cross-Validation

A popular way to choose the bandwidth is by *cross-validation* (CV). We present a particular variant called *leave-one-out cross-validation*, proposed by Rudemo in [157]. The idea is to choose a to minimize the following notion of risk called the *mean integrated squared error*

$$\mathscr{R}(a) := \mathbb{E}\left[\int (\hat{f}_a(x) - f(x))^2 dx \right],$$

where the expectation is with respect to the sample that underlies the estimate $\hat{f}_a$. Based on this risk, Rudemo proposes the following choice of bandwidth

$$\hat{a} := \arg\min_{a>0} Q(a),$$

$$Q(a) := \int \hat{f}_a(x)^2 dx - \frac{2}{n-1} \sum_{i=1}^{n} \hat{f}_a(X_i).$$

Problem 16.84 Relate $\mathbb{E}[Q(a)]$ to $\mathscr{R}(a)$.

R corner This is very close to how the R function bw.ucv selects the bandwidth. A more faithful implementation may be found in the function h.ucv in the kedd package.

16.10.6 Monotonic Density Estimation

Kernel density estimation is, as we saw, founded on the implicit assumption that the underlying density has a certain degree of smoothness. An alternative is to assume that the density has a certain *shape*. We present the simplest, and most famous example, where the underlying density is supported on $\mathbb{R}_+$ and assumed to be monotone (and therefore non-increasing).

It so happens that there is a MLE for this model, proposed by Ulf Grenander (1923–2016) [83]. The likelihood is here defined as

$$\arg\max_{f \in \mathcal{F}} \prod_{i=1}^{n} f(x_i),$$

where $x_1, \ldots, x_n$ denote the observations and $\mathcal{F}$ denotes the class of monotone densities on $\mathbb{R}_+$. It turns out to be maximized by the first derivative of the least concave majorant of the empirical distribution function, which is a non-increasing piecewise-constant function.

R corner The grenander function in the fdrtool package computes this estimator.

16.11 Additional Problems

Problem 16.85 In R, for $n \in \{10, 10^2, 10^3\}$, generate n points from the uniform distribution on $[0, 1]$. Draw the empirical distribution function (solid line) and overlay the actual distribution function (dashed line). Do it several times for each n to get a feel for the randomness. Repeat with the standard normal distribution.

Problem 16.86 In R, write a function which behaves as ecdf but returns the piecewise linear variant of Remark 16.10 instead. In addition, write another function which behaves as plot.ecdf. To each plot of Problem 16.85, add (in a different color) a graph of this variant.

Problem 16.87 In R, generate a sample of size $n = 10$ from the standard normal distribution. Plot the quantile function, type 1, 4, 5, in red, green, and blue, respectively. Add the underlying quantile function (given by qnorm). Make sure to use a fine grid, say, 10^3 points covering $[0, 1]$, for otherwise visual artifacts will result. Add dotted vertical lines at k/n for $k = 0, 1, \ldots, n$, and dotted horizontal lines at the data points. Repeat with $n = 100$.

Problem 16.88 In R, generate a sample of size $n \in \{10^2, 10^3\}$ from the standard normal distribution. Plot the histogram in various ways, each time overlaying the density. Do it several times in each setting to get a feel for the randomness. Try different methods for choosing the bins.

Problem 16.89 Show that the variant (16.9) is unbiased. However, 'unbiased' does not mean 'better', and indeed, show that for the estimation of the variance in the normal location-scale family, (16.7) has smaller MSE than (16.9).

Problem 16.90 Suppose that P_θ is the exponential distribution with rate $\theta > 0$, and that we have an iid sample of size n, $X_1, \ldots, X_n$, sampled from this distribution. Suppose we want to estimate $\varphi := \theta^{1/2}$.

(i) Show that the MLE for θ is $\bar{X}^{-1}$, where $\bar{X}$ is the sample mean, so that the MLE for φ is $\bar{X}^{-1/2}$.

(ii) Compute its bias by numerical integration. [Note that the family is a scale family, so that the bias under θ can be obtained from the bias under $\theta = 1$ by rescaling.]

(iii) In R, for $n \in \{10, 20, 30, 50\}$, under $\theta = 1$, estimate that quantity by Monte Carlo, using $B = 10^4$ replicates.

(iv) Now estimate that quantity by bootstrap based on $B = 10^4$ replicates. Do this for various choices of n. Repeat a few times to get a feel for the randomness and compare with the value obtained by numerical integration.

Problem 16.91 Repeat Problem 16.90 with the variance instead of the bias.

Problem 16.92 (Kolmogorov distribution) In [108], Kolmogorov derived the null distribution of the test statistic (16.15). He showed that, based on an iid sample of size n from a continuous distribution F, the empirical distribution function $\widehat{F}_n$ as a random function satisfies

$$\lim_{n \to \infty} \mathbb{P}\left(\sup_{x \in \mathbb{R}} \left| \widehat{F}_n(x) - F(x) \right| \geq t / \sqrt{n} \right) = 2 \sum_{k \geq 1} (-1)^{k-1} \exp(-2k^2 t),$$

for all $t \geq 0$. In R, perform simulations to probe the accuracy of this limit for various choices of sample size n.

Problem 16.93 (A failure of the bootstrap) Consider the situation where $X_1, \ldots, X_n$ are iid normal with mean θ and variance 1. It is desired to provide a confidence interval for $|\theta|$. A natural estimator is the MLE, which is $|\bar{X}|$. The bootstrap confidence interval of Section 16.6.3 is based on estimating the distribution of $|\bar{X}| - |\theta|$ by the bootstrap distribution of $|\bar{X}^*| - |\bar{X}|$. It happens to fail when $\theta = 0$. (This can be explained by the fact that the absolute value function is not smooth at the origin.)

(i) Compute the distribution function of $\sqrt{n}(|\bar{X}| - |\theta|)$. Specialize to the case where $\theta = 0$ and draw it.

(ii) In R, generate a sample of size $n = 10^6$ and estimate the bootstrap distribution $\sqrt{n}(|\bar{X}^*| - |\bar{X}|)$ using $B = 10^3$ Monte Carlo replicates. Add the resulting empirical distribution function to the plot.

Problem 16.94 In R, do the following. Generate a sample of size $n \in \{10, 20, 50, 100, 200, 500, 1000\}$ from the standard normal distribution (so that $\mu = 0$ and $\sigma = 1$). Set the confidence level at 0.95.

(i)   Compute the Student confidence interval. This is the gold standard if the sample is known to be normal and the variance is unknown.

(ii)   Compute the bootstrap confidence interval using the function implemented in Problem 16.43.

(iii)   Compute the bootstrap Studentized confidence interval using the function implemented in Problem 16.44.

Each time, record whether the true mean is in the interval and measure the length of the interval. After doing that 200 times, for each of the three confidence interval constructions, display the fraction of times the interval contained the true mean and plot a histogram of its length.

Problem 16.95 Derive a bound that is sub-exponential in t for the probability that appears in Problem 16.68.

Problem 16.96 (Student test) Consider a normal experiment where we observe a realization of $X_1, \ldots, X_n$, assumed to be an iid sample from $\mathcal{N}(\mu, \sigma^2)$. Both parameters are unknown. Our goal is to test $\mu = \mu_0$ for some given $\mu_0 \in \mathbb{R}$. The *Student test* (aka *t-test*) rejects for large values of $|T|$ where $T := (\bar{X} - \mu_0)/S$, with $\bar{X}$ being the sample mean and S being the sample standard deviation.

(i)   Show that there is a constant c_n such that, under the null hypothesis, $c_n T$ has the Student distribution with $n - 1$ degrees of freedom.

(ii)   Show that this test corresponds to the LRT under the present model.

Problem 16.97 Propose and study, analytically or via computer simulations, a test for goodness-of-fit based on the characteristic function, that is, based on a test statistic comparing the sample characteristic function (i.e., the characteristic function of the empirical distribution) and the null characteristic function (i.e., the characteristic function of the null distribution).

Problem 16.98 In R, do the following. Generate a sample of size n from some exponential distribution (say, with rate $\lambda = 1$, although this is inconsequential). Plot a histogram. Compute the MLE for λ and overlay the corresponding density. Then compute the Grenander's estimator and overlay the corresponding density (in a different color). Repeat several times for several choice of sample size n.

Problem 16.99 (Log-concave densities) A function $f: \mathbb{R} \to \mathbb{R}_+$ is said to be *log-concave* if $\log f$ is concave. Any normal distribution is log-concave, for example. Just as for monotonic densities, the class of log-concave densities admits an MLE, which also turns out to be piecewise linear. This estimator is available in the package longcondens. In R, do the following. Generate a sample of size n from some normal distribution (say, standard normal, although this is inconsequential). Plot a histogram. Compute the MLE among normal distributions and overlay the corresponding density. Then compute the MLE among log-concave distributions and overlay the corresponding density (in a different color). Repeat several times for several choice of sample size n.

Problem 16.100 Verify as many statements in [38] as you can, in particular those in dimension one.

Problem 16.101 (A uniform law of large numbers) The Glivenko–Cantelli Theorem is an example of a uniform law of large numbers. Here is another example, due to Jennrich [103]. Assume that X is a random variable on some probability space, and that Θ is a compact parameter space. Consider a (measurable) function $g: \mathbb{R} \times \Theta \to \mathbb{R}$ such that $\theta \mapsto g(x, \theta)$ is continuous for all $x \in \mathbb{R}$; and $|g(x, \theta)| \le h(x)$ for all x and all θ, with $\mathbb{E}[h(X)] < \infty$. Then, if (X_i) are iid copies of X,

$$\sup_{\theta \in \Theta} \left| \frac{1}{n} \sum_{i=1}^{n} g(X_i, \theta) - \mathbb{E}[g(X, \theta)] \right| \xrightarrow{\text{P}} 0, \quad \text{as } n \to \infty.$$

Prove this result.

Problem 16.102 In the setting of Section 16.9, with the same notation, consider the following estimator for the survival function

$$\widetilde{S}(t) := \frac{\#\{i : X_i > t\}}{\#\{i : X_i > t\} + D_t}. \tag{16.25}$$

Show that $\widetilde{S}(t)$ is unbiased for $S(t)$. Add the corresponding estimates to Figure 16.5 as dash lines. [Thus, while the Kaplan–Meier estimator given in (16.20) is the MLE, the estimator given in (16.25) can be seen as the MME for the survival function.]

17

Multiple Numerical Samples

In Chapter 16, we considered an experiment resulting in an iid real-valued sample from an unknown distribution and the task was to infer this distribution or some of its features. Here, we consider a setting where either two or more real-valued samples are observed, and the goal is to compare the underlying distributions that generated the samples or some of their features.

Randomized clinical trials (Section 11.2.5) are a rich source of examples of such experiments. Although it is somewhat more complicated, the methodology parallels that of Chapter 16. As in Chapter 15, permutation tests play a central role in goodness-of-fit testing.

Example 17.1 (Weight loss maintenance) The paper [183] reports on a "two-phase trial in which 1032 overweight or obese adults with hypertension, dyslipidemia, or both, who had lost at least 4 kg during a 6-month weight loss program (Phase 1) were randomized to a weight-loss maintenance intervention (Phase 2)." There were three intervention groups: monthly personal contact, unlimited access to an interactive technology, or self-directed (which served as control). Subjects were followed for a total of 30 months.

Example 17.2 (Manual v. automated blood pressure measurement) The paper [135] reports on a clinical trial where the objective was to "compare the quality and accuracy of manual office blood pressure and automated office blood pressure". It was a cluster randomized trial where 67 medical practices in eastern Canada were randomized to measuring blood pressure either manually or using a BpTRU device. (The awake ambulatory blood pressure was used as gold standard.)

We start with two groups to ease the presentation, and then extend the narrative to multiple groups. We typically leave the dependence on the sample sizes implicit to keep the notation light.

17.1 Inference about the Difference in Means

We have two samples, $X_{1,1}, \ldots, X_{n_1,1}$ assumed to be iid from P_1 and, independent of that, $X_{1,2}, \ldots, X_{n_2,2}$ assumed to be iid from P_2. The sample sizes are therefore n_1 and n_2, respectively, and we let $n := n_1 + n_2$ denote the total sample size.

Assume that P_1 has mean μ_1 and that P_2 has mean μ_2. We are interested in the difference

$$\delta := \mu_1 - \mu_2.$$

The plug-in estimator is the difference of the sample means, or in formula

$$D := \bar{X}_1 - \bar{X}_2,$$

where

$$\bar{X}_j := \frac{1}{n_j} \sum_{i=1}^{n_j} X_{i,j}.$$

In particular, D is unbiased for δ.

In the following, we present various ways of building a confidence interval based on that estimator.

17.1.1 Normal Confidence Interval

Assume that P_1 has variance σ_1^2 and that P_2 has variance σ_2^2. Note that D has variance

$$\gamma^2 := \frac{\sigma_1^2}{n_1} + \frac{\sigma_2^2}{n_2}.$$

Problem 17.3 Prove, using the Central Limit Theorem, that $(D - \delta)/\gamma$ is asymptotically standard normal as $n_1, n_2 \to \infty$. Is this still true if one of the sample sizes remains constant?

This normal limit implies that, if z_u denotes the u-quantile of the standard normal distribution,

$$\delta \in \left[D - z_{1-\alpha/2}\, \gamma, D - z_{\alpha/2}\, \gamma \right],$$

with probability converging to $1 - \alpha$ as $n_1, n_2 \to \infty$.

Typically, σ_1^2 and σ_2^2 are unknown, in which case this interval is not a confidence interval. We obtain a bona fide confidence interval by plugging the sample variances, S_1^2 and S_2^2, in place of the variances, where

$$S_j^2 := \frac{1}{n_j} \sum_{i=1}^{n_j} (X_{i,j} - \bar{X}_j)^2,$$

so that γ is estimated by

$$G := \left[S_1^2/n_1 + S_2^2/n_2 \right]^{1/2}.$$

Then by Slutsky's theorem, $(D - \delta)/G$ is asymptotically standard normal as the sample sizes diverge to infinity, and this implies that

$$\delta \in \left[D - z_{1-\alpha/2} G, D - z_{\alpha/2} G \right],$$

with probability converging to $1 - \alpha$ as $n_1, n_2 \to \infty$. (This interval is a confidence interval since it can be computed from the data.)

Student Confidence Interval

As in the one-sample setting, using quantiles from the Student distribution is the common practice, despite the fact that even with normal samples the Student distribution is only an approximation. (In that case, the sample variance is defined as in (16.9).) The computation of the number of degrees of freedom identifying the Student distribution is a bit involved. This derivation is due to Welch [204] and for this reason the test sometimes bares his name. A simpler but more conservative choice for the number of degrees of freedom is $\min(n_1, n_2) - 1$.

R corner The Student–Welch confidence interval, as well as the corresponding test, can be computed using the function t.test.

17.1.2 Bootstrap Studentized Confidence Interval

There are analogue s to the bootstrap confidence intervals presented in Section 16.6.3 and Section 16.6.4 in the setting of one sample. The Studentized variant is typically preferred as it tends to mc more accurate, so this is the one we focus on.

Problem 17.4 Before reading further, derive the analogue to the bootstrap confidence interval of Section 16.6.3 in the present setting of two samples.

The basic idea is the same: go to the bootstrap world to estimate the distribution of $T := (D - \delta)/G$. Assume that the data, $(x_{i,j})$, have been observed. As usual, a star indicates a quantity in the bootstrap world. In particular, P_j^* denotes the empirical distribution for Group j. A bootstrap sample is thus $(X_{i,j}^*)$, where

$$X_{i,j}^* \sim \mathsf{P}_j^*, \quad \text{independent.} \tag{17.1}$$

Its sample mean and variance will be denoted by $\bar{X}_j^*$ and S_j^*. In the bootstrap world we can define the analogue of the various quantities that are needed for the analysis:

$$\mu_j^* := \bar{x}_j, \qquad \delta^* := \mu_1^* - \mu_2^* = \bar{x}_1 - \bar{x}_2,$$

and

$$D^* := \bar{X}_1^* - \bar{X}_2^*, \qquad G^* := \left[S_1^{*2}/n_1 + S_2^{*2}/n_2 \right]^{1/2},$$

and $T^* := (D^* - \delta^*)/G^*$. The bootstrap distribution of T^* is used as an estimate of the distribution of T. Let t_u^* be the u-quantile of the bootstrap distribution of T^*. Then

$$\delta \in \left[D - t_{1-\alpha/2}^* G, D - t_{\alpha/2}^* G \right], \tag{17.2}$$

with approximate probability $1 - \alpha$.

As usual, an analytically derivation of the bootstrap distribution of T^* is impractical and one resorts to Monte Carlo simulation to estimate it.

Problem 17.5 In R, write a function that takes in the data, the confidence level, and number of Monte Carlo replicates, and returns the bootstrap confidence interval (17.2) estimated by Monte Carlo.

17.2 Inference about a Parameter

In this section, we consider the task of comparing a parameter of interest, denoted φ (e.g., standard deviation). Let $\varphi_j = \varphi(\mathsf{P}_j)$. We are interested in building a confidence interval for

$$\delta := \varphi_1 - \varphi_2.$$

Remark 17.6 One might be interested, instead, in the ratio φ_1/φ_2. Almost invariably, φ is a non-negative parameter in this case, and if so, one can simply take the logarithm and the problem becomes that of estimating the difference $\log \varphi_1 - \log \varphi_2$.

Problem 17.7 Suppose that $I = [A, B]$ is a $(1 - \alpha)$-confidence interval for $\log \varphi_1 - \log \varphi_2$. Turn that into a $(1 - \alpha)$-confidence interval for φ_1/φ_2.

17.2.1 Naive Approach

A naive, yet very reasonable approach consists in building a confidence interval for each parameter based on the corresponding sample, for example, using a bootstrap approach as described in Section 16.7, and then combining these intervals to obtain a confidence interval for the difference.

Problem 17.8 Let $I_1 = [A_1, B_1]$ be a $(1 - \alpha/2)$-confidence interval for φ_1 based on the 1st sample, and let $I_2 = [A_2, B_2]$ be a $(1 - \alpha/2)$-confidence interval for φ_2 based on the 2nd sample. Show that $[A_1 - B_2, B_1 - A_2]$ is a $(1 - \alpha)$-confidence interval for δ.

Although this approach is clearly valid, it can be conservative, in the usual sense that, although the level of confidence may be satisfied, the resulting interval is relatively wide.

17.2.2 Bootstrap Confidence Interval

We now present the two-sample analogue of the approach described in Section 16.6.3. We assume that φ can be defined for a discrete distribution so that we may use the empirical bootstrap. A reasonable estimator for δ is the plug-in estimator, which may be written as

$$D := \varphi(\mathsf{P}_1^*) - \varphi(\mathsf{P}_2^*).$$

Beyond mere estimation, the construction of a confidence interval necessitates knowledge of the distribution of $D - \delta$. We estimate this distribution by bootstrap.

Having observed the data $(x_{i,j})$, we travel to the bootstrap world. There, bootstrap samples are generated as in (17.1), and the quantities of interest are

$$\delta^* := \varphi(\mathsf{P}_1^*) - \varphi(\mathsf{P}_2^*),$$

and

$$D^* := \varphi(\mathsf{P}_1^{**}) - \varphi(\mathsf{P}_2^{**}),$$

where P_j^{**} is the empirical distribution of $X_{1,j}^*, \ldots, X_{n_j,j}^*$.

We estimate the distribution of $D - \delta$ by the bootstrap distribution of $D^* - \delta^*$. As usual, computing this bootstrap distribution in closed form is impractically and we resorts to Monte Carlo simulation to estimate it.

Remark 17.9 A bootstrap Studentized confidence interval can also be derived, but the method is in general more complex and computationally more intensive.

Problem 17.10 In R, write a function that takes in the two samples as numerical vectors, the desired confidence level, and a number of Monte Carlo replicates, and returns an appropriate bootstrap confidence interval for the difference in medians.

Remark 17.11 In the two-sample setting, there does not seem to be a distribution-free confidence interval for the difference in medians that would mimic the one developed in Section 16.3.2 in the one-sample setting. (That is, unless one is willing to assume that the two underlying distributions are translates of each other, as in (17.6).)

17.3 Goodness-of-Fit Testing

Consider the problem of goodness-of-fit testing where we test whether the two samples come from the same distribution,

$$\mathcal{H}_0 : \mathsf{P}_1 = \mathsf{P}_2. \tag{17.3}$$

We considered this problem in Section 15.2 in the discrete setting.

Problem 17.12 (Naive approach) A naive, although reasonable approach, consists in computing a $(1-\alpha/2)$-confidence band for F_j as in Section 16.8.2, denoted $\mathcal{B}_j$, and then rejecting the null if $\mathsf{F}_1 \in \mathcal{B}_2$ or $\mathsf{F}_2 \in \mathcal{B}_1$. Show that this yields a test at level α.

17.3.1 Kolmogorov–Smirnov Test

We present a more direct approach based on rejecting for large values of $\Delta(\widehat{\mathsf{F}}_1, \widehat{\mathsf{F}}_2)$, where Δ is a measure of dissimilarity between distribution functions. We saw some examples in Section 16.8.1. We focus on the supremum norm (16.15), which is one of the most popular choices

$$\Delta(\widehat{\mathsf{F}}_1, \widehat{\mathsf{F}}_2) = \sup_x \left| \widehat{\mathsf{F}}_1(x) - \widehat{\mathsf{F}}_2(x) \right|. \tag{17.4}$$

Proposition 17.13. *The distribution of* $\Delta(\widehat{\mathsf{F}}_1, \widehat{\mathsf{F}}_2)$ *when both samples are drawn from the same distribution, say* F_0, *does not depend on* F_0 *as long as it is continuous. The distribution does depend on the sample sizes* (n_1, n_2).

Problem 17.14 Prove Proposition 17.13.

There are recursive formulas for computing that distribution, which are valid when there are no ties in the data.

R corner Such recursive formulas are implemented in the function ks.test, although for larger sample sizes, the function relies on the asymptotic distribution, which is known in closed form.

When there are no ties in the data, the p-value may be obtained by Monte Carlo simulation, which consists in drawing the two samples independently from the uniform distribution in $[0, 1]$ (or any other continuous distribution). Alternatively, this can be done by permutation (Section 17.3.2).

Problem 17.15 (Random walk) There is a close connection with the simple random walk process of Example 9.14, particularly when the sample sizes are equal ($n_1 = n_2$). In that case, provide an interpretation of $\Delta(\widehat{F}_1, \widehat{F}_2)$ in terms of a simple random walk.

17.3.2 Permutation Distribution

Define the concatenated sample

$$X_i = \begin{cases} X_{i,1} & \text{if } i \leq n_1; \\ X_{i-n_1,2} & \text{if } i > n_1. \end{cases}$$

Thus, if we denote the two samples by $x_1 = (x_{i,1} : i = 1, \ldots, n_1)$ and $x_2 = (x_{i,2} : i = 1, \ldots, n_2)$, the concatenated sample is $x = (x_1, x_2)$ and is of length $n_1 + n_2 = n$.

Let Π denote the group of permutations of $\{1, \ldots, n\}$. For a permutation π, let x_π denote the corresponding vector, meaning $x_\pi = (x_{\pi_1}, \ldots, x_{\pi_n})$ if $\pi = (\pi_1, \ldots, \pi_n)$. The *permutation distribution* of a statistic T conditional on x is the uniform distribution on $(T(x_\pi) : \pi \in \Pi)$. If a test rejects for large values of T, the corresponding p-value is defined as in (15.4).

Problem 17.16 Assume that the samples come from the same continuous distribution. Show in that case that the distribution of the Kolmogorov–Smirnov statistic coincides with its permutation distribution. Show that the same is true of any test statistic based on a test of randomness (Section 15.7).

Remark 17.17 Nominally, there are $n! = (n_1 + n_2)!$ permutations. In fact, some of these are equivalent, as any permutation within a group does not change the underlying distribution, even under an alternative. (Recall that each sample is assumed to be iid.) The number of non-equivalent permutations is

$$\binom{n}{n_1} = \binom{n}{n_2} = \frac{n!}{n_1! \, n_2!}.$$

Even then, the number of such permutations is often impractically large, and the permutation p-value is typically estimated by Monte Carlo simulation.

Remark 17.18 (Re-randomization) In the context of a randomized trial where, say, individuals are assigned to either of two groups at random to receive one of two treatments, under the null distribution, all the permutations are equally likely, and obtaining a p-value by permutation is an example of re-randomization testing (Section 22.1.1). That said, the permutation p-value is valid regardless, and can be motivated by conditional inference (Section 22.1).

17.3.3 Rank Tests

Let $r_{i,j}$ denote the *rank* of $x_{i,j}$ in increasing order when the two samples are combined. Ties, if present, can be broken in any number of ways, for example, at random, or by giving to all the tied observations their average rank. (In principle, ties do not arise when the underlying distributions are continuous, but in practice they are common.)

R corner The function rank offers a number of ways for breaking ties, including these two.

Problem 17.19 Prove that

$$\sum_{j=1}^{2} \sum_{i=1}^{n_j} r_{i,j} = n(n+1)/2,$$

so that the set of ranks for Group 1 determines the set of ranks for Group 2.

A *rank statistic* is any function of the ranks.

Problem 17.20 Show that the Kolmogorov–Smirnov statistic (17.4) is a rank statistic.

Problem 17.21 (Rank tests are permutation tests) Show that under the null hypothesis where both samples come from the same distribution, if that distribution is continuous or the ranks are broken at random, the concatenated vector of ranks, $(R_{i,j})$, is uniformly distributed over all permutations of $\{1, \ldots, n_1 + n_2\}$. In particular, the p-value of a rank test coincides with its permutation p-value.

Wilcoxon Rank-Sum Test

This test is based on the sum of the ranks from Group 1, meaning

$$r_1 := \sum_{i=1}^{n_1} r_{i,1}. \tag{17.5}$$

In the two-sided setting of (17.3), we reject for large and small values of this test statistic. The test was proposed by Frank Wilcoxon (1892–1965) in the 1940s, and remains very popular.

Problem 17.22 Show that the null distribution of R_1 is symmetric about its mean, namely, $n_1(n + 1)/2$. Deduce that the two-sided rank-sum test corresponds to rejecting for large value of $(r_1 - n_1(n + 1)/2)^2$.

Remark 17.23 The Wilcoxon rank-sum test is sometimes presented as a test for comparing medians. This is *not* correct in general, the reason being that, as any other permutation test (Problem 17.21), the rank-sum test is a goodness-of-fit test. By this we mean that it results in a valid p-value (in the sense of (12.12)) when the distributions are the same under the null. When only the medians are the same under the null, the p-value is not guaranteed to be valid. That being said, testing for the equality of medians is equivalent to goodness-of-fit testing when the underlying distributions are assumed to be translates of each other, meaning

$$\mathsf{F}_1 = \mathsf{F}_2(\cdot - \mu) \text{ for some } \mu. \tag{17.6}$$

In that case, obviously, the distributions are the same if and only if their medians are the same.

Remark 17.24 There are advantages and disadvantages to using ranks. The main disadvantage is a loss in sensitivity, which may result in a loss of power. This loss is typically quite mild. The two main advantages are in terms of computation and robustness.

- *Computation and tabulation*   Assuming that ties are broken at random, the null distribution of R_1 only depends on (n_1, n_2), which can therefore be pre-tabulated. (This can done using recursive formulas for smaller sample sizes and asymptotic calculations for larger sample sizes.)
- *Robustness*   Using ranks offers some protection against gross outliers. For example, consider the permutation test based on the difference in sample means. Assume that one value has been corrupted and is now larger than the sum of the absolute values of all the other observations. Then the p-value will be approximately $1/2$, whether the null hypothesis is true or not. If one uses ranks, the effect of this corruption is minimal.

17.3.4 Patterns and Tests of Randomness

A *pattern* is obtained by ordering the observations in the combined sample and labeling each observation according to the group it belongs to. For example, the data

$$\text{Group 1:} \quad 2.5 \quad 0.9 \quad -1.0 \quad -1.6 \quad 0.7$$
$$\text{Group 2:} \quad -1.2 \quad -0.5 \quad -1.4$$

results in the following pattern:

$$1 \quad 2 \quad 2 \quad 1 \quad 2 \quad 1 \quad 1 \quad 1$$

Once the pattern is computed on a particular dataset, tests of randomness (Section 15.7) can be used for goodness-of-fit testing.

Problem 17.25 Suppose that F_1 and F_2 are continuous. Show that any pattern is equally likely if and only if $F_1 = F_2$.

Therefore, when ties are broken at random, we can rely on the p-value returned by the test of randomness applied to the pattern. (In any case, we can rely on the permutation p-value.)

Remark 17.26 A pattern provides the same information as the ranks modulo the ordering within each group, but this within-group ordering is irrelevant for inference because each group is assumed to be iid.

17.4 Multiple Samples

We now consider the more general situation where g groups of observations are available and need to be compared. The observations from Group j are denoted $(X_{i,j} : i = 1, \ldots, n_j)$ and assumed to be iid from some distribution denoted P_j. As usual, the samples are assumed to be independent of each other. We let $n := n_1 + \cdots + n_g$ denote the total sample size.

Remark 17.27 On a computer, multiple-sample is typically stored an array with two variables, 'values' and 'group label'. For example, the data

$$\text{Group 1:} \quad 2.2 \quad 0.8 \quad 1.0$$
$$\text{Group 2:} \quad 1.1 \quad 0.3 \quad 1.6$$
$$\text{Group 3:} \quad 0.3 \quad 0.8$$

may be equivalently written

$$\text{Values:} \quad 2.2 \quad 0.8 \quad 1.0 \quad 1.1 \quad 0.3 \quad 1.6 \quad 0.3 \quad 0.8$$
$$\text{Group:} \quad 1 \quad 1 \quad 1 \quad 2 \quad 2 \quad 2 \quad 3 \quad 3$$

17.4.1 All Pairwise Comparisons

A reasonable approach is to perform all pairwise comparisons using a method for the comparison of two groups. If we are performing a test, for example, this leads us to do so for every pair, so that $\binom{g}{2}$ tests are performed in total. Such *multiple testing* situations are discussed in more detail in Chapter 20.

R corner The function pairwise.t.test performs all the pairwise Student–Welch tests, while pairwise.wilcox.test performs all the pairwise Wilcoxon rank-sum tests. A number of ways to correct for multiple testing are offered.

Remark 17.28 (Many-to-one comparisons) Another approach consists in performing a *many-to-one comparison*, where all the 'treatment' groups are compared to a 'control' group, the latter serving as benchmark. Treatment groups are *not* compared to each other. (For more on this, see the classic book by Miller [131].)

In the remainder of this section, we focus on various forms of *global testing*, by which we mean testing whether there is any difference between the groups. (We develop this further in Section 20.2.)

17.4.2 Testing for a Difference in Means

Let μ_j denote the mean of P_j, and consider testing for the equality in means

$$\mathcal{H}_0 : \mu_1 = \cdots = \mu_g.$$

This generalizes the testing problem of Section 17.1, where we considered the case $g = 2$. The methods presented there have analogue s.

F-test

This test was proposed by Fisher in the 1920s and later modified by Welch [203] to settings where the groups do not necessarily have the same variance. This test, which we will call the Fisher–Welch test, generalizes the Student–Welch test and, in particular, also relies on the Central Limit Theorem. Its form and derivation are rather complex and will not be given here.

R corner This test is implemented in the R function oneway.test.

Bootstrap Tests

A bootstrap p-value may be obtained in a way which is analogous to the one presented in Section 17.1.2. In particular, the resampling is within groups.

As for the choice of a test statistic, one possibility is to choose the *treatment sum-of-squares*, defined as

$$\sum_{j=1}^{p} n_j(\bar{x}_j - \bar{x})^2, \tag{17.7}$$

where $\bar{x}_j$ is the average for Group j and $\bar{x}$ is the overall average. For a Studentized version, use the Fisher–Welch test statistic instead.

We assume that we reject for large values of a test statistic T. Importantly, we need to place ourselves under the null distribution before bootstrapping. We do this by centering each group, which effectively makes the groups have the same mean (equal to 0). Having observed the data, let $\mathsf{P}_{0,j}^*$ denote the empirical distribution for Group j after centering. A bootstrap sample is thus $X^* = (X_{i,j}^*)$, where

$$X_{i,j}^* \sim \mathsf{P}_{0,j}^*, \quad \text{independent.}$$

Then the bootstrap p-value is the probability that $T^* := T(X^*) \geq t := T(x)$, as usual. This p-value is typically estimated by Monte Carlo simulation.

Problem 17.29 In R, write a function that takes in the values and group labels, and the number of bootstrap samples to be generated, and returns the Monte Carlo estimate for the bootstrap p-value of the Fisher–Welch statistic (Although we did not provide an analytic form for this statistic, it can be computed using the function oneway.test.)

17.4.3 Goodness-of-Fit Testing

As in Section 17.3, suppose we are interested in comparing the distributions that generated the groups in the sense of testing

$$\mathcal{H}_0 : \mathsf{P}_1 = \cdots = \mathsf{P}_g.$$

Permutation Tests

As in Section 17.3.2, a calibration by permutation is particularly appealing, and can be motivated from re-randomization perspective (Remark 17.18).

Suppose we reject for large values of T. This could be the treatment sum-of-squares or the Fisher–Welch statistic, or any other. We concatenate the groups as we did in Section 17.3.2, obtaining $x = (x_1, \dots, x_g)$, where $x_j := (x_{1,j}, \dots, x_{n_j,j})$ are the observations from Group j. Let Π be the group of permutations of $\{1, \dots, n\}$ and for $\pi \in \Pi$, let x_π denote the dataset permuted according to π. The corresponding p-value is defined as in (15.4), and it is a valid p-value under the null hypothesis $\mathcal{H}_0$ above.

Problem 17.30 In R, write a function that takes in the values and group labels and the number of permutations to be generated, and returns a Monte Carlo permutation p-value for the Fisher–Welch test statistic.

Rank Tests

As in Section 17.3.3, using ranks is a viable option. Let $r_{i,j}$ be the rank of $x_{i,j}$ in increasing order in the combined sample. A direct extension of the Wilcoxon rank-sum test, in particular in view of Problem 17.22, consists in rejecting for large values of

$$\sum_{j=1}^{g} \left(r_j - n_j(n+1)/2\right)^2, \tag{17.8}$$

where $r_j := \sum_{i=1}^{n_j} r_{i,j}$ is the rank sum for Group j.

Problem 17.31 Show that, indeed, this reduces to the two-sided rank-sum test when $g = 2$.

Problem 17.32 In R, write a function that implements the test based on (17.8). The function takes as input the data and the number of permutations to be drawn and returns the corresponding estimated p-value.

The most popular rank test in the present setting, however, is a. test due to William Kruskal (1919–2005) and W. Allen Wallis (1912–1998), which rejects for large values of the treatment sum-of-squares (17.7) computed on the ranks, namely

$$\sum_{j=1}^{g} \frac{1}{n_j} \left(r_j - n_j(n+1)/2\right)^2. \tag{17.9}$$

Problem 17.33 Show that the Kruskal–Wallis test equivalently rejects for large value of $\sum_{j=1}^{g} r_j^2/n_j$. [Use the fact that $r_1 + \cdots + r_g = n(n+1)/2$.]

Remark 17.34 The actual Kruskal–Wallis test statistic is an affine function of (17.9) that makes the resulting statistic have, under the null hypothesis, the chi-squared distribution with $g-1$ degrees of freedom in the large-sample limit where $\min_j n_j \to \infty$.

R corner This test is implemented in the function kruskal.test, which returns a p-value based on the limiting distribution.

Problem 17.35 In R, write a function that takes in the data and a number of Monte Carlo replicates, and returns a Monte Carlo estimate of the permutation p-value for the Kruskal–Wallis statistic.

17.5 Further Topics

17.5.1 Two-Sample Median Test

Despite its name, the *median test* is for goodness-of-fit, meaning a test for (17.3). It works as follows. We consider two groups, of sizes n_1 and n_2, as before. Let M denote the sample median of all the observations combined. Let T denote the number of observations from Group 1 that exceed M. The two-sided variant of the test rejects for large and small values of T.

Problem 17.36 Assume that ties are broken at random.

(i)   Assume the $n = n_1 + n_2$ is even. Show that, under the null hypothesis stated in (17.3), T has the hypergeometric distribution with parameters $(n_1, n/2, n/2)$.

(ii) Assume that n is odd. What is the distribution of T under the null?

Problem 17.37 In R, write a function that implements this test.

17.5.2 Consistency

Consider the two-sample setting. We say that a testing procedure is universally consistent if, when $P_1 \neq P_2$, at any level $\alpha > 0$ the corresponding test has power converging to 1 has $n_1, n_2 \to \infty$.

Problem 17.38 Show that the Kolmogorov–Smirnov testing procedure is universally consistent. [Use the fact that the empirical distribution function is consistent for the underlying distribution function.]

Proposition 17.39. *The Wilcoxon rank-sum test procedure is* not *universally consistent.*

The Wilcoxon rank-sum test is, however, consistent in a shift model (17.6). More generally, the rank-sum test is consistent when, under the alternative, P_1 stochastically dominates P_2, or vice versa (assuming the two-sided version of the test). We say that P, with distribution function F, *stochastically dominates* Q, with distribution function G, if

$$\bar{F}(t) \geq \bar{G}(t), \quad \text{for all } t \in \mathbb{R}.$$

This means that, if X has distribution F and Y has distribution G, then $\mathbb{P}(X > t) \geq \mathbb{P}(Y > t)$ for all t.

17.6 Additional Problems

Problem 17.40 (Kolmogorow–Smirnov vs Wilcoxon) In R, perform some simulations to compare the power of the (one-sided) Kolmogorov–Smirnov test and that of the Wilcoxon test. Consider the case where both samples are of same size $m \in \{20, 50, 100\}$. The first group comes from $\mathcal{N}(0, 1)$. In the first situation, the second group comes from $\mathcal{N}(\theta, 1)$. In the second situation, the second group comes from $\mathcal{N}(0, 1 + \theta)$. In each case $\theta > 0$ is chosen carefully on a grid to make the setting interesting, showing the power going from the level, set at $\alpha = 0.01$, to close to 1. Repeat each setting 10^3 times. For each (situation, m), in the same plot draw the power curve for each test as a function of θ. Use different colors and add a legend.

Problem 17.41 (Permutation vs rank tests) In R, perform some simulations to compare the power of the permutation test based on the difference in sample means and the corresponding rank test, which is none other than the rank-sum test. Specifically, consider the case where both samples are of same size $m \in \{20, 50, 100\}$. The first group comes from $\mathcal{N}(0, 1)$, while the second group comes from $\mathcal{N}(\theta, 1)$. In each case $\theta > 0$ is chosen carefully to make the setting interesting, showing the power going from $\alpha := 0.01$ to close to 1. Repeat each setting (situation, m, θ) $B = 10^3$ times. For each (situation, m), in the same plot, draw the power curve for each test as a function of θ.

Problem 17.42 Strictly speaking, a permutation p-value is only valid for the null hypothesis that the underlying distributions are the same. How does it behave when it is used when testing for a parameter? Consider a two sample setting where we test for a difference in means. We choose as test statistic the difference in sample means, which we calibrate by permutation. The first group comes from $\mathcal{N}(0, 1)$, while the second group comes from $\mathcal{N}(0, 3)$. Clearly, we are under the null hypothesis (same means), and the distributions are different. Assume both groups are of same size m, and the permutation p-value is based on a number B of Monte Carlo replicates. By varying m and B, assess the accuracy of the permutation p-value. Offer some brief comments, and possibly some elements of explanation for that behavior.

Problem 17.43 The two tests, based on (17.8) and (17.9) respectively, coincide when the design is balanced in the sense that the group sizes are identical. How do they compare when the group sizes are not the same? Perform some simulations to investigate that.

Problem 17.44 (Bootstrap goodness-of-fit tests) In goodness-of-fit testing, permutation is typically considered the calibration of choice, in large part because the permutation p-value is valid, regardless of the group sizes. However, a bootstrap approach is also possible. Explain, when testing whether two groups come from the same distribution, how you would obtain a p-value by bootstrap.

Problem 17.45 (Mann–Whitney test) This test is based on the test statistic

$$U := \sum_{j=1}^{2} \sum_{i=1}^{n_j} \{X_{i,1} > X_{j,2}\}.$$

Show that, at least when there are no ties, $U = R_1 - n_1(n_1+1)/2$, where R_1 is rank-sum for Group 1 (17.5). [Hence, the Mann-Whitney test is equivalent to the Wilcoxon rank-sum test.]

Problem 17.46 (Student test) Consider a normal experiment where we observe independent realizations of $X_1, \ldots, X_m$, assumed an iid sample from $\mathcal{N}(\mu, \sigma^2)$, and of $Y_1, \ldots, Y_n$, assumed an iid sample from $\mathcal{N}(\xi, \sigma^2)$. Importantly, the two normal distributions are assumed to have the *same variance*. All three parameters are unknown. Our goal is to test the null hypothesis that the means are equal, $\mu = \xi$. The *Student test* rejects for large values of $|T|$ where

$$T := \frac{\bar{X} - \bar{Y}}{S}, \tag{17.10}$$

with $\bar{X}$ and $\bar{Y}$ being the sample means and S being the pooled sample standard deviation.

(i)   Show that there is a constant $c_{m,n}$ such that $c_{m,n}T$ has the Student distribution with $m + n - 2$ degrees of freedom under the null.

(ii) Show that this test corresponds to the LRT under the present model.

Problem 17.47 (Student–Welch test) In the same setting, the *Student–Welch test* rejects for large values of $|T|$ where

$$T := \frac{\bar{X} - \bar{Y}}{\sqrt{S_X^2/m + S_Y^2/n}}, \tag{17.11}$$

with $\bar{X}$ and $\bar{Y}$ being the sample means and S_X and S_Y being the sample standard deviations. It turns out that T does not have a Student distribution under the null hypothesis. Does this test correspond to the LRT under the present model?

Problem 17.48 (Fisher test) Consider a normal experiment where we observe independent realizations of $(X_{i,j} : i = 1, \ldots, n_j)$ assumed an iid sample from $\mathcal{N}(\mu_j, \sigma^2)$, for $j = 1, \ldots, g$. Importantly, the normal distributions are assumed to have the same variance. All parameters are unknown. Our goal is to test the null hypothesis that $\mu_1 = \cdots = \mu_g$. The *F-test* rejects for large values of F where

$$F := \frac{\sum_{j=1}^{g} n_j (\bar{X}_j - \bar{X})^2}{S^2},$$

with $\bar{X}_j$ being the sample mean for the group j, $\bar{X}$ the pooled sample mean, and S the pooled sample standard deviation. This is the multiple-sample analogue of (17.10).

(i) Show that there is a constant $c_{m,n}$ such that the distribution of $c_{m,n}F$ under the null hypothesis is the Fisher distribution with $g - 1$ and $n - g$ degrees of freedom, where n is the total sample size.

(ii) Show that this test corresponds to the LRT under the present model.

(There is a Welch version of this test which is the multi-sample analogue of (17.11). The exact form of the test statistic is rather complicated [203].)

Problem 17.49 (Energy statistics) Cramér [66] proposed in [34] the following dissimilarity for comparing two distribution functions

$$\Delta(\mathsf{F}, \mathsf{G})^2 := \int_{-\infty}^{\infty} \left(\mathsf{F}(x) - \mathsf{G}(x) \right)^2 dx.$$

(Note the difference with the Cramér–von Mises dissimilarity (16.16).)

(i) Letting X, X' be iid from F and (independently) Y, Y' be iid from G, show that

$$\Delta(\mathsf{F}, \mathsf{G})^2 = \mathbb{E}[|X - Y|] - \tfrac{1}{2}\mathbb{E}[|X - X'|] + \tfrac{1}{2}\mathbb{E}[|Y - Y'|].$$

(ii) Use this to provide an explicit expression for the sample equivalent, meaning when applying the dissimilarity to the empirical distribution functions of the two samples under consideration. [68]

(iii) Implement this as a test in R. (Calibration is by permutation based on a specified number of Monte Carlo replicates.)

(iv) Is the test distribution-free?

---

[68] The resulting statistic is an example of what Székely and collaborators call *energy statistics*. See [184] for a survey of the relevant literature which puts this early proposal by Cramér in context.

Problem 17.50 The test of Problem 17.49 was expressed there in terms of distribution functions. However, it turns out it also has a relatively simple expression in terms of characteristic functions. Indeed, show that

$$\Delta(\mathsf{F},\mathsf{G})^2 = \frac{1}{2\pi} \int_{-\infty}^{\infty} \frac{|\varphi_\mathsf{F}(t) - \varphi_\mathsf{G}(t)|^2}{t^2}\, dt,$$

where φ_F and φ_G are the characteristic functions of F and G, respectively.

Problem 17.51 Censoring is also common with multiple-sample numerical data. In fact, we already saw an instance of that in Example 16.63, where an analysis would lead one to compare the group survival functions. Propose a Kolmogorov–Smirnov-type test based on the Kaplan–Meier estimator given in (16.20), specifying how you obtain the p-value. In R, write a function that implements that test and apply it to the data of Example 16.63.

Problem 17.52 (The log-rank test) Continuing with censored data, consider two groups (as in Example 16.63). The goal remains the same, which is, to compare the underlying distributions. Let $D_{t,j}$ (resp. $Y_{t,j}$) denote the number of events (resp. of subjects at risk) at time t in Group j. Define $D_t = D_{t,1} + D_{t,2}$ and $Y_t = Y_{t,1} + Y_{t,2}$. Argue that, under the null hypothesis of identical distributions, $D_{t,1} \sim \mathrm{Hyper}(D_t, Y_{t,1}, Y_{t,2})$ when conditioning on $(D_t, Y_{t,1}, Y_{t,2})$. Derive the corresponding mean and variance, denoted $E_{t,1}$ and $V_{t,1}$ below. The *log-rank test* is based on the statistic

$$\frac{\sum_t (D_{t,1} - E_{t,1})}{\left(\sum_t V_{t,1}\right)^{1/2}}.$$

In R, implement this test, with a p-value obtained by permutation, and apply it to Example 16.63. Traditionally, the p-value is derived from the standard normal distribution based on a normal approximation. Can you justify this approximation?

18

Multiple Paired Numerical Samples

In this chapter, we consider experiments where the variables of interest are paired. Importantly, we assume that these variables are directly comparable. (This is in contrast with Chapter 19 and Chapter 21, where the variables can be measurement in different units.)

Crossover trials are important examples of such experiments. Other examples include the following.

Example 18.1 (Judge panel) In the food industry, in particular, it is common to ask individuals to rate the taste of different products, typically of the same type. For example, in [28], 12 experienced wine tasters were asked to rate 78 wines on a variety of characteristics.

Example 18.2 (Father-son heights) Karl Pearson collected data on the heights of 1078 fathers and their (adult) sons [142]. This dataset is discussed at length in [66] and, in fact, a scatterplot of the data is on the cover of that book, and reproduced in Figure 18.1.

18.1 Two Paired Variables

We start with a setting where we observe a bivariate numerical sample $(X_1, Y_1), \ldots, (X_n, Y_n)$. The pairs are assumed iid from some unknown distribution. We assume that the X and Y variables can be compared directly.

Taking the example of a crossover trial comparing two treatments (one of them could be a placebo), when there is no difference between treatments it is assumed that X and Y are *exchangeable*. When testing for a difference in treatment, we are thus testing the following null hypothesis

$$\mathcal{H}_{0,*} : (X, Y) \sim (Y, X).$$

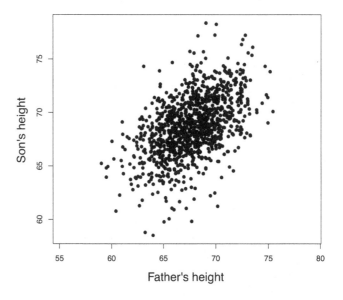

Figure 18.1 A scatterplot of the data described in Example 18.2.

18.1.1 Testing for Symmetry

There is a one-to-one correspondence

$$(X, Y) \leftrightarrow (U, Z) := (X + Y, X - Y),$$

and the null hypothesis can be equivalently expressed as follows:

$$\mathcal{H}_{0,*} : (U, Z) \sim (U, -Z).$$

It is tempting to drop U and focus on Z, and doing so[69] leads to testing the following null hypothesis of *symmetry*

$$\mathcal{H}_0 : Z \sim -Z. \tag{18.1}$$

The rest of this section is dedicated to testing this hypothesis based on an iid sample $Z_1, \ldots, Z_n$. (Unless restricted further, the underlying distribution is simply assumed to be in the family of all distributions on the real line.)

---

[69] This can be justified on the basis of invariance considerations [110, Sec 6.8].

18.1.2 Symmetric Distributions

A random variable Z is said to be *symmetric* about μ if

$$\mathbb{P}(Z < z) = \mathbb{P}(Z > \mu - z), \quad \text{for all } z \in \mathbb{R}.$$

Problem 18.3 Assuming that Z has continuous distribution function F, show that this is equivalent to

$$F(z) = 1 - F(\mu - z), \quad \text{for all } z \in \mathbb{R}.$$

In particular, F is symmetric about 0 if

$$F(z) = 1 - F(-z),$$

that is, if $z \mapsto F(z) - 1/2$ is odd.

Problem 18.4 Suppose that F has a piecewise continuous density f. Show that F is symmetric about μ if and only if

$$f(z) = f(\mu - z), \text{ at any continuity point } z.$$

In particular, if f is continuous, F is symmetric about 0 if and only f is even. (In any case, f is essentially even.)

Problem 18.5 Show that if F is symmetric about μ, then μ is necessarily a median of F, and also its mean if it has a mean.

Although symmetry can be about any point on the real line, in what follows we assume that point to be the origin. This is the most important case, in part because it is motivated by (18.1), and also because it can be considered without loss of generality. In particular, in what follows, by 'symmetric' we mean 'symmetric about 0'.

Problem 18.5 justifies the application of tests for the median, such as the sign test (Section 16.3.3), as well as tests for the mean, for example a bootstrap test (Section 16.6.5). If such a test rejects, it is evidence against the (null) hypothesis of symmetry. However, such a test cannot be universally consistent since a distribution can be asymmetrical and yet have zero median, or zero mean, or both.

Problem 18.6 Construct a distribution that is asymmetrical and has median and mean both equal to 0. One avenue is to consider a Gaussian mixture of the form $p\mathcal{N}(a, 1) + (1 - p)\mathcal{N}(b, \sigma^2)$, where $p \in [0, 1]$, $a, b \in \mathbb{R}$, and $\sigma^2 > 0$ are chosen to satisfy the requirements. Another avenue is to consider a distribution with finite support. In that case, what is the minimum support size needed to satisfy the requirements?

The procedures presented below are based on the following characterizations of symmetry.

Problem 18.7 Let Z be a random variable. Show that the following assertions are equivalent:

(i) Z is symmetric;
(ii) $\mathrm{sign}(Z)$ is symmetric and independent of $|Z|$;
(iii) $\mathbb{P}(Z > 0) = \mathbb{P}(Z < 0)$ and $Z \mid Z > 0$ and $-Z \mid Z < 0$ share the same distribution.

Problem 18.7 can be used to motivate the comparison of the positive part of the sample, meaning $\{Z_i : Z_i > 0\}$, with the negative part of the sample, meaning $\{-Z_i : Z_i < 0\}$, as one would for two different groups. The techniques developed in Section 17.3 for that purpose are particularly relevant. Following this logic leads to two well-known methods for testing for symmetry that we present next.

Remark 18.8 (Zero values) Since the values that are exactly 0 do not carry any information on the asymmetry of the underlying distribution, it is common to simply drop these values before applying a procedure. (This is an example of conditional inference.) This is what we do in what follows, and, although it changes the sample size, we redefine n as the sample size after removing these observations.

18.1.3 A Test based on Sign Flips

Let $z = (z_1, \ldots, z_n)$ denote the observed sample, and for a sign vector $\varepsilon = (\varepsilon_1, \ldots, \varepsilon_n) \in \{-1, 1\}^n$, let $z_\varepsilon = (\varepsilon_1 z_1, \ldots, \varepsilon_n z_n)$. Suppose that we reject for large values of a test statistic $T(z)$. A popular choice is

$$T(z) = \left| \sum_{i=1}^{n} z_i \right|. \tag{18.2}$$

Having observed $Z = z$, the p-value is

$$\mathrm{pv}_T(z) = \frac{\#\{\varepsilon : T(z_\varepsilon) \geq T(z)\}}{2^n}. \tag{18.3}$$

The denominator is the cardinality of the set of sign vectors $\{-1, 1\}^n$. Thus, this is the proportion of sign vectors that lead to a value of the test statistic that is at least as extreme as the one observed.

Proposition 18.9. *This quantity is a valid p-value in the sense that it satisfies* (12.12), *meaning that*

$$\mathbb{P}(\mathsf{pv}_T(\boldsymbol{Z}) \le \alpha) \le \alpha, \quad \text{for all } \alpha \in [0, 1],$$

when the underlying distribution is symmetric.

In practice, computing the p-value (18.3) is quickly intractable. This is because the number of sign vectors of interest is equal to 2^n, which grows exponentially ($> 10^{30}$ when $n = 100$). As usual, one resorts to Monte Carlo simulation to estimate this p-value, by repeatedly drawing sign vectors uniformly at random.

Problem 18.10 In R, write a function that takes as input the sample and a number of Monte Carlo replicates, and returns the p-value estimated by Monte Carlo for the test statistic (18.2).

Remark 18.11 Although the test is built on a permutation test, it is not a permutation test per se. Nothing is being permuted. However, its construction can be motivated by conditional inference (Section 22.1): we condition on the absolute values, $|Z_1|, \ldots, |Z_n|$, which a priori do not carry any information on whether the underlying distribution is symmetric.

Problem 18.12 How is the test above, based on the statistic (18.2) and returning the p-value (18.3) different from the permutation test, based on the absolute value of the difference in sample means, applied to compare the positive and negative samples? [The two tests are almost the same, but not quite identical.]

18.1.4 Wilcoxon Signed-Rank Test

We now turn to the test that results from comparing the distributions of $Z \mid Z > 0$ and $-Z \mid Z < 0$ using the Wilcoxon rank-sum test presented in Section 17.3.3. This leads one to use the test statistic

$$\sum_{i=1}^{n} r_i\{z_i > 0\},$$

where r_i is the rank of $|z_i|$ among $|z_1|, \ldots, |z_n|$. In the two-sided situation, we reject for large and small values of this statistic, as we did for the two-sided rank-sum test.

Problem 18.13 Verify that this is indeed the resulting test statistic when using the rank-sum test to compare the distributions of $Z \mid Z > 0$ and $-Z \mid Z < 0$.

R corner The function wilcox.test computes the signed-rank test when provided with a numerical vector, and the rank-sum test when provided with two numerical vectors.

Problem 18.14 Show that it is equivalent to use the following statistic

$$\sum_{i=1}^{n} r_i \operatorname{sign}(z_i) = \sum_{z_i > 0} r_i - \sum_{z_i < 0} r_i. \tag{18.4}$$

(In this form, it is clear that this is the rank variant of the test of Section 18.1.3.)

Problem 18.15 When the underlying distribution is symmetric, and assuming in addition that it is continuous or that ties among ranks are broken at random, show that the signed-rank statistic in the form of (18.4) has the distribution of $\sum_{i=1}^{n} i \varepsilon_i$ with[70] $\varepsilon_1, \ldots, \varepsilon_n$ iid uniform in $\{-1, 1\}$.

Whether the underlying distribution is continuous or not, and whether ties are broken at random or not, an approach by conditional inference remains available: it consists in computing a p-value by fixing the ranks while sampling sign vectors uniformly at random.

Sign Pattern

In Section 17.3.4, we saw that any (reasonable) rank test for the two-sample setting is based on the pattern defined by the two samples combined. The situation is analogous here. Indeed, any (reasonable) test for symmetry based on the ranks and signs is based on the *sign pattern* given by ordering the absolute values and then listing the signs in that order. For example, the following observations

$$1.3 \quad 2.1 \quad 2.5 \quad -1.4 \quad 1.0 \quad 0.4 \quad -3.5 \quad -1.0 \quad 0.2$$

yield the following sign sequence

$$+ \quad + \quad + \quad - \quad + \quad - \quad - \quad + \quad +$$

---

[70] The uniform distribution on $\{-1, 1\}$ is sometimes called the *Rademacher distribution*.

18.2 Multiple Paired Variables

We now consider the more general case of p paired variables. The sample is of size n, and denoted

$$(X_{1,1}, \ldots, X_{1,p}), \ldots, (X_{n,1}, \ldots, X_{n,p}).$$

Let $X_i = (X_{i,1}, \ldots, X_{i,p})$ denote the ith observation, which is a vector of length p here. We assume these n observations to be iid from some unknown distribution, and we also assume that all the variables are directly comparable. This is the case in Example 18.1, where $X_{i,j}$ is the rating of ith wine by the jth judge.

These observations are typically gathered in a n-by-p data matrix, $(X_{i,j})$. See Table 18.1.

Table 18.1 *Prototypical data matrix in the context of a crossover clinical trial.*

	Treatment 1	Treatment 2	$\cdots$	Treatment p
Subject 1	$X_{1,1}$	$X_{1,2}$	$\cdots$	$X_{1,p}$
Subject 2	$X_{2,1}$	$X_{2,2}$	$\cdots$	$X_{2,p}$
$\vdots$	$\vdots$	$\vdots$		$\vdots$
Subject n	$X_{n,1}$	$X_{n,2}$	$\cdots$	$X_{n,p}$

Taking the example of a crossover clinical trial, the goal is to assess whether the treatments are different. This is again modeled by testing

$$\mathcal{H}_0 : X_1, \ldots, X_p \text{ are exchangeable.}$$

18.2.1 Permutation Tests

A calibration by permutation is particularly attractive in the present context, since permutations are at the core of the definition of exchangeability. In fact, in the context of a crossover trial, this corresponds to re-randomization testing (Section 22.1.1).

Compared to a completely randomized design, the permutation in the context of a crossover trial or any other repeated measures design is done differently. Indeed, here the permutation is within subject. In particular, permuting across subjects (as is done in the context of a completely randomized design) is not appropriate.

Let T be a test statistic whose large values are evidence against the null. Having observed $(x_1, \ldots, x_n)$, a calibration by permutation is done as follows. Let Π_0 be the group of permutations of $\{1, \ldots, p\}$ and let $\Pi := \Pi_0^{\times n}$, which also a group in the algebra sense of the term. The permutations in Π are the valid permutations in the present context.

Problem 18.16 Show that $|\Pi| = (p!)^n$.

Let $t := T(x_1, \ldots, x_n)$ denote the observed value of the statistic. For $\pi = (\pi_1, \ldots, \pi_n) \in \Pi$, with $\pi_i := (\pi_{i,1}, \ldots, \pi_{i,p}) \in \Pi_0$, let t_π denote the value of the statistic when it is applied to the same data except permuted by π, meaning $t_\pi = T(\pi_1(x_1), \ldots, \pi_n(x_n))$ where $\pi_i(x_i) := (x_{\pi_{i,1}}, \ldots, x_{\pi_{i,p}})$. The permutation p-value is then defined as in (15.4). In practice, it is typically estimated by Monte Carlo simulation based on a number of permutations that are sampled independently and uniformly at random from Π.

Problem 18.17 Show that the permutation p-value and its Monte Carlo estimate, seen as random variables, are both valid in the sense of (12.12). [Use the conclusions of Problem 8.53.]

Problem 18.18 In R, write a function that takes in as input the data matrix and a number of Monte Carlo replicates, and returns the estimated permutation p-value for the treatment sum-of-squares defined in (17.7). [Note that there are p groups here, with sample size $n_j = n$ for all j.]

18.2.2 Rank Tests

Using ranks is particularly appealing in settings where the measurements across subjects are not easy to compare. This is for example the case when the measurements are subjective evaluations. A prototypical example is that of a judge panel experiment (Example 18.1) as the judges may be more or less liberal in the use of the full appraisal scale. Another important example is that of crossover trials where what is measured is the improvement of symptoms on a visual analogue scale (VAS). Such subjective evaluations are notoriously difficult to compare. The use of ranks disregards the scale implicitly (and often unconsciously) used by a subject, and focuses on the subject's ranking instead. Thus, ranks are computed within subjects.

In detail, let $r_{i,j}$ be the rank of $x_{i,j}$ among $x_{i,1}, \ldots, x_{i,p}$ and let $r_j = \sum_{i=1}^{n} r_{i,j}$ be the rank sum for Treatment j. (Although we are using the same notation, these ranks are defined differently than in Section 17.4.3.)

Remark 18.19 Rank tests are special cases of permutation tests, as their null distribution is the permutation distribution (at least when the ties are broken at random).

Friedman Test

Like the Kruskal–Wallis test, this test uses as test statistic the treatment sum-of-squares applied to the ranks defined above. It was proposed by Milton Friedman (1912–2006) in [74].

Problem 18.20 Show that using the treatment sum-of-squares is equivalent to using $\sum_{j=1}^{p} r_j^2$.

Remark 18.21 The actual Friedman test statistic involves a standardization that makes the resulting statistic have, under the null hypothesis, the chi-squared distribution with $p - 1$ degrees of freedom in the large-sample limit where $n \to \infty$.

R corner The function friedman.test, which implements that test, uses the limiting distribution to compute the p-value.

18.3 Additional Problems

Problem 18.22 Consider a strictly increasing and continuous distribution function F. Derive a necessary and sufficient condition on the corresponding quantile function F^{-1} (a true inverse in this case) for F to be symmetric about a given $\mu \in \mathbb{R}$.

Problem 18.23 (Sign-flip vs signed-rank) Perform some simulations to compare the power of the sign-flip test of Section 18.1.3 and the Wilcoxon signed-rank test of Section 18.1.4.

Problem 18.24 (Smirnov test) The *Smirnov test for symmetry* is based on comparing the distributions of $Z \mid Z > 0$ and $-Z \mid Z < 0$ using the two-sample Kolmogorov–Smirnov test.

(i)   Write down a simple expression for the test statistic.

(ii)   In R, write a function that takes in the data and a number of Monte Carlo replicates, and returns the Monte Carlo estimate for the p-value (18.3).

Problem 18.25 (Hodges–Lehmann estimator) Suppose we have a numerical sample $z_1, \ldots, z_n$. For $\theta \in \mathbb{R}$, let $r_i(\theta)$ denote the rank of $|z_i - \theta|$ among $\{|z_j - \theta| : j = 1, \ldots, n\}$.

(i)   Show that

$$\sum_i r_i(\theta)\,\text{sign}(z_i - \theta) = \sum_{i \le j} \text{sign}\left(\tfrac{1}{2}(z_i + z_j) - \theta\right).$$

From this, argue that the confidence interval which is derived (as in Section 12.4.7) from the Wilcoxon signed-rank test for the null hypothesis that $\theta = \theta_0$ (at some arbitrary level α) is symmetric about

$$\text{Med}\left(\tfrac{1}{2}(z_i + z_j) : i \le j\right).$$

This statistic is the *Hodges–Lehmann estimator*.

(ii)   Assume the sample is generated iid from some distribution. Determine the large-sample limit of this statistic. This limit quantity defines the (location) parameter of the underlying distribution that the statistic is consistent for as an estimator. Show that this parameter is the median when the distribution is symmetric about a unique median.

19

Correlation Analysis

We consider an experiment which results in a sample of paired numerical variables, $(X_1, Y_1), \ldots, (X_n, Y_n)$. The paired variables are generically denoted by (X, Y), and the general goal addressed in this chapter is that of quantifying the strength of association between X and Y. By association we mean dependence. Contrary to the setting of Chapter 18, and, in particular, Section 18.1, here X and Y can be measurements of completely different kinds.[71]

Example 19.1 The study described in [106] evaluates the impact of the time in detention on the mental health of asylum seekers in the United States in terms of (self-reported) symptoms of anxiety, depression, and post-traumatic stress disorder. Focusing on just one symptom, say anxiety, the data are of the form $(T_1, A_1), \ldots, (T_n, A_n)$, where T_i denotes the time spent in detention and A_i the level of anxiety for Individual i. (There were $n = 70$ individuals interviewed for this study.)

A *correlation analysis* amounts to quantifying the level of association between X and Y. A more detailed description of this association is the goal of *regression analysis*, which is the topic of Chapter 21.

19.1 Testing for Independence

Suppose we want to test

$$\mathcal{H}_0 : X \text{ and } Y \text{ are independent.}$$

This is, in some sense, the most extreme form of non-association between two variables. Most of this chapter is dedicated to *testing for independence*.

---

[71] We assume throughout that neither X nor Y are constant, for otherwise a correlation analysis is not relevant as we effectively only have one numerical sample in that case.

19.1.1 Tests based on Binning

In Section 15.6, we saw how to test for independence between (paired) discrete variables. In principle, the tools developed there could be used when the variables are numerical, but only after binning.

Problem 19.2 (Independence tests based on binning) Based on the methodology introduced in Section 15.6 for discrete variables, propose at least one test for independence that is applicable to numerical variables. The general idea is to first bin the numerical variables, thus obtaining discrete variables, and then apply an independence test for discrete variables. Implement that test procedure in R. [Notice the parallel with Problem 16.61.]

19.1.2 Permutation Tests

In the present context, a permutation consists in permuting one coordinate, say, the Y coordinate (without loss of generality), while leaving the other variable fixed. Doing this leaves a null distribution unchanged while breaking any dependence under an alternative.

Suppose we reject for large values of a statistic T. Let t denote the observed value of the statistic, meaning $t = T((x_1, y_1), \ldots, (x_n, y_n))$. Letting Π be the group of permutations of $\{1, \ldots, n\}$, for $\pi = (\pi_1, \ldots, \pi_n) \in \Pi$, let t_π denote the value of the statistic when applied to the data permuted by π, meaning $t_\pi = T((x_1, y_{\pi_1}), \ldots, (x_n, y_{\pi_n}))$. Based on this, the permutation p-value is defined as in (15.4). Since $|\Pi| = n!$ can be quite large, this p-value is typically estimated by Monte Carlo simulation.

Problem 19.3 Show that the permutation p-value and its Monte Carlo estimate, seen as random variables, are both valid in the sense of (12.12). [Use the conclusions of Problem 8.53.]

Problem 19.4 (Bootstrap tests) Although a calibration by permutation is favored, in large part because the permutation p-value is valid regardless of the sample size, a calibration by bootstrap is also possible (and reasonable). Propose a way to do so in the present context.

In the remainder of this chapter, we give several examples of test statistics that are commonly used for the purpose of testing for independence. The alternative set is very large, comprising all distributions on $\mathbb{R}^2$ that are *not* the product of their marginals, and there is no test that is uniformly best. Instead, each of the following tests is designed for certain alternatives.

19.2 Affine Association

The variables X and Y are in perfect *affine association* if they are affine functions of each other, for example,

$$Y = aX + b, \text{ for some } a, b \in \mathbb{R}.$$

Such a perfect association is extremely rare in real applications. Even in settings governed by the laws of physics the relation is not exact, for example, because of various factors including measurement precision and error.

Example 19.5 (Boiling temperature in the Himalayas) In [64], James Forbes reports data collected by Joseph Hooker on the boiling temperature of water at different elevations in the Himalayas. The variables are boiling temperature (degrees Fahrenheit) and barometric pressure (inches of mercury). Although the laws of physics predict an affine relationship, this is not exactly the case in this dataset, although it is an excellent model.

19.2.1 Pearson Correlation

The correlation, defined in (7.8), was seen to measure affine association between paired random variables. This motivates the use of the sample correlation, defined as the correlation of the empirical distribution.

Problem 19.6 Show that the sample correlation is given by

$$r := \frac{\sum_{i=1}^{n} (x_i - \bar{x})(y_i - \bar{y})}{\sqrt{\sum_{i=1}^{n} (x_i - \bar{x})^2} \sqrt{\sum_{i=1}^{n} (y_i - \bar{y})^2}}.$$

This is often called the *Pearson sample correlation*.

Problem 19.7 Show that $r \in [-1, 1]$, and $|r| = 1$ if and only if there are $a, b \in \mathbb{R}$ such that $y_i = ax_i + b$ for all i. (We are assuming that the x_i are not constant.)

Problem 19.8 (Consistency of the sample correlation) Let $\{(X_i, Y_i) : i \geq 1\}$ be iid bivariate numerical with correlation ρ. Let R_n denote the sample correlation of $(X_1, Y_1), \ldots, (X_n, Y_n)$. Show that $R_n \to_P \rho$ as $n \to \infty$.

The Pearson correlation test, which in its two-sided variant rejects for large values of $|R|$, is *not* universally consistent, essentially because there are bivariate distributions with $\rho = 0$ that are not the product of their marginals.

Problem 19.9 Suppose that X is uniform in $[-1, 1]$ and define $Y = X^2$. Show that the distribution of (X, Y) has correlation $\rho = 0$. Generalize this result as much as you can (within reason).

Proposition 19.10. *If X and Y are independent and normal,*

$$T := \frac{R\sqrt{n-2}}{\sqrt{1-R^2}}$$

has the Student distribution with $n-2$ degrees of freedom. In general, if X and Y are independent and have finite 2nd moments,

$$T = T_n \xrightarrow{\mathcal{L}} \mathcal{N}(0, 1), \quad as\ n \to \infty.$$

R corner The function cor computes, by default, the Pearson correlation, while the function cor.test implements, by default, the Pearson correlation test, albeit returning a p-value computed based on Proposition 19.10.

Problem 19.11 Detail how Proposition 19.10 is used to produce a p-value. (Note that the null hypothesis, in this case, can be that the variables are independent, or more generally, that they have zero Pearson correlation. In either case, the p-value is approximate, except in the exceedingly rare situation where the underlying distribution is known to be bivariate normal.)

Problem 19.12 In R, write a function that takes in the dataset and a number of Monte Carlo replicates, and returns the estimated permutation p-value for the Pearson correlation.

Remark 19.13 If one obtains a p-value for R, or equivalently, T, by permutation, then it is much safer to take the null hypothesis to be that the two variables X and Y are independent. The paper [48] shows that a calibration by permutation is not appropriate when testing the null hypothesis that X and Y are uncorrelated.

19.3 Monotonic Association

The variables X and Y are in perfect *monotonic association* if they are monotonic functions of each other, for example,

$$Y = g(X), \text{ for some monotone function } g.$$

As before, and for similar reasons, perfect monotonic association is extremely rare in practice.

Example 19.14 (Antoine's equation) For a pure liquid, the vapor pressure (p) and temperature (t) are related, to first order, by *Antoine's equation*

$$\log(p/p_0) = t_0/t,$$

where p_0 and t_0 are constants. Note that p is a monotone function of t in this model. Actual data does not fit the equation perfectly, but comes very close to that. (See [98] for more details in the case of mercury, including a discussion of more refined equations.)

Rank Pattern

The two most popular tests for monotonic association, which we introduce below, are based on the *rank pattern*, which is given by ranking the Y_i among themselves, and then listing these ranks according to increasing values of the X_i. For example, the following data

$$X: \quad -1.0 \quad -1.3 \quad 0.8 \quad 1.1 \quad -0.4 \quad -0.3 \quad 0.9 \quad -0.8$$
$$Y: \quad 0.2 \quad 0.0 \quad 1.6 \quad 0.1 \quad 0.6 \quad -0.6 \quad 1.0 \quad -0.7$$

yield the following rank pattern

$$3 \quad 5 \quad 1 \quad 6 \quad 2 \quad 8 \quad 7 \quad 4$$

Problem 19.15 Suppose that ties, if present, are broken at random. Prove that the rank pattern is uniformly distributed among the permutations of $(1, \ldots, n)$ when X and Y are independent.

19.3.1 Spearman Correlation

The Spearman correlation is the rank variant of the Pearson correlation. We start with the sample version. Let a_i denote the rank of x_i within $x_1, \ldots, x_n$ and b_i denote the rank of y_i within $y_1, \ldots, y_n$. The *Spearman sample correlation* is the Pearson sample correlation of $(a_1, b_1), \ldots, (a_n, b_n)$.

Problem 19.16 Show that this is a rank statistic.

Problem 19.17 Show that the Spearman sample correlation can be written

$$r_s = 1 - \frac{6}{n^3 - n} \sum_{i=1}^{n} (a_i - b_i)^2.$$

Problem 19.18 Show that $r_s \in [-1, 1]$, and $= 1$ (resp. $= -1$) if and only if there is a non-decreasing (resp. non-increasing) function g such that $y_i = g(x_i)$ for all i.

We first defined the sample version of the Spearman correlation, the reason being that it is easy to motivate. One may wonder if there is a corresponding feature of the underlying distribution. Coming from another angle, is there a result analogous to Problem 19.8 here?

Proposition 19.19. *Let* $\{(X_i, Y_i) : i \geq 1\}$ *be iid bivariate numerical. Let* $R_{s,n}$ *denote the Spearman correlation of* $(X_1, Y_1), \ldots, (X_n, Y_n)$. *Then* $R_{s,n} \to_P \rho_s$ *as* $n \to \infty$, *where*

$$\rho_s := 3 \mathbb{E} \left[\operatorname{sign} \left((X_1 - X_2)(Y_1 - Y_3) \right) \right].$$

(ρ_s is sometimes called *Spearman's* ρ.)

The Spearman correlation test, which in its two-sided variant rejects for large values of $|R_s|$, is not universally consistent, essentially because there are bivariate distributions with $\rho_s = 0$ that are not the product of their marginals (see Problem 19.9).

R corner The function cor can be used to compute the Spearman correlation, while the function cor.test can be used to perform the Spearman correlation test. The p-value is computed analytically up to a certain sample size, and after that the large-sample null distribution is used. (It turns out that the second part of Proposition 19.10 applies to R_s.)

19.3.2 Kendall Correlation

The *Kendall sample correlation* is defined as

$$r_{\text{K}} := \frac{2}{n(n-1)} \sum_{1 \leq i < j \leq n} \sum \operatorname{sign} \left((x_j - x_i)(y_j - y_i) \right).$$

Problem 19.20 Show that this is a rank statistic.

Problem 19.21 Show that $r_{\text{K}} \in [-1, 1]$, and $= 1$ (resp. $= -1$) if and only if there is a non-decreasing (resp. non-increasing) function g such that $y_i = g(x_i)$ for all i.

Here too, this statistic estimates a feature of the underlying distribution.

Proposition 19.22. *Let* $\{(X_i, Y_i) : i \geq 1\}$ *be iid bivariate numerical. Let* $R_{\text{K},n}$ *denote the Kendall correlation of* $(X_1, Y_1), \ldots, (X_n, Y_n)$. *Then* $R_{\text{K},n} \to_P \rho_{\text{K}}$ *as* $n \to \infty$, *where*

$$\rho_{\text{K}} := \mathbb{E} \left[\operatorname{sign} \left((X_1 - X_2)(Y_1 - Y_2) \right) \right].$$

(ρ_K is sometimes denoted by τ and called *Kendall's τ*.)

The Kendall correlation test, which in its two-sided variant rejects for large values of $|R_\text{K}|$, is not universally consistent, essentially because there are bivariate distributions with $\rho_\text{K} = 0$ that are not the product of their marginals (see Problem 19.9).

R corner The function cor can be used to compute the Kendall correlation, while the function cor.test can be used to perform the Kendall correlation test. The p-value is computed analytically up to a certain sample size, and after that the large-sample null distribution is used. (It turns out that R_K is asymptotically normal.)

Problem 19.23 Show that, under the null hypothesis of independence, R_K has mean zero and variance given by $(4n + 10)/9n(n - 1)$.

19.4 Universal Tests for Independence

We saw that none of the correlation tests is universally consistent. This is because they focus on features that are not characteristic of independence. We present below approaches that can lead to universally consistent tests, which do so by looking at the entire distribution through its distribution function, its density, or its characteristic function.

19.4.1 Tests based on the Distribution Function

Recall the definition of the distribution function of a random vector given in (6.2). For (X, Y) bivariate numerical, it is defined as

$$F_{X,Y}(x,y) = \mathbb{P}(X \le x, Y \le y).$$

In Proposition 6.3, we saw that it characterizes the underlying distribution. In particular, the following is true.

Problem 19.24 X and Y are independent if and only if

$$F_{X,Y}(x,y) = F_X(x)F_Y(y), \quad \text{for all } x, y \in \mathbb{R}.$$

Prove this claim.

In view of this, it becomes natural to consider test statistics of the form

$$\Delta\big(\widehat{F}_{X,Y}, \widehat{F}_X \otimes \widehat{F}_Y\big), \tag{19.1}$$

where Δ denotes a measure of dissimilarity between distribution functions as considered in Section 16.8.1, while $\widehat{F}_X$, $\widehat{F}_Y$, and $\widehat{F}_{X,Y}$ denote the empirical distribution functions of the X, Y, and (X, Y) samples, respectively. In particular,

$$\widehat{F}_{X,Y}(x,y) := \frac{1}{n} \sum_{i=1}^{n} \{X_i \le x, Y_i \le y\}.$$

For example, the analogue of the Kolmogorov–Smirnov test rejects for large values of

$$\sup_{x,y \in \mathbb{R}} \left| \widehat{F}_{X,Y}(x,y) - \widehat{F}_X(x)\widehat{F}_Y(y) \right|.$$

Problem 19.25 Show that this statistic is a function of the ranks (so that the resulting test is a rank test).

Problem 19.26 Argue that this test is universally consistent.

Hoeffding [96] proposed, instead, the analogue of the Cramér–von Mises test, except in reverse, as it rejects for large values of $\Delta(\widehat{F}_X \otimes \widehat{F}_Y, \widehat{F}_{X,Y})$, with the Δ defined in (16.16).

Problem 19.27 The statistic (19.1), with the same Δ, appears to be an equally fine choice. Perform some numerical experiments to compare these two choices.

19.4.2 Tests based on the Density

Tests for independence based on binning the observations, as studied in Problem 19.2, can interpreted as tests based on the density function.

Problem 19.28 In parallel with Section 19.4.1, but this time in analogy with Section 16.8.3, propose a class of tests for independence based on the density function. Speculate on whether the tests you propose are universally consistent, or not. Implement your favorite test among these in R, and perform some simulations to assess its power.

19.4.3 Tests based on the Characteristic Function

As we saw in Remark 7.57, the characteristic function of a random vector, (X, Y), is defined as

$$\varphi_{X,Y}(s,t) := \mathbb{E}[\exp(\iota(sX + tY))].$$

We also saw there that a distribution on $\mathbb{R}^2$ is characterized by its characteristic function.

Problem 19.29 Show that X and Y are independent if and only if

$$\varphi_{X,Y}(s,t) = \varphi_X(s)\varphi_Y(t), \quad \text{for all } s,t \in \mathbb{R}.$$

Problem 19.30 Propose a class of tests for independence based on the characteristic function. Speculate on whether the tests you propose are universally consistent, or not. Implement your favorite test among these in R, and perform some simulations to assess its power.

Distance Covariance

In [185], Székely, Rizzo, and Bakirov propose a test for independence based on pairwise distances, which in fact turns out to be based on the characteristic function.

Based on data $(x_1, y_1), \ldots, (x_n, y_n)$, define

$$a_{ij} = |x_i - x_j|, \quad a_i = \frac{1}{n}\sum_{j=1}^{n} a_{ij}, \quad a = \frac{1}{n^2}\sum_{i=1}^{n}\sum_{j=1}^{n} a_{ij},$$

and $u_{ij} = a_{ij} - a_i - a_j + a$. Similarly, define

$$b_{ij} = |y_i - y_j|, \quad b_i = \frac{1}{n}\sum_{j=1}^{n} b_{ij}, \quad b = \frac{1}{n^2}\sum_{i=1}^{n}\sum_{j=1}^{n} b_{ij},$$

and $v_{ij} = b_{ij} - b_i - b_j + b$. The test rejects for large values of the sample *distance covariance* defined as

$$\frac{1}{n^2}\sum_{i=1}^{n}\sum_{j=1}^{n} u_{ij}v_{ij}. \tag{19.2}$$

Being a test for independence, a p-value is typically obtained by permutation – via Monte Carlo simulation in practice.

To see how the test is based on the characteristic function, let $\widehat{\varphi}_x$ denote the empirical characteristic function based on the sample x, meaning the characteristic function of $\widehat{\mathbb{P}}_x$, and define $\widehat{\varphi}_y$ as well as $\widehat{\varphi}_{x,y}$ analogously.

Problem 19.31 Show that, when computed on $x = (x_1, \ldots, x_n)$,

$$\widehat{\varphi}_x(s) = \frac{1}{n}\sum_{j=1}^{n}\exp(\iota s x_j).$$

Prove that the sample characteristic function is pointwise consistent for the characteristic function, meaning that $\widehat{\varphi}_{X_n}(s) \to_P \varphi_X(s)$ as $n \to \infty$, for all $s \in \mathbb{R}$, where $X_n = (X_1, \ldots, X_n)$ are iid copies of a random variable X. Repeat all this with the joint characteristic function, $\widehat{\varphi}_{x,y}$.

Proposition 19.32. *The statistic* (19.2) *is equal to*

$$\frac{1}{\pi^2} \int_{-\infty}^{\infty} \int_{-\infty}^{\infty} \frac{\left|\widehat{\varphi}_{X,Y}(s,t) - \widehat{\varphi}_X(s)\widehat{\varphi}_Y(t)\right|^2}{s^2 t^2} \, ds \, dt$$

In view of this result, it is not too hard to believe that the sample distance covariance is consistent for the *distance covariance* of (X, Y), defined as

$$\frac{1}{\pi^2} \int_{-\infty}^{\infty} \int_{-\infty}^{\infty} \frac{\left|\varphi_{X,Y}(s,t) - \varphi_X(s)\varphi_Y(t)\right|^2}{s^2 t^2} \, ds \, dt.$$

And from this it is not too hard to argue that the distance covariance test is universally consistent.

Problem 19.33 The distance covariance is intimately related to the energy statistic of Problem 17.49. Can you see that? [This is analogous to how the Hoeffding test is related to the Cramér–von Mises test.]

19.5 Further Topics

19.5.1 When One Variable is Categorical

In this chapter, so far, we have focused on the situation where both variables are numerical. This comes after Section 15.6, where we addressed the situation where both are categorical. Suppose now that one of the variables, say X, is categorical while the other variable, Y, is numerical.

To derive tests for independence, we make a connection with goodness-of-fit testing. Indeed, regardless of the nature of the variables, testing for the independence of X and Y is equivalent to testing the null hypothesis that the conditional distribution of Y given X is the same as the (marginal) distribution of Y. Rephrased, testing for independence in the present context is thus equivalent to testing

$$\mathcal{H}_0 : Y \mid X = x \text{ is distributed as } Y, \quad \text{for all } x.$$

This is clearly a goodness-of-fit testing problem where the groups are given by the different values that X takes. The only difference with the setting of Section 17.3 and Section 17.4.3 is that here the group sizes are random. However, conditional on $X_1, \ldots, X_n$, the setting is exactly that of goodness-of-fit testing, and the methods presented in these sections are directly applicable.

Remark 19.34 Note that proceeding with goodness-of-fit testing after conditioning on X is another example of conditional inference (Section 22.1).

20

Multiple Testing

In a wide range of real-life situations, not one but several, even many hypotheses are to be tested, and not accounting for multiple inference can lead to a grossly incorrect analysis. In this chapter, we look closely at this important issue, describing some pitfalls and presenting remedies that 'correct' for this multiplicity. We also discuss related issues having to do with the publication of scientific papers.

Example 20.1 (Genetics) In genetics an important line of research revolves around discovering how an individual's genetic material influences his/her health. In particular, biologists have developed ways to measure how 'expressed' a gene is, and a typical experiment for understanding what genes are at play in a given disease can be described as follows. A number of subjects with the disease, and a number of subjects without the disease, are recruited. For each individual in the study, the expression levels of certain genes (m of them) are measured. For each gene, a test comparing the two groups is performed, so that m tests are performed in total [36]. In practice, for human subjects, m is on the order of 10^4. Experiments focusing on single nucleotide polymorphisms (SNP's) instead of genes result in an even larger number of tests, on the order of 10^5.

Example 20.2 (Surveillance) In a surveillance setting, a signal is observed over time and the task is to detect a change in the signal of particular relevance. The signal can be almost anything and the change is typically in terms of features that are deemed important for the task at hand. Practical examples include the detection of fires from satellite images [113] and the detection of epidemics (aka *syndromic surveillance*) based on a variety of data such as transcripts from hospital emergency visits and pharmacy sales of over-the-counter drugs [91, 92]. In such settings, a test is applied at every location/time point.

Example 20.3 (Functional MRI) Functional magnetic resonance imaging (fMRI) can be used as a non-invasive technique for understanding what regions of the human brain are active when performing a certain task [118]. In an experiment involving a single subject, a person's brain is observed over time while the individual is performing several tasks – one of them might be 'no task' or 'at rest' and serve as control. The goal is then to identify which parts of the brain are most active during a particular task. The identification is typically done by performing a test for each voxel comparing its activation level during the different tasks. (A voxel represents a small unit of space on the order of a few cubic millimeters.) There are on the order of 10^6 voxels.

When confronted with the task of testing a number of (null) hypotheses we talk of *multiple testing*. For now, consider a simplified situation where m null hypotheses, denoted $\mathcal{H}_1, \ldots, \mathcal{H}_m$, need to be tested. (Note that, in the present setting, $\mathcal{H}_1$ is a null hypothesis and *not* an alternative hypothesis.) We apply a test to each null hypothesis $\mathcal{H}_j$, resulting in a p-value denoted P_j. Assume for simplicity that the p-values are independent and that each P_j is uniform in $[0, 1]$ under $\mathcal{H}_j$, that is,

$$P_1, \ldots, P_m \text{ are independent,} \tag{20.1}$$

$$P_j \sim \text{Unif}(0, 1) \text{ under } \mathcal{H}_j, \text{ for } j = 1, \ldots, m. \tag{20.2}$$

Here are two aspects of the situation that illustrate the underlying difficulties:

- Suppose that we proceed as usual, choosing a level $\alpha \in (0, 1)$ and rejecting $\mathcal{H}_j$ if $P_j \leq \alpha$. Then, even if all the hypotheses are true, on average there are αm rejections (all incorrect). In settings where very many tests are performed (m is very large), choosing α to be the usual 0.05 or 0.01 leads to an impractically large number of rejections. Take Example 20.1, where (say) $m = 10^4$ tests are performed. Then rejecting at level $\alpha = 0.05$ leads to 500 rejections on average, even when all the hypotheses are true (meaning that no gene is truly differentially expressed when comparing the two conditions).
- The smallest p-value can be quite small even if all the hypotheses are true. Indeed, $\min_j P_j$ has expectation $1/(m + 1)$ in that case.

Problem 20.4 Verify these assertions.

20.1 Testing Multiple Hypotheses

In what follows, we postulate a statistical model $(\Omega, \Sigma, \mathcal{P})$ as in Chapter 12, where Ω is the sample space containing all possible outcomes, Σ is the class of events of interest, and $\mathcal{P}$ is a family of distributions on Σ. We assume as before that $\mathcal{P}$ is parameterized as $\mathcal{P} = \{\mathbb{P}_\theta : \theta \in \Theta\}$.

Within this framework, we consider a situation where m null hypotheses need to be tested, with the jth null hypothesis being

$$\mathcal{H}_j : \theta_* \in \Theta_j,$$

for some given $\Theta_j \subset \Theta$. (Recall that θ_* denotes the true value of the parameter.) The alternative to $\mathcal{H}_j$ will simply the negation of $\mathcal{H}_j$ and denoted $\mathcal{H}_j^c$ in congruence with the fact that $\mathcal{H}_j^c : \theta_* \in \Theta_j^c = \Theta \setminus \Theta_j$.

Recall that a test is applied to each null hypothesis, resulting in a total of m p-values, denoted $P_1, \ldots, P_m$. What tests are used obviously depends on the situation. In Example 20.1, for instance, the rank-sum test could be applied to each hypothesis. We always assume that each p-value P_j is valid in the usual sense that it satisfies (12.12), meaning here that

$$\text{Under } \mathcal{H}_j : \quad \mathbb{P}(P_j \le \alpha) \le \alpha, \ \forall \alpha \in (0,1). \tag{20.3}$$

Remark 20.5 For the sake of conciseness, we focus on methods for multiple testing that are based on the p-values. Such methods are all based on the ordered p-values, denoted $p_{(1)} \le \cdots \le p_{(m)}$. We will let $\mathcal{H}_{(j)}$ denote the hypothesis associated with $p_{(j)}$.

Normal Sequence Model

The *normal sequence model* provides a stylized mathematical framework within which methods can be studied. Although it is too simple to accurately model real-life situations, it is nevertheless relevant, in part because common test statistics used in practice are approximately normal in large samples.

The model is as follows. We observe $Y_1, \ldots, Y_m$, independent, with $Y_j \sim \mathcal{N}(\theta_j, 1)$. We are interested in testing $\mathcal{H}_1, \ldots, \mathcal{H}_m$, where

$$\mathcal{H}_j : \theta_j = 0.$$

The problem can be consider one-sided, in which case $\mathcal{H}_j^c : \theta_j > 0$; or two-sided, in which case $\mathcal{H}_j^c : \theta_j \ne 0$.

Assuming the one-sided setting, it makes sense to reject $\mathcal{H}_j$ for large values of Y_j, since doing so is optimal for that particular hypothesis (Theorem 13.26). The corresponding p-value is $P_j := 1 - \Phi(Y_j)$, where Φ denotes the standard normal distribution function.

20.2 Global Null Hypothesis

The *global null hypothesis* (aka *complete null hypothesis*) is defined as

$$\mathcal{H}_0 : \text{``the hypotheses } \mathcal{H}_1, \ldots, \mathcal{H}_m \text{ are all true''},$$

or, equivalently,

$$\mathcal{H}_0 : \theta_* \in \Theta_0 := \bigcap_{j=1}^{m} \Theta_j.$$

In most situations, a null hypothesis represents "business as usual". We will assume this is the case throughout. Then the global null hypothesis represents "there is nothing at all going on". Although one is typically interested in identifying the false hypotheses, testing the global null hypothesis might be relevant in some applications, for example, in surveillance settings (Example 20.2).

The global null hypothesis is just a null hypothesis. We present below some commonly used tests, all based on the available p-values. Such tests are sometimes called *combination tests*.

Remark 20.6 Since a small p-value provides evidence against the hypothesis it is associated with, all the tests below are one-sided.

Fisher Test

This test rejects for large values of

$$T(p_1, \ldots, p_m) := -2 \sum_{j=1}^{m} \log p_j.$$

The test statistic was designed that way because its null distribution is stochastically dominated by the chi-squared distribution with $2m$ degrees of freedom. (This explains the presence of the factor 2.)

Problem 20.7 Assuming that (20.1)–(20.2) hold, show that, under the global null, T has the chi-squared distribution with $2m$ degrees of freedom.

Liptak–Stouffer Test

This test rejects [115] for large values of

$$T(p_1, \ldots, p_m) := \frac{1}{\sqrt{m}} \sum_{j=1}^{m} \Phi^{-1}(1 - p_j). \tag{20.4}$$

The test statistic was designed that way because its null distribution is stochastically dominated by the standard normal distribution.

Problem 20.8 Assuming that (20.1)–(20.2) hold, show that, under the global null, T has the standard normal distribution.

Problem 20.9 Consider the normal sequence model. First, express the test statistic as a function of $y_1, \ldots, y_m$. Then, setting the level to some given $\alpha \in (0, 1)$, provide a sufficient condition for the test to have power tending to 1. [Use Chebyshev's inequality.]

Tippett–Šidák Test

This test [166] rejects for small values of

$$T(p_1, \ldots, p_m) := \min_{j=1,\ldots,m} p_j.$$

Problem 20.10 Assuming that (20.1)–(20.2) hold, derive the distribution of T under the global null.

Problem 20.11 Repeat Problem 20.9 with this test. [This time, use Boole's inequality together with the fact that $1 - \Phi(x) \leq \phi(x)/x$, where ϕ denotes the density of the standard normal distribution.]

Simes Test

This test [168] rejects for small values of

$$T(p_1, \ldots, p_m) := \min_{j=1,\ldots,m} m\, p_{(j)}/j.$$

Proposition 20.12. *Assuming that (20.1)–(20.2) hold, T has the uniform distribution in $[0, 1]$ under the global null.*

Problem 20.13 Prove Proposition 20.12, and perform some simulations in R to numerically confirm it.

Tukey Test

Better known as the *higher criticism* test [50], it comes from applying the one-sided Anderson–Darling procedure (16.17) to test the hypothesis that the p-values are iid uniform in $[0, 1]$ – that is, the global null hypothesis under (20.1)–(20.2).

Problem 20.14 Show that the test rejects for large values of

$$T(p_1, \ldots, p_m) := \max_{j=1,\ldots,m} \frac{j/m - p_{(j)}}{\sqrt{p_{(j)}(1 - p_{(j)})}}.$$

Under the global null, T has a complicated distribution, but asymptotically ($m \to \infty$) it becomes of Gumbel type (after a proper standardization) [101].

Problem 20.15 In R, write a function that computes the statistic T. Then, using that function, write another one hc.test that returns a p-value for the test based on a specified number of Monte Carlo replicates.

A general recipe for designing a combination test is to apply a test for uniformity in its proper one-sided version to the p-values (playing the role of sample).

Problem 20.16 Show that the Berk–Jones test (Section 16.10.1) for the present setting rejects for small values of

$$T(p_1, \ldots, p_m) := \min_{j=1,\ldots,m} \text{Prob}\big(\text{Beta}(j, m - j + 1) \le p_{(j)}\big).$$

Repeat Problem 20.15 with this test (name your function aappropriately).

20.3 Multiple Tests

Testing the global null hypothesis amounts to weighing the evidence that one or several hypotheses are false. However, even if we reject, we do not know what hypotheses are doubtful. We now turn to the more ambitious goal of identifying the false hypotheses (if there are any). We will call a procedure for this task a *multiple test*.

While a test is a function of the data with values in $\{0, 1\}$, with '1' indicating a rejection, a multiple test in the context of Remark 20.5 is a function of the p-values with values in $\{0, 1\}^m$, with '1' in the jth component indicating a rejection of $\mathcal{H}_j$. Thus, a multiple test is of the form

$$\begin{array}{ccc} [0, 1]^m & \longrightarrow & \{0, 1\}^m \\ \boldsymbol{p} = (p_1, \ldots, p_m) & \longmapsto & \varphi(\boldsymbol{p}) = (\varphi_1(\boldsymbol{p}), \ldots, \varphi_m(\boldsymbol{p})) \end{array}$$

Seen as a function on $[0, 1]^m$, φ_j is a test for $\mathcal{H}_j$, but possibly based on all the p-values instead of just p_j.

For $\theta \in \Theta$, let $h_j(\theta) = \{\theta \in \Theta_j\}$, so that $h_j(\theta) = 0$ if $\mathcal{H}_j$ is true, and $= 1$ if it is false. Also, let $m_0(\theta) = \#\{j : \theta \in \Theta_j\}$, which is the number of true hypotheses, and $m_1(\theta) = \#\{j : \theta \notin \Theta_j\}$, which is the number of false hypotheses, that is, under θ. Note that $m_0(\theta) + m_1(\theta) = m$.

For a given multiple test φ, define the following quantities:[72]

$$N_{0|0}(\varphi, \theta) = \#\{j : \varphi_j = 0 \text{ and } h_j(\theta) = 0\},$$
$$N_{1|0}(\varphi, \theta) = \#\{j : \varphi_j = 1 \text{ and } h_j(\theta) = 0\},$$
$$N_{0|1}(\varphi, \theta) = \#\{j : \varphi_j = 0 \text{ and } h_j(\theta) = 1\},$$
$$N_{1|1}(\varphi, \theta) = \#\{j : \varphi_j = 1 \text{ and } h_j(\theta) = 1\}.$$

(These are summarized in Table 20.1.) In particular, $N_{1|0}(\varphi, \theta)$ is the number of Type I errors and $N_{0|1}(\varphi, \theta)$ the number of Type II errors made by the multiple test φ when the true value of the parameter is θ, and the total number of errors made by the multiple test is given by

$$N_{1|0}(\varphi, \theta) + N_{0|1}(\varphi, \theta) = \#\{j : \varphi_j \neq h_j(\theta)\},$$

and the total number of rejections is given by

$$R(\varphi) = N_{1|0}(\varphi, \theta) + N_{1|1}(\varphi, \theta) = \#\{j : \varphi_j = 1\}.$$

These counts are all functions of the p-values and, with the exception of $R(\varphi)$, of the true value of the parameter.

Table 20.1 *The counts below summarize the result of applying a multiple test φ when the true value of the parameter is θ.*

	No rejection	Rejection	Total
True Null	$N_{0|0}(\varphi, \theta)$	$N_{1|0}(\varphi, \theta)$	$m_0(\theta)$
False Null	$N_{0|1}(\varphi, \theta)$	$N_{1|1}(\varphi, \theta)$	$m_1(\theta)$
Total	$m - R(\varphi)$	$R(\varphi)$	m

For a single hypothesis, the accepted modus operandi is to control the level and, within that constraint, design a test that maximizes the power (as much as possible). We introduce some notion of level and power for multiple tests below. These apply to a given multiple test φ, although this is left implicit in some places.

---

[72] A different notation is typically used in the literature, stemming from the influential paper [9], but that choice of notation is not particularly mnemonic. Instead, we follow [77].

20.3.1 Notions of Level for Multiple Tests

Family-Wise Error Rate (FWER)

For a long time, this was the main notion of level for multiple tests. It is defined as the probability of making at least one Type I error or, using the notation of Table 20.1,

$$\mathsf{FWER}(\varphi) = \sup_{\theta \in \Theta} \mathbb{P}_\theta\big(N_{1|0}(\varphi, \theta) \geq 1\big).$$

A multiple test φ controls the FWER at α if

$$\mathsf{FWER}(\varphi) \leq \alpha.$$

Problem 20.17 Assuming that (20.1)–(20.2) hold, derive the FWER for the multiple test defined by

$$\varphi_j = \{P_j \leq \alpha\},$$

which is the multiple test that ignores multiple testing.

False Discovery Rate (FDR)

This notion is more recent. It was suggested in the mid 1990s by Benjamini and Hochberg [9]. It is now the main notion of level used in large-scale multiple testing problems. It is defined as the expected proportion of incorrect rejections among all rejections or, using the notation of Table 20.1,

$$\mathsf{FDR}(\varphi) = \sup_{\theta \in \Theta} \mathbb{E}_\theta\left(\frac{N_{1|0}(\varphi, \theta)}{R(\varphi) \vee 1}\right).$$

A multiple test φ controls the FDR at α if

$$\mathsf{FDR}(\varphi) \leq \alpha.$$

(The name comes from the fact that, in most settings, a rejection indicates a *discovery*.)

Problem 20.18 As notions of level for multiple tests, the FDR is always less severe then the FWER. Indeed, show that in any situation – meaning any model, any set of null hypotheses, and any value of the parameter within that model – and for any multiple test φ,

$$\mathsf{FDR}(\varphi) \leq \mathsf{FWER}(\varphi).$$

20.3.2 Notions of Power for Multiple Tests

Notions of power for multiple tests can be defined by analogy with the notions of level presented above.

Problem 20.19 Define the power equivalent of FWER. (This quantity does not seem to have a name in the literature.)

False Non-Discovery Rate (FNR)

To define the power equivalent of FDR, one possibility is

$$\mathsf{FNR}(\varphi) = \sup_{\theta \in \Theta} \mathbb{E}_\theta \left(\frac{N_{0|1}(\varphi, \theta)}{(m - R(\varphi)) \vee 1} \right).$$

This definition [77] leads to a quantity that could look artificially small when m_0/m is close to 1, common in practice, in which case the following variant might be preferred:

$$\mathsf{FNR}(\varphi) = \sup_{\theta \in \Theta} \frac{\mathbb{E}_\theta(N_{0|1}(\varphi, \theta))}{m_1(\theta) \vee 1}.$$

20.4 Methods for FWER Control

For a given $\theta \in \Theta$, define

$$\mathcal{T}_\theta = \{j : \theta \in \Theta_j\}, \tag{20.5}$$

which is the subset of indices corresponding to true null hypotheses.

Tippett Multiple Test

This multiple test rejects $\mathcal{H}_j$ if $p_j \leq c_\alpha$, where c_α is such that

$$\sup_{\theta \in \Theta} \mathbb{P}_\theta \left(\min_{j \in \mathcal{T}_\theta} P_j \leq c_\alpha \right) \leq \alpha. \tag{20.6}$$

Note that c_α is a valid critical value for the Tippett test for the global null.

Proposition 20.20. *The Tippett multiple test controls the FWER as desired.*

Proof A Type I error occurs if the multiple test rejects some $\mathcal{H}_j$ with $j \in \mathcal{T}_\theta$. This happens with probability

$$\mathbb{P}_\theta \left(\exists j \in \mathcal{T}_\theta : \varphi_j = 1 \right) = \mathbb{P}_\theta \left(\min_{j \in \mathcal{T}_\theta} P_j \leq c_\alpha \right) \leq \alpha,$$

using (20.6) at the end. $\qquad\qquad\square$

Problem 20.21 (Šidák multiple test) Assuming that (20.1) holds, show that the inequality (20.6) is valid with $c_\alpha = 1 - (1 - \alpha)^m$.

Problem 20.22 (Bonferroni multiple test) Show that, under all circumstances, the inequality (20.6) holds with $c_\alpha = \alpha/m$. The resulting multiple testing procedure is known as the *Bonferroni correction*.

Holm Multiple Test

This multiple test [97] rejects $\mathcal{H}_{(j)}$ if

$$p_{(k)} \leq \alpha/(m - k + 1) \text{ for all } k \leq j.$$

Problem 20.23 Show that a brute force implementation based on this description requires on the order of $O(m^2)$ basic operations. In fact, the method can be implemented in order $O(m)$ basic operations after ordering the p-values. Describe such an implementation.

Remark 20.24 (Step down methods) This is a *step-down* procedure as it moves from the most significant to the least significant p-value.

Proposition 20.25. *The Holm multiple test controls the FWER as desired.*

Proof   Recall (20.5) and let $j_0 = \arg\min_{j \in \mathcal{T}_\theta} p_{(j)}$. Note that j_0 is a function of the p-values. Since $|\mathcal{T}_\theta| = m_0(\theta)$, necessarily,

$$j_0 \leq m - m_0(\theta) + 1. \tag{20.7}$$

The procedure makes an incorrect rejection if and only if it rejects $\mathcal{H}_{(j_0)}$, which happens exactly when

$$p_{(j)} \leq \alpha/(m - j + 1) \text{ for all } j \leq j_0,$$

which in particular implies that

$$p_{(j_0)} \leq \alpha/(m - j_0 + 1) \leq \alpha/m_0(\theta),$$

by (20.7). But

$$\mathbb{P}_\theta\big(P_{(j_0)} \leq \alpha/m_0(\theta)\big) = \mathbb{P}_\theta\Big(\min_{j \in \mathcal{T}_\theta} P_j \leq \alpha/m_0(\theta)\Big)$$

$$\leq \sum_{j \in \mathcal{T}_\theta} \mathbb{P}_\theta(P_j \leq \alpha/m_0(\theta))$$

$$\leq \sum_{j \in \mathcal{T}_\theta} \alpha/m_0(\theta) = \alpha,$$

using the union bound and then the fact that the p-values are valid in the sense of (20.3).   □

Problem 20.26 Prove (20.7).

Problem 20.27 How would you change Holm's procedure if you knew that the p-values where independent?

Problem 20.28 (Holm vs Bonferroni) Show that Holm's procedure is always preferable to Bonferroni's, in the (strongest possible) sense that any hypothesis that Bonferroni's rejects Holm's also rejects.

Hochberg Multiple Test

This multiple test rejects $\mathcal{H}_{(j)}$ if

$$p_{(k)} \leq \alpha/(m - k + 1) \text{ for some } k \geq j.$$

Problem 20.29 Repeat Problem 20.23 but for the Hochberg multiple test.

Remark 20.30 (Step up methods) This is a *step-up* procedure as it moves from the least significant to the most significant p-value.

Proposition 20.31. *The Hochberg multiple test controls the FWER as desired when the p-values are independent.*

Problem 20.32 (Hochberg vs Holm) Prove that, if the p-values are independent, Hochberg's multiple test is more powerful than Holm's in the (strongest possible) sense that any hypothesis that Holm's rejects Hochberg's also rejects.

Remark 20.33 (Hommel multiple test) There is another procedure, Hommel's, that is more powerful than Hochberg's. However, it is a bit complicated to describe, and we do not detail it here.

20.5 Methods for FDR Control

Benjamini–Hochberg Multiple Test

This multiple rejects $\mathcal{H}_{(j)}$ if

$$p_{(k)} \leq k\alpha/m \text{ for some } k \geq j.$$

It was the first (and still the main) method used to control the FDR at the desired level, although this only happens under appropriate conditions.

Problem 20.34 Let r denote the number of rejections when this method is applied to a particular situation. Show that the method rejects $\mathcal{H}_j$ if and only if $p_j \leq r\alpha/m$.

Proposition 20.35. *The Benjamini–Hochberg multiple test above controls the FDR at αC_m, in general, where $C_m := 1 + 1/2 + \cdots + 1/m$, and at α if the p-values are independent.*

Proof sketch We only prove[73] the first part, and only when each p-value is uniform in $[0, 1]$ under its respective null. Let φ denote the multiple test and let $R(\varphi)$ denote the number of rejections when applied to a particular situation. Define $A_l = ((l-1)\alpha/m, l\alpha/m]$. We have

$$\frac{\{\varphi_j = 1\}}{R(\varphi) \vee 1} = \sum_{r=1}^{m} \{p_j \le r\alpha/m\} \frac{\{R(\varphi) = r\}}{r}$$

$$= \sum_{r=1}^{m} \sum_{l=1}^{r} \{p_j \in A_l\} \frac{\{R(\varphi) = r\}}{r}$$

$$= \sum_{l=1}^{m} \{p_j \in A_l\} \frac{\{R(\varphi) \ge l\}}{R(\varphi)}$$

$$\le \sum_{l=1}^{m} \frac{\{p_j \in A_l\}}{l}.$$

Thus,

$$\mathbb{E}_\theta \left(\frac{N_{1|0}(\varphi, \theta)}{R(\varphi) \vee 1} \right) = \mathbb{E}_\theta \left(\frac{\sum_{j \in \mathcal{T}_\theta} \{\varphi_j = 1\}}{R(\varphi) \vee 1} \right)$$

$$\le \sum_{j \in \mathcal{T}_\theta} \sum_{l=1}^{m} \frac{1}{l} \mathbb{P}_\theta(P_j \in A_l)$$

$$= \sum_{j \in \mathcal{T}_\theta} \sum_{l=1}^{m} \frac{1}{l} \frac{\alpha}{m} \le \alpha C_m.$$

In the last line we used the fact that $|\mathcal{T}_\theta| = m_0(\theta) \le m$. □

Problem 20.36 Show that $C_m < \log m + 1$.

One way to arrive at the Benjamini–Hochberg procedure is as follows. Consider the multiple test that rejects $\mathcal{H}_j$ when $p_j \le t$. With some abuse of notation, let $N_{1|0}(t)$ and $R(t)$ denote the corresponding number of Type I errors and total number of rejections, and define $F_t = N_{1|0}(t)/(R(t) \vee 1)$. Ideally, we would like to choose t largest such that $F_t \le \alpha$. However, $N_{1|0}(t)$ cannot be computed solely based on the p-values as it depends on knowing which hypotheses are true.

---

[73] We learned of this proof from Emmanuel Candès.

The idea is to replace it by an estimate. Since we assume the p-values to be valid (20.3), we have

$$\mathbb{E}_\theta\big(N_{1|0}(t)\big) \leq m_0(\theta)t \leq mt.$$

If we replace $N_{1|0}(t)$ by mt, we effectively estimate F_t by $\hat{F}_t := mt/(R(t) \vee 1)$.

Problem 20.37 Let $\hat{t} = \max\{t : \hat{F}_t \leq \alpha\}$. Show that the multiple test that rejects $\mathcal{H}_j$ when $p_j \leq \hat{t}$ is Benjamini–Hochberg's.

20.6 Meta-Analysis

Meta-analysis is a branch of Statistics/Epidemiology whose goal is to gather evidence from multiple studies in order to reach stronger conclusions on a particular issue (e.g., the effectiveness of a particular class of treatments for a particular medical condition). For this, a number of tools have been developed, and we present a few here.

Example 20.38 (Bone density and fractures) The study [125] is a "meta-analysis of prospective cohort studies published between 1985 and 1994 with a baseline measurement of bone density in women and subsequent follow up for fractures". The stated purpose of this analysis was to "determine the ability of measurements of bone density in women to predict later fractures". Combined, the studies comprised "eleven separate study populations with about 90000 person years of observation time and over 2000 fractures in total".

Example 20.39 (Alcohol consumption) The paper [179] presents a meta-analysis of 87 studies on the relationship between alcohol consumption and all-cause mortality. Some previous studies had concluded that consuming a small amount of alcohol (1–2 drinks per day) was associated with a slightly longer lifespan. The authors of [179] argue that this association can be explained in large part by the classification of former drinkers (who might have stopped drinking because of health issues) as abstainers.

Remark 20.40 It goes without saying that there are meta-analyses of meta-analyses [41].

It is often the case that several studies examine the same effect, and it is rather tempting to use all this information combined to boost the power of the statistical inference. This is possible under restrictive assumptions. In particular, the studies have to be comparable.

The bulk of the effort in a meta-analysis goes, in fact, to deciding which studies to include. Although there are some guidelines [94, Ch 5], ultimately, this triaging of studies requires domain-specific knowledge.

Otherwise, in terms of methods for inference, the meta-analysist makes use of various tests, including combination tests (Section 20.2). Some, mostly ad hoc, methods have been developed for detecting the presence of *publication bias*, defined in [49] as "any tendency on the parts of investigators or editors to fail to publish study results on the basis of the direction or strength of the study findings". Most of these methods, like the popular *funnel plot*, are based on the so-called *small study effect*, which is the empirically-observed fact that small studies are more prone to publication bias compared to larger studies, presumably because large studies are better funded and better executed, and also cannot remain unpublished as easily.

Example 20.41 (FDA-registered anti-depressant studies) The authors in [190] obtained "reviews from the US Food and Drug Administration (FDA) for studies of 12 antidepressant agents involving 12564 patients." There were 74 FDA-registered studies in total. Among the 38 studies viewed by the FDA as having positive results, all but one were published. Among the 36 studies viewed by the FDA as having negative or questionable results, 22 were not published and 11 were published, but presented in a in a way that conveyed a positive outcome.

Example 20.42 (Minimum wage studies) The article [52] revisits Card and Krueger's meta-analysis of the employment effects of minimum wages by looking at 64 minimum-wage studies (in the US). The article concludes that "The minimum-wage effects literature is contaminated by publication selection bias [...]. Once this publication selection is corrected, little or no evidence of a negative association between minimum wages and employment remains." – confirming Card and Krueger's original findings.

20.6.1 Cochran–Mantel–Haenszel Test

It is not uncommon for studies in the medical field (e.g., clinical trials) to result in a 2-by-2 table. This happens, for example, with a completely randomized design on two treatments and a binary outcome of 'success' or 'failure'. The *Cochran–Mantel–Haenszel* (CMH) *test* is applied when examining a number of such studies. The goal is to determine whether there is a treatment effect or not, and (optionally) to specify the direction of the effect when it is determined that there is an effect.

Example 20.43 (Low protein diets in chronic renal insufficiency) In [65], a meta-analysis is undertaken to better assess the impact that low protein diets have on chronic renal insufficiency. A total of 46 clinical trials where examined, from which 6 were selected (5 European and 1 Australian, between 1982 and 1991). This amounted to a combined sample of size 890 subjects with mild to severe chronic renal failure. Among these, 450 patients received a low protein diet (treatment) and 440 a control diet. Assignment was at random in all trials. Each subject was followed for at least one year. The main outcome was renal death (start of dialysis or death) during the study. Table 20.2 provides a summary in the form of six contingency tables (one for each study).

Table 20.2 *The following data are taken from Table 3 in [65].*

	Diet	Survived	Died
Study 1	Control	95	15
	Treatment	110	8
Study 2	Control	2	7
	Treatment	5	5
Study 3	Control	194	32
	Treatment	209	21
Study 4	Control	8	17
	Treatment	14	11
Study 5	Control	25	13
	Treatment	30	4
Study 6	Control	21	11
	Treatment	21	12

The simplest approach is arguably to *collapse* of all these 2-by-2 tables into a single 2-by-2 table, followed by applying one of the tests seen in Sections 15.2, 15.3, or 15.5. Some meta-analysts, however, are reluctant to do that because, although the studies are supposed to be comparable, they are invariably performed on samples from different populations, and collapsing the tables could dilute the strength of association in some of the studies. (Remember Simpson's paradox.)

For notation, assume there are m studies and let the contingency table resulting from jth study be as follows:

	Success	Failure
Treatment	a_j	b_j
Control	c_j	d_j

The one-sided CMH test is based on rejecting for large values of the total number of successes in the treatment group, namely $\sum_{j=1}^{m} a_j$.

Problem 20.44 In which direction is the test one-sided? How would you define a two-sided CMH test?

Problem 20.45 (Normal approximation) The classical version of the test relies on a normal approximation for calibration. Specify this normal approximation. Can you justify this normal approximation when the results from the studies are independent?

Problem 20.46 How would you calibrate the test by simulation on a computer?

Problem 20.47 Compare this with the test that collapses the tables into a single table. (This test is based on the same statistic. What distinguishes the tests is in how the p-value is computed.)

It is rather natural to approach this problem from a multiple testing perspective. After all, we are testing a global null.

Problem 20.48 Can you propose a combination test based on Fisher's exact test (Section 15.5) applied to each table? [Recall Remark 15.32.]

20.6.2 File Drawer Problem

Rosenthal [155] refers to publication bias as the *file drawer problem*. This was the 1970s, so that manuscripts were written on paper, and an unpublished paper would remain hidden in a file drawer somewhere.

Rosenthal considers a setting where some published papers address the same general question (formalized as a null hypothesis) and report on a p-value obtained from a test of significance performed on independently collected data. Having access to all these published papers, Rosenthal asks the question: How many papers addressing the same question (each performing a test on a separate dataset) would have to be left unpublished, stored away in a file drawer, to offset the combined significance resulting from the published papers?

Remark 20.49 Ghostwriting in the sciences pose the opposite problem to the file drawer problem. See, e.g., [186].

Example 20.50 (Antipsychotic medication) As reported on in [163], "Atypical antipsychotic medications are widely used to treat delusions, aggression, and agitation in people with Alzheimer's disease and other dementia; however, concerns have arisen about the increased risk for cerebrovascular adverse events, rapid cognitive decline, and mortality with their use." The announced objective in this article is "To assess the evidence for increased mortality from atypical antipsychotic drug treatment for people with dementia." This is done via a meta-analysis of 15 clinical trials. Among these, 9 trials were unpublished, and all but 1 were conducted by a pharmaceutical company. The authors say that the trials were of quality comparable to those that were published, and speculate that "the likely reasons for the delays in publication were that most did not show statistically significant results on their primary efficacy outcomes."

Assume that we have available m studies testing for the presence of the same effect (e.g., effectiveness of a particular drug compared to a placebo), with Study j resulting in a p-value denoted p_j. The Liptak–Stouffer (LS) combination test seen in (20.4) rejects for large values of $y := \sum_{j=1}^{m} z_j$, where $z_j := \Phi^{-1}(1 - p_j)$ is the z-score associated with p_j. (Φ denotes the standard normal distribution function.) The test is significant at level γ if $y/\sqrt{m} \geq \Phi^{-1}(1 - \gamma)$.

Rosenthal's Method

Suppose the LS test is significant at level γ. In an effort to answer his own question, Rosenthal then computes the *fail-safe number*,[74] defined as

$$\hat{n} := \min\left\{ n : y/\sqrt{m + n} < \Phi^{-1}(1 - \gamma) \right\}.$$

To motivate this definition, let $z_{m+1}, \ldots, z_{m+n}$ denote the z-scores corresponding to the studies that have remained unpublished. Let $y' = \sum_{j=m+1}^{m+n} z_j$. If we had access to all studies, published and unpublished, we would base our inference on $\sum_{j=1}^{m+n} z_j = y + y'$. Specifically, we would fail to reject at level γ when

$$(y + y')/\sqrt{m + n} < \Phi^{-1}(1 - \gamma).$$

In light of this, the fail-safe number is based on replacing the unobservable y' with 0, which is the mean of the z_j when there is no effect and no selection.

---

[74] Rosenthal does not make a connection with the Liptak–Stouffer test in his original paper. Also, we are working with p-values and transforming them into z-scores. Rosenthal works directly with z-scores and assumes a two-sided situation. This leads to a different definition for the fail-safe number.

In a variant, Iyengar and Greenhouse [100] suggest replacing 0 above with the expected value of a standard normal random variable conditional on not exceeding $\Phi^{-1}(1 - \alpha)$, where α is the common level of significance a study is typically required to achieve in order to be published. Note that α may be different from γ.

Remark 20.51 It has been observed that, among published articles reporting on empirical findings involving at least one test, there is a discernible transition at the classical significance level of $\alpha = 0.05$ [152, 78].

Gleser and Olkin's Method

Gleser and Olkin [80] take a more principled approach, based on a worst-case scenario in which, out of a total of $m + n$ studies, each yielding a p-value as above, we only get to observe the smallest m. The goal remains to estimate n assuming that there is no effect. Thus, denoting $p_{(1)} \leq \cdots \leq p_{(m+n)}$ the ordered p-values, we only get to observe $p_{(1)}, \ldots, p_{(m)}$.

Assuming that (20.1) and (20.2) hold, when there is no effect we have a completely specified model, with likelihood

$$\frac{(m + n)!}{n!}(1 - q_m)^n \{0 \leq q_1 \leq \cdots \leq q_m \leq 1\},$$

where we wrote q_j in place of $p_{(j)}$ for clarity. (This plays the role of null model in the present context.)

Problem 20.52 Show that $p_{(m)}$ is sufficient for n and derive its distribution.

Problem 20.53 Show that the maximum likelihood estimator is

$$\hat{n}_{\text{MLE}} := \lfloor m(1 - p_{(m)})/p_{(m)} \rfloor.$$

In fact, Gleser and Olkin prefer to use an unbiased estimator.

Problem 20.54 Show that the following estimator is unbiased

$$\hat{n} := \frac{m(1 - p_{(m)}) - 1}{p_{(m)}}.$$

(As it turns out, this estimator is the only unbiased estimator of n which is a function of the sufficient statistic $p_{(m)}$.)

20.7 Further Topics

20.7.1 Adjusted P-Values

An *adjusted p-value* is such that, when it is below α (the desired FWER or FDR level) the corresponding null hypothesis is rejected. Take, for example, Bonferroni's multiple test meant to control the FWER at α. The corresponding Bonferroni adjusted p-values are defined as

$$p_j^{\text{Bonf}} := (m\, p_j) \wedge 1.$$

And indeed, the Bonferroni multiple test with parameter α rejects $\mathcal{H}_j$ if and only if $p_j^{\text{Bonf}} \le \alpha$.

Problem 20.55 Show that the Holm adjusted p-value for $\mathcal{H}_j$ can be defined as

$$p_j^{\text{Holm}} := \max\left\{ (m - k + 1)p_k : k \le j \right\} \wedge 1,$$

in the sense that the Holm multiple test with parameter α rejects $\mathcal{H}_j$ if and only if $p_j^{\text{Holm}} \le \alpha$.

Problem 20.56 Derive the Hochberg adjusted p-values.

Problem 20.57 Derive the Benjamini–Hochberg adjusted p-values. (These are called *q-values* in [180].)

R corner The multiple tests presented here are implemented in the function p.adjust, which returns the adjusted p-values. The default multiple test is Holm's, which is the safest since it applies regardless of the dependence structure of the p-values, and it is more powerful than Bonferroni's (Problem 20.28).

20.8 Additional Problems

Problem 20.58 (Comparing global tests) Perform some numerical experiments to compare the tests for the global null presented in Section 20.2 in the normal sequence model.

Problem 20.59 (Comparing multiple tests for FWER control) Perform some numerical experiments to compare the multiple tests available in p.adjust for FWER control. Do this in the setting of the normal sequence model.

Problem 20.60 (*k*-FWER) For a multiple test φ, and for $k \geq 1$ integer, the *k-FWER* is defined as

$$\mathsf{FWER}_k(\varphi) = \sup_{\theta \in \Theta} \mathbb{P}_\theta\big(N_{1|0}(\varphi, \theta) \geq k\big).$$

Note that the 1-FWER coincides with the FWER. In general, k stands for the number of Type I errors that the researcher is willing to tolerate.

(i) (Bonferroni) Show that the multiple test that rejects $\mathcal{H}_j$ if $P_j \leq \alpha k/m$ controls the k-FWER at α.

(ii) (Tippett) More generally, how would you change the definition of c_α given in (20.6) to control the k-FWER at the desired level?

Problem 20.61 (marginal FDR) For a multiple test φ, the *marginal false discovery rate (mFDR)* is defined as

$$\mathsf{mFDR}(\varphi) = \sup_{\theta \in \Theta} \frac{\mathbb{E}_\theta[N_{1|0}(\varphi, \theta)]}{\mathbb{E}_\theta[R(\varphi)]}.$$

(i) Show that the mFDR is the probability that a null hypothesis, chosen uniformly at random among those that were rejected, is true. (Thus, in looser terms, the mFDR is the probability that a claimed discovery is actually false.)

(ii) Show that, in general, the mFDR cannot be controlled at any level strictly less than 1.

Problem 20.62 Relate the CMH test of Section 20.6.1 to the log-rank test of Problem 17.52.

Problem 20.63 Suppose you receive a promotional email from a investment manager offering his predictions on a particular stock, specifically whether the price of the stock will increase or decrease by the end of the week. For ten weeks straight, his predictions are correct. If this were a scam, explain how it would work. (Adapted from [89].)

21

Regression Analysis

Beyond quantifying the amount of association between (necessarily paired) variables, as was the goal in Chapter 19, *regression analysis* aims at describing that association and/or at predicting one of the variables based on the other ones. Partly because the task is so much more ambitious, the literature on the topic is vast. Our treatment in this chapter is necessarily very limited in scope but provides some essentials. For more on regression analysis, we recommend [90] or the lighter version [102].

We consider an experiment that results in paired observations, assumed to be iid from an underlying unknown distribution and denoted

$$D := \{(X_1, Y_1), \ldots, (X_n, Y_n)\}. \tag{21.1}$$

Throughout

$$d := \{(x_1, y_1), \ldots, (x_n, y_n)\} \tag{21.2}$$

will denote a realization of D. We also let $|d|$ denote the size of the sample d, also denoted by n above and in what follows.

At this stage, the variables x and y can be of any type. Suppose we set as a general goal that of predicting y as a function of x. In that case, x be will called the *predictor variable* and y will called the *response variable*.

There are two main motives for performing a regression analysis.

• *Modeling*   Here the main purpose is to build a model that describes how the response variable varies as a function of the predictor variable. A simple, parsimonious model is often desirable to ease interpretability. Modeling building is important, most notably, in fundamental sciences like Physics, Chemistry, or Biology, where gaining a functional understanding of how the variables that drive a system are related is a center of focus. An example of that is described in Example 19.14 where the vapor pressure of a pure liquid is related to its temperature.

- *Prediction* Here the main purpose is to predict the value of the response variable given the predictor variable. Examples of applications where this is needed abound in engineering and a broad range of industries (insurance, finance, marketing, etc.). For example, in the insurance industry, when pricing a policy, the predictor variables encapsulate the available information about what is being insured, and the response variable is a measure of risk that the insurance company would take if underwriting the policy. In this context, a procedure is evaluated based solely on its performance at predicting that risk, and can otherwise be very complicated and have no simple interpretation since all relevant computations are done on a computer. (In fact, very large computer clusters in modern applications [23].)

We focus here on the goal of prediction because it is simpler, its scope is broader in terms of applications, and it is easier to formalize mathematically. Also, the framework of prediction is not limited by the assumption that there is a 'true model' that needs to be uncovered.

Example 21.1 (Real estate prices) A number of real estate websites, besides listing properties that are currently on the market (for which the asking price is set by the sellers), also estimate the price of properties that are not currently for sale, using proprietary regression models that take in all the available information on these properties (prediction variable) and returns an estimated value (response variable). For a residential property, the prediction variable may include square footage, number of bedrooms, number of bathrooms, location, etc.

Example 21.2 (MNIST dataset) The special case where the response is categorical is most often called *classification* instead of regression. The MNIST dataset is a dataset that researchers have used for many years for comparing procedures for classification. Each observation is a 28×28 grey level pixel image of a handwritten digit, which is labeled accordingly. The main goal is to recognize the digits. This can be cast as a classification task where each observation is of the form (x, y) with $x \in \mathbb{R}^p$ with $p = 28 \times 28 = 784$ and $y \in \{0, 1, \ldots, 9\}$. Importantly, y is categorical here. Indeed, the fact that y is a digit, and therefore a number, is irrelevant, as the order between the digits is not pertinent to the classification task.

21.1 Prediction

Stating a prediction problem amounts to specifying the class of possible distributions for (X, Y), as well as a functional quantifying the error made by a procedure.

21.1.1 Loss and Risk

Assume that X takes values in $\mathcal{X}$ and Y takes values in $\mathcal{Y}$. Most of the time, both $\mathcal{X}$ and $\mathcal{Y}$ are subsets of Euclidean spaces. Choose a function $\mathcal{L} : \mathcal{Y} \times \mathcal{Y} \to \mathbb{R}_+$ meant to measure dissimilarity. This function $\mathcal{L}$ is referred to as the *loss function*. For a function $f : \mathcal{X} \to \mathcal{Y}$, define its *risk* (aka *expected loss* or *prediction error*) as

$$\mathcal{R}(f) = \mathbb{E}\left[\mathcal{L}(Y, f(X))\right]. \tag{21.3}$$

Thus, $\mathcal{R}(f)$ quantifies the average loss, measured in terms of $\mathcal{L}$, when predicting Y by $f(X)$. Note that f has to be measurable. Henceforth, we let $\mathcal{M}$ denote the class of measurable functions from $\mathcal{X}$ to $\mathcal{Y}$. (As usual, $\mathcal{X}$ and $\mathcal{Y}$ are implicitly equipped with σ-algebras.)

Example 21.3 (Numerical response) In the important case where $\mathcal{Y} = \mathbb{R}$, $\mathcal{L}$ is very often chosen of the form $\mathcal{L}(y, y') = |y - y'|^\gamma$ for some $\gamma > 0$. Popular choices in that family of losses include

$$\begin{aligned} \textit{squared error loss} \quad & \mathcal{L}(y, y') = (y - y')^2, \\ \textit{absolute loss} \quad & \mathcal{L}(y, y') = |y - y'|. \end{aligned}$$

Example 21.4 (Categorical response) Another important example is where $\mathcal{Y}$ is a discrete set, which arises when the response is categorical, i.e., in a classification setting. A popular choice of loss function in that case is

$$\textit{0-1 loss} \quad \mathcal{L}(y, y') = \{y \neq y'\}.$$

21.1.2 Regression Estimators

A *regression estimator* is of the form

$$\hat{f} : \mathcal{D} \longrightarrow \mathcal{M}$$
$$d \longmapsto \hat{f}_d$$

where $\mathcal{D} := \bigcup_{n \geq 1} (\mathcal{X} \times \mathcal{Y})^n$ represents the space where the data (21.2) resides.

The action of applying $\hat{f}$ to data $\boldsymbol{d}$ to obtain $\hat{f}_{\boldsymbol{d}}$ is referred to as *fitting* or *training* the regression estimator $\hat{f}$ on the data $\boldsymbol{d}$, and the resulting estimate, $\hat{f}_{\boldsymbol{d}}$, is then referred to as the fitted or trained estimator. The result is a (measurable) function from $\mathcal{X}$ to $\mathcal{Y}$, which is meant to predict Y from future observations of X.

An estimator being a random function, we use its *expected risk*, or *generalization error*, to quantify its performance, which for an estimator $\hat{f}$ is defined as

$$\overline{\mathcal{R}}_n(\hat{f}) = \mathbb{E}[\mathcal{R}(\hat{f}_{\boldsymbol{D}})], \tag{21.4}$$

where the expectation is with respect to a dataset $\boldsymbol{D}$ of size n. (Besides n, this quantity also depends on the distribution of (X, Y), but this dependency is left implicit.)

21.1.3 Regression Functions

A more ambitious goal than just finding a function with low risk is to approach a minimizer, meaning an element of

$$\mathcal{F}_* := \arg\min_f \mathcal{R}(f),$$

when this set is not empty. Any element of $\mathcal{F}_*$ is called a *regression function*. In many cases of interest, $\mathcal{F}_*$ is (essentially) a singleton.

It helps to work conditional on X, because of the following.

Problem 21.5 (Conditioning on X) Show that

$$\inf_f \mathbb{E}\left[\mathcal{L}(Y, f(X))\right] \geq \mathbb{E}\left(\inf_{y' \in \mathcal{Y}} \mathbb{E}\left[\mathcal{L}(Y, y') \mid X\right]\right).$$

Deduce that any function f satisfying

$$f(x) \in \arg\min_{y' \in \mathcal{Y}} \mathbb{E}\left[\mathcal{L}(Y, y') \mid X = x\right], \quad \text{for all } x, \tag{21.5}$$

minimizes the risk (21.3).

Problem 21.6 (Mean regression) Consider a setting where $\mathcal{Y} = \mathbb{R}$. Assume that Y has a 2nd moment and take $\mathcal{L}$ to be the squared error loss. Show that the minimum in (21.5) is uniquely attained at $y' = \mathbb{E}[Y \mid X = x]$, so that the risk (21.3) is minimized by

$$f_*(x) := \mathbb{E}[Y \mid X = x]. \tag{21.6}$$

[Use Problem 7.87.]

Problem 21.7 (Median regression) Consider a setting where $\mathcal{Y} = \mathbb{R}$. Assume that Y has a 1st moment and take $\mathcal{L}$ to be the absolute loss. Show that the minimum in (21.5) is attained at any median of $Y \mid X = x$, and only there. In the special case where, for all x, $Y \mid X = x$ has a unique median, the risk (21.3) is thus minimized by

$$f_*(x) := \mathrm{Med}(Y \mid X = x).$$

[Use Problem 7.88.]

Problem 21.8 (Classification with 0-1 loss) Consider a classification setting with 0-1 loss. Show that the minimum in (21.5) is attained at any y' maximizing $y \mapsto \mathbb{P}(Y = y \mid X = x)$, so that the risk (21.3) is minimized by any function f_* satisfying

$$f_*(x) \in \arg\max_{y \in \mathcal{Y}} \mathbb{P}(Y = y \mid X = x). \tag{21.7}$$

Such a function is called a *Bayes classifier*.

21.2 Local Methods

The methods that follow are said to be *local*, and this is in the sense that the value of the estimated function at some point $x \in \mathcal{X}$ is computed based on the observations (x_i, y_i) with x_i in a neighborhood of x.

Let δ denote a dissimilarity on $\mathcal{X}$, so that $\delta(x, x')$ is a measure of how dissimilar $x, x' \in \mathcal{X}$ are. 'Local' is henceforth understood in the context of $\mathcal{X}$ equipped with the dissimilarity δ.

Example 21.9 (Euclidean metric) When $\mathcal{X}$ is a Euclidean space, it is most common to use the Euclidean metric, meaning that $\delta(x, x') = \|x - x'\|$, with $\|\cdot\|$ denoting the Euclidean norm.

There are two main types of neighborhood used in practice:

- *Ball neighbors*   The h-ball neighbors of $x \in \mathcal{X}$ are indexed by

$$I_d^h(x) := \{i : x_i \in \mathcal{B}_h(x)\}, \quad \text{where } \mathcal{B}_h(x) := \{x' \in \mathcal{X} : \delta(x', x) \le h\}.$$

- *Nearest neighbors*   The k-nearest neighbors of $x \in \mathcal{X}$ are indexed by

$$J_d^k(x) := \{i : \delta(x_i, x) \text{ among } k \text{ smallest } \delta(x_j, x)\}.$$

We mostly work with an h-ball neighborhood, as it is easier to understand and to handle – even though nearest neighbors are equally, if not more, popular in practice.

We assume throughout that the distribution of X has support $\mathcal{X}$, in the sense that $\mathbb{P}(X \in \mathcal{B}_h(x)) > 0$ for every $x \in \mathcal{X}$ and every $h > 0$.

Problem 21.10 In that case, show that for every $x \in \mathcal{X}$,

$$\min_{i=1,\dots,n} \delta(X_i, x) \xrightarrow{\text{P}} 0, \quad \text{as } n \to \infty. \tag{21.8}$$

21.2.1 Local Methods for Regression

Consider the setting where $\mathcal{Y} = \mathbb{R}$ and the loss is the squared error loss, so that the regression function is the conditional expectation given in (21.6). The methods that we present below aim directly at estimating f_*.

Local Average

Computing the regression function as given in (21.6) is impossible without access to the distribution of $Y \mid X$, which is unknown. A local average approach attempts to estimate this function by making two approximations:

- *Conditioning on a neighborhood*   While in (21.6) the conditioning is on $X = x$, we approximate this by conditioning on a neighborhood. Using a ball neighborhood, the approximation is

$$\mathbb{E}[Y \mid X \in \mathcal{B}_h(x)] \approx \mathbb{E}[Y \mid X = x]. \tag{21.9}$$

  when h is small. This approximation is reasonable when the regression function f_* is continuous, and can indeed be shown to be valid under mild assumptions (Problem 21.53).
- *Averaging*   As we often do, we estimate an expectation with an average, yielding the approximation

$$\frac{1}{|I_D^h(x)|} \sum_{i \in I_D^h(x)} Y_i \approx \mathbb{E}[Y \mid X \in \mathcal{B}_h(x)].$$

By the Law of Large Numbers, the approximation is valid when the number of data points in the neighborhood, $|I_D^h(x)|$, is large.

The *local average* estimator combines these two approximations to take the form

$$\hat{f}_d^h(x) := \frac{1}{|I_d^h(x)|} \sum_{i \in I_d^h(x)} y_i. \tag{21.10}$$

The tuning parameter h is often called the *bandwidth*.

Kernel Regression

Kernel regression is a form of weighted local average, and as such includes (21.10) as a special case. Choose a non-increasing function $Q: \mathbb{R}_+ \to \mathbb{R}_+$, and for $h > 0$, define

$$K_h(x', x) = Q(\delta(x', x)/h).$$

The function Q is sometimes referred to as the kernel function, and most often chosen compactly supported or fast-decaying.

The *Nadaraya–Watson estimate*[75] is defined as

$$\hat{f}_d^h(x) := \sum_{i=1}^{n} w_{i,h}(x) y_i, \tag{21.11}$$

where

$$w_{i,h}(x) := \frac{K_h(x_i, x)}{\sum_{j=1}^{n} K_h(x_j, x)}.$$

R corner The Nadaraya–Watson kernel regression estimate can be computed using the function ksmooth. Several choices of kernel function are offered.

Remark 21.11 Kernel regression is analogous to kernel density estimation (Section 16.10.5).

Local Linear Regression

While a kernel regression estimate is built by fitting a constant locally (Problem 21.57), *local linear regression* is based on fitting an affine function locally. For this to make sense, we need to assume that $\mathcal{X}$ is a Euclidean space, and we assume that δ is a norm for concreteness.

Assuming the regression function f_* is differentiable, we have the Taylor expansion

$$f_*(x') \approx f_*(x) + \nabla f_*(x)^\top (x' - x),$$

the approximation being accurate to first order when $\delta(x', x)$ is small. Having noticed that $x' \mapsto f_*(x) + \nabla f_*(x)^\top (x' - x)$ is an affine function, its coefficients are estimated in a neighborhood of x (since the approximation is only valid near x).

---

[75] Named after Èlizbar Nadaraya (1936–) and Geoffrey Watson (1921–1998).

In more detail, having chosen a kernel function Q and a bandwidth $h > 0$, for $x \in \mathcal{X}$, define $\left(a_d^h(x), b_d^h(x)\right)$ to be the solution to

$$\min_{(a,b)} \sum_{i=1}^{n} \mathsf{K}_h(x_i, x)\left(y_i - a - b^\top(x_i - x)\right)^2.$$

The intercept, $a_d^h(x)$, is meant to estimate $f_*(x)$, while the slope, $b_d^h(x)$, is meant to estimate $\nabla f_*(x)$. The local linear regression estimate is simply the intercept, namely

$$\hat{f}_d^h(x) = a_d^h(x), \text{ as computed above.}$$

Figure 21.1 illustrates an application of local linear regression to synthetically generated data.

R corner The function loess implements local linear regression.

Remark 21.12 Loader [116] proposes a local linear density estimation method based on a Taylor expansion of the logarithm of the density.

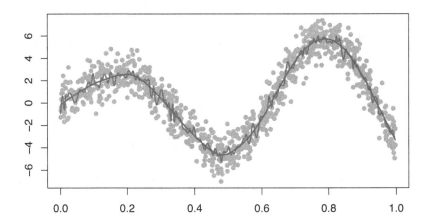

Figure 21.1 An example of application of local linear regression. The model that was generated is the following: $Y_i = f_*(X_i) + Z_i$, with $X_1, \ldots, X_n$ iid uniform in $[0,1]$, $f_*(x) = (1 + 10x - 5x^2)\sin(10x)$, and (independently) $Z_1, \ldots, Z_n$ iid standard normal. Local linear regression was applied with two different values of h, resulting in a rough curve and a smooth curve, with the latter coming very close to the function f_* (and virtually indistinguishable from it in the current plot).

21.2.2 Local Methods for Classification

Consider the setting where $\mathcal{Y}$ is discrete and the loss is the 0-1 loss, so that the regression function is the Bayes classifier given in (21.7). The methods that we present below aim directly at estimating f_*.

Local Majority Vote

The arguments that lead to the local average of (21.10) can be adapted to the present setting, starting from (21.7) instead of (21.6). The end result is the following classifier

$$\hat{f}_d^h(x) \in \arg\max_{y \in \mathcal{Y}} \sum_{i \in I_d^h(x)} \{y_i = y\}. \tag{21.12}$$

In words, the classifier, at a given x, returns the most common category in the neighborhood of x.

Problem 21.13 Detail the arguments leading to (21.12) following those that lead to (21.10).

Nearest-Neighbor Classifier

The expected risk of a classifier $\hat{f}$ at a point $x \in \mathcal{X}$, when $\hat{f}$ is fitted to data D, is

$$\mathbb{P}(Y \neq \hat{f}_D(x) \mid X = x),$$

where the expectation with respect to $Y \mid X = x$ and the data D. Here, (X, Y) represents a 'future' observation not available in D – and in fact independent of D. In Problem 21.8, we saw that this is bounded from below by the risk of the Bayes classifier (21.7), which at x is equal to

$$1 - \max_{y \in \mathcal{Y}} \mathbb{P}(Y = y \mid X = x).$$

Local majority vote based on nearest neighbors has some universality property, in the sense that its risk comes close to that of the Bayes classifier under mild assumptions.

Proposition 21.14 (Nearest neighbor classifier). *In the present setting, assume that for all $y \in \mathcal{Y}$, the function $x \mapsto g(y \mid x) := \mathbb{P}(Y = y \mid X = x)$ is continuous on $\mathcal{X}$. Then, as the sample size increases, the limiting expected risk of the nearest neighbor classifier is, pointwise, at most twice the risk of the Bayes classifier.*

Proof sketch Fix $x \in \mathcal{X}$, and with some abuse of notation let $\hat{f}_n$ denote the nearest neighbor classifier based on a sample of size n. Specifically,

$\hat{f}_n(x) = Y_{i_n(x)}$, where $i_n(x) \in \{1, \ldots, n\}$ indexes one the data points closest to x (with ties, if any, broken in a systematic way). Then (21.8) implies that

$$\delta(X_{i_n(x)}, x) \xrightarrow{\text{P}} 0, \quad \text{as } n \to \infty. \tag{21.13}$$

The expected risk of $\hat{f}_n$ at x is

$$\mathbb{P}(Y \neq Y_{i_n(x)} \mid X = x)$$
$$= 1 - \mathbb{P}(Y = Y_{i_n(x)} \mid X = x)$$
$$= 1 - \sum_{y \in \mathcal{Y}} \mathbb{P}(Y = y, Y_{i_n(x)} = y \mid X = x)$$
$$= 1 - \sum_{y \in \mathcal{Y}} g(y \mid x) \mathbb{P}(Y_{i_n(x)} = y),$$

using the independence of the generic observation (X, Y) and the data (and the fact that $Y_{i_n(x)}$ is a function of the data). We also have

$$\mathbb{P}(Y_{i_n(x)} = y) = \mathbb{E}\left[\mathbb{P}(Y_{i_n(x)} = y \mid X_{i_n(x)})\right]$$
$$= \mathbb{E}\left[g(y \mid X_{i_n(x)})\right]$$
$$\to g(y \mid x), \quad \text{as } n \to \infty, \tag{21.14}$$

by (21.13) combined with the continuity of $x \mapsto g(y \mid x)$ and dominated convergence (Proposition 8.11). Hence,

$$\mathbb{P}(Y \neq \hat{f}_n(x) \mid X = x) \to 1 - \sum_{y \in \mathcal{Y}} g(y \mid x)^2, \quad \text{as } n \to \infty.$$

We then conclude with Problem 21.16. $\qquad \square$

Problem 21.15 Prove the convergence (21.14).

Problem 21.16 For any probability vector (p_j), show that

$$1 - \sum_j p_j^2 \leq 2(1 - \max_j p_j).$$

Remark 21.17 This performance result may be impressive given the simplicity of the 1-nearest neighbor classifier, but the result is very much asymptotic in nature.

21.2.3 Curse of Dimensionality

We assume in this section that $\mathcal{X}$ is Euclidean and that the dissimilarity δ derives from a norm on $\mathcal{X}$. In this context, the local methods presented in Section 21.2 are better suited for when $\mathcal{X}$ has small dimension. (In fact, a dimension as low as $\dim(\mathcal{X}) = 5$ is already a stretch in practice.) This is because the space is mostly empty of data points unless the sample size is exponential in the dimension. This phenomenon, in regression, is called the *curse of dimensionality*.

For a concrete example, take $\mathcal{X} = [0,1]^p$, which is a 'nice' compact domain of $\mathbb{R}^p$. Assume furthermore that X has the uniform distribution on $\mathcal{X}$. Fix $h \in (0, 1/2)$. Then, in the setting where the data are as in (21.1), the chances that a Euclidean ball centered at $x \in \mathcal{X}$ of radius h is empty of data points are given by

$$
\begin{aligned}
\mathbb{P}\big(\forall i : X_i \notin \mathcal{B}_h(x)\big) &= \mathbb{P}\big(X_1 \notin \mathcal{B}_h(x)\big)^n \\
&= \big[1 - \mathbb{P}\big(X_1 \in \mathcal{B}_h(x)\big)\big]^n \\
&\geq \big(1 - (2h)^p\big)^n \\
&\to 1, \quad \text{when } n(2h)^p \to 0.
\end{aligned}
$$

(Note that the inequality is very conservative.) Taking h to be fixed, the condition on n and p holds, for example, when $p \gg \log n$.

We conclude that, when the dimension is a little more than logarithmic in the sample size, the ball neighborhood of any given point is likely empty of data points, which is, of course, problematic for any local method.

Problem 21.18 In the same setting, compute as precisely as you can the minimum sample size n such that the probability that a Euclidean ball of radius h is empty of data points is at most $1/2$. Do this for $p = 1, \ldots, 10$.

21.3 Empirical Risk Minimization

The empirical risk is the risk computed on the empirical distribution. It can be used to produce an estimator by minimization over an appropriately chosen function class. Although such an estimator is typically less 'local', it may nevertheless suffer from the curse of dimensionality (Problem 21.59). This is not the case for estimators based on linear models, an important family of function classes.

21.3.1 Empirical Risk

Given data (21.1), the *empirical risk* of a function f is defined as

$$
\widehat{\mathcal{R}}_D(f) := \frac{1}{n} \sum_{i=1}^{n} \mathcal{L}(Y_i, f(X_i)). \tag{21.15}
$$

Problem 21.19 Show that the empirical risk is an unbiased and consistent estimate for the risk, in the sense that, for any function $f \in \mathcal{M}$, $\mathbb{E}[\widehat{\mathcal{R}}_D(f)] = \mathcal{R}(f)$, and

$$\widehat{\mathcal{R}}_D(f) \xrightarrow{\text{P}} \mathcal{R}(f), \quad \text{as } |D| \to \infty. \tag{21.16}$$

21.3.2 Empirical Risk Minimization

With the empirical risk estimating the risk, it is rather natural to aim at minimizing the empirical risk. This is done over a carefully chosen subclass $\mathcal{F} \subset \mathcal{M}$.

Assuming a minimizer over $\mathcal{F}$ exists, and that some other measurability issues are taken care of, *empirical risk minimization* (ERM) over the class $\mathcal{F}$ amounts to returning a minimizer of the empirical risk over $\mathcal{F}$, namely

$$\hat{f}_D^{\mathcal{F}} \in \arg\min_{f \in \mathcal{F}} \widehat{\mathcal{R}}_D(f).$$

Thus, a function class $\mathcal{F} \subset \mathcal{M}$ defines an estimator via ERM minimization.

We say that ERM is *consistent*[76] for the class $\mathcal{F}$ when

$$\mathcal{R}(\hat{f}_D^{\mathcal{F}}) \xrightarrow{\text{P}} \inf_{f \in \mathcal{F}} \mathcal{R}(f), \quad \text{as } |D| \to \infty. \tag{21.17}$$

Importantly, this consistency is not implied by (21.16). Instead, what is needed is a uniform consistency over $\mathcal{F}$.

Problem 21.20 Show that (21.17) holds when

$$\sup_{f \in \mathcal{F}} \left| \widehat{\mathcal{R}}_D(f) - \mathcal{R}(f) \right| \xrightarrow{\text{P}} 0, \quad \text{as } |D| \to \infty.$$

21.3.3 Interpolation and Inconsistency

Consider the case where $\mathcal{X} = \mathbb{R}$ and $\mathcal{Y} = \mathbb{R}$. For simplicity, assume that X has a continuous distribution, so that the X_i are distinct with probability one. We work with squared error loss, meaning $\mathcal{L}(y, y') = (y - y')^2$.

Having observed the data, for a function f, the empirical risk (21.15) takes the form

$$\widehat{\mathcal{R}}_d(f) = \frac{1}{n} \sum_{i=1}^{n} (y_i - f(x_i))^2.$$

---

[76] This notion of consistency is with respect to the underlying distribution. If it holds regardless of the underlying distribution, then ERM is said to be *universally consistent*.

Clearly, this risk is non-negative and equal to 0 if and only if $y_i = f(x_i)$ for all $i = 1, \ldots, n$, meaning that the function f *interpolates* the data points. Consequently, if for any sample size n there is a function in $\mathcal{F}$ that interpolates the data points, then $\widehat{\mathcal{R}}_d(\hat{f}_d^{\mathcal{F}}) = 0$. If, at the same time, $\inf_{f \in \mathcal{F}} \mathcal{R}(f) > 0$, then ERM cannot be consistent.

Problem 21.21 Suppose, without loss of generality, that Y as support $\mathcal{Y}$. Take a loss $\mathcal{L}$ such that $\mathcal{L}(y, y') = 0$ if and only if $y = y'$. Show that $\inf_{f \in \mathcal{F}} \mathcal{R}(f) = 0$ if and only if

$$\mathbb{P}(Y = f(X)) = 1, \quad \text{for some } f \in \mathcal{F}.$$

Overfitting

We find it desirable to choose a function class for which ERM is consistent, for otherwise it is difficult to know what ERM does. When ERM is not consistent, we say that it *overfits*, and from our discussion above, we know that this happens, for example, when the function class is so 'expressive' that interpolation is possible.

Problem 21.22 In R, generate data according to the model of Figure 21.1. In an effort to perform ERM on the class of all polynomials, interpolate the data points by Lagrange interpolation, which is available via the package polynom. Produce a scatterplot with the fitted polynomial overlaid (as done in that same figure for local linear regression). Repeat for increasing values of n to get a sense of how (wildly) the estimated function behaves.

21.3.4 Linear Models

Linear function classes, that is, classes of functions which have the structure of a linear space, have been popular for decades. This is because of their simplicity, their expressive power, and the fact that ERM is relatively easy to compute. Throughout, we assume that the linear class has finite dimension.

Linear Regression

Assume that the response is numerical, meaning that $\mathcal{Y} = \mathbb{R}$. Given a set of functions, $f_1, \ldots, f_m : \mathcal{X} \to \mathbb{R}$, we may consider linear combinations, meaning functions of the form

$$f(x) = a_1 f_1(x) + \cdots + a_m f_m(x),$$

for arbitrary reals $a_1, \ldots, a_m$.

Example 21.23 (Polynomial regression) Polynomials of degree at most k form such a class. In dimension one, meaning when $\mathcal{X} = \mathbb{R}$, we can choose $f_j(x) = x^{j-1}$ for $j = 1, \ldots, k+1$.

Proposition 21.24. *Assume that the loss is as in Example 21.3, and that each f_j has finite risk. Then ERM is consistent for the linear class defined by $f_1, \ldots, f_m$.*

Problem 21.25 Prove Proposition 21.24, possibly under some additional assumptions, based on Problem 16.101.

Problem 21.26 (Least squares) ERM with the squared error loss is implemented via the *method of least squares*, defined by the following optimization problem

$$\text{minimize } \sum_{i=1}^{n} \left(y_i - a_1 f_1(x_i) - \cdots - a_m f_m(x_i) \right)^2$$

over $a_1, \ldots, a_m \in \mathbb{R}$.

Show that this optimization problem can be reduced to solving an $m \times m$ linear system.

Linear Classification

Assume that the response is binary, so that we may take $\mathcal{Y}$ to be $\{-1, 1\}$ without loss of generality. Given a set of functions, $g_1, \ldots, g_m \colon \mathcal{X} \to \mathbb{R}$, we may consider the class of functions of the form

$$f(x) = \text{sign}(a_1 g_1(x) + \cdots + a_m g_m(x)),$$

for arbitrary reals $a_1, \ldots, a_m$.

Proposition 21.27. *Working with the 0-1 loss, ERM is consistent for any such class.*

It turns out that implementing ERM with the 0-1 loss is in general quite difficult due to the fact that the sign function is not smooth. For this reason, a *surrogate loss* is sometimes chosen. Such a loss is defined not just on $\mathcal{Y}$, but on $\mathbb{R}$. Let $\mathcal{S} \colon \mathbb{R} \times \mathbb{R} \to \mathbb{R}_+$ be such a loss function. In general, a class $\mathcal{G}$ of functions $g \colon \mathcal{X} \to \mathbb{R}$ defines a class $\mathcal{F}$ of functions $f \colon \mathcal{X} \to \{-1, 1\}$ of the form $f(x) = \text{sign}(g(x))$, for some $g \in \mathcal{G}$. ERM for such a class $\mathcal{F}$ with the surrogate loss $\mathcal{S}$ proceeds by first minimizing the empirical risk over $\mathcal{G}$, yielding

$$\hat{g}_d \in \arg\min_{g \in \mathcal{G}} \sum_{i=1}^{n} \mathcal{S}(y_i, g(x_i)), \tag{21.18}$$

and then returning $\hat{f}_d = \text{sign}(\hat{g}_d)$.

Examples of surrogate losses used in practice include

exponential loss	$\mathcal{S}(y,z) = \exp(-yz),$
logistic loss	$\mathcal{S}(y,z) = \log(1 + \exp(-yz)),$
hinge loss	$\mathcal{S}(y,z) = (1 - yz) \vee 0.$

Remark 21.28 The surrogate losses above are all convex, which is desirable in the context of a linear class $\mathcal{G}$ since in that case (21.18) is a convex optimization problem. With $\mathcal{G}$ being a linear class, ERM with the logistic loss is also called *logistic regression*, and ERM with the hinge loss corresponds to *support vector machines*.

Under some conditions, it turns out that ERM with one of these surrogates losses leads to consistency with respect to the 0-1 loss. This is the case, for example, when the class is linear and, importantly, when the minimum risk is achieved over that class (i.e., when the linear class contains a Bayes classifier) [19, Sec 4.2].

21.4 Selection

The local methods presented in Section 21.2 depend on a choice of bandwidth, while empirical risk minimization depends on a choice of function class. In general, when having to choose among various regression estimators, one would want to compare their expected risk (21.4). However, this is not an option, as the expected risk is based on the underlying distribution, which is known. Instead, we substitute the expected risk with an estimate.

Remark 21.29 (Beyond the empirical risk) Even when consistent for the risk (21.17), the empirical risk is typically not useful for comparing various regression estimators. This is because the empirical risk favors expressiveness or richness, and as a consequence leads to choosing an estimate that interpolates the data points when this is possible.

21.4.1 Data Splitting

The main motive for splitting the data is to separate the two operations of fitting and risk assessment, by having them use disjoint parts of the data.

Recall that d denotes the data as in (21.2). Let $t \subset d$ and $v \subset d$ denote the *training set* and *validation set*, respectively. These are chosen disjoint, namely $t \cap v = \varnothing$. Let $\mathscr{C}$ denote a set of estimators to be compared. For each estimator in that set, $\hat{f} \in \mathscr{C}$, we do the following:

- *Fitting*   Fit the estimator on the training set, obtaining $\hat{f}_t$.
- *Assessment*   Compute the average loss of the fitted estimator on the validation set, obtaining $\widehat{\mathscr{R}}_v(\hat{f}_t)$.

Having done this, we choose the estimator among those in $\mathscr{C}$ that has the smallest estimated prediction error, obtaining

$$\hat{f}^{\mathscr{C}} := \arg\min_{\hat{f} \in \mathscr{C}} \widehat{\mathscr{R}}_v(\hat{f}_t). \tag{21.19}$$

We call this estimator the *selected estimator*. (The selection process was based on t and v, but we leave this implicit.)

Remark 21.30 The selected estimator is typically fitted on the entire dataset, resulting in $\hat{f}_d^{\mathscr{C}}$, which is in turn used for prediction.

Problem 21.31 Show that, for a given estimator $\hat{f}$, we have

$$\mathbb{E}[\widehat{\mathscr{R}}_V(\hat{f}_T)] = \overline{\mathscr{R}}_m(\hat{f}),$$

if the training set is of size m. How does $\overline{\mathscr{R}}_m(\hat{f})$ compare with $\overline{\mathscr{R}}_n(\hat{f})$ (which is arguably what we would like to estimate)?

Test Set

The use of a set separate from the training and validation sets becomes necessary if it is of interest to estimate the prediction error of the selected estimator, namely $\hat{f}^{\mathscr{C}}$ of (21.19). This set is called the *test set*. The reason the training and validation sets cannot be used for that purpose is because they were used to arrive at $\hat{f}^{\mathscr{C}}$.

Let $s \subset d$ denote the test set. It is disjoint from the training and validation sets, meaning that $s \cap (t \cup v) = \varnothing$. The risk estimate for the selected estimator is obtained by fitting the selected estimator on the training and validation sets, and then computing the average loss on the test set, obtaining $\widehat{\mathscr{R}}_s(\hat{f}_{t \cup v}^{\mathscr{C}})$.

21.4.2 Cross-Validation

Data splitting is often seen as being wasteful in the sense that the dataset is subdivided into even smaller subsets (training and validation sets) with each subset playing only one role. The methods we present next mimic data splitting while attempting to make better use of the available data. These are all variants of *cross-validation* (CV), arguably the most popular approach for comparing estimators in the context of regression.

For a comprehensive review of cross-validation (both for regression and density estimation), we refer the reader to the survey [5].

k-Fold Cross-Validation

Let $k \in \{2, \ldots, n\}$ and partition the dataset d into k subsets of (roughly) equal size, denoted $d_1, \ldots, d_k$. In a nutshell, in the jth round, the jth subset plays the role of validation set while the others together play the role of training set, resulting in a risk estimate. The final risk estimate is the average of these k estimates. See Table 21.1 for an illustration.

In more detail, take one of the estimators, $\hat{f} \in \mathscr{C}$. We first fit the estimator on $t_l := d \setminus d_l$, obtaining $\hat{f}_{t_l}$. Then we compute the average loss on d_l. Finally, we average these k risk estimates, obtaining

$$\frac{1}{k} \sum_{l=1}^{k} \widehat{\mathscr{R}}_{d_l}(\hat{f}_{t_l}). \tag{21.20}$$

Remark 21.32 The choices $k = 5$ and $k = 10$ appear to be among the most popular in practice.

Problem 21.33 Compute the expectation of the risk estimate (21.20).

Table 21.1 *5-fold cross-validation for a given estimator $\hat{f}$. The j-th row illustrates the j-th round where the j-th data block (highlighted) plays the role of validation set. Each round results in a risk estimate and the average of all these risk estimates (5 of them here) is the cross-validation risk estimate for $\hat{f}$.*

	Subset 1	Subset 2	Subset 3	Subset 4	Subset 5
Round 1	**Validate**	Train	Train	Train	Train
Round 2	Train	**Validate**	Train	Train	Train
Round 3	Train	Train	**Validate**	Train	Train
Round 4	Train	Train	Train	**Validate**	Train
Round 5	Train	Train	Train	Train	**Validate**

Leave-q-Out Cross-Validation

Pushing the rationale behind CV to its extreme leads to having each subset of observations of size q play the role of validation set in turn.

Problem 21.34 Write down the leave-q-out CV risk estimate directly in formula, taking care of defining any mathematical symbols that you use.

Because there are $\binom{n}{q}$ subsets of size q out of n observations total, this procedure is computationally prohibitive for almost all choices of q – even with $q = 2$ or $q = 3$ the procedure may be too costly.

Although leave-q-out CV is typically not computationally tractable, Monte Carlo simulation is possible. Indeed, each subset of size q yields a risk estimate and the leave-q-out CV risk estimate is the average of all these risk estimates. The Monte Carlo approach consists in drawing a certain number of subsets of size q at random, computing their associated risk estimates, and returning their average.

The special choice $q = 1$, *leave-one-out cross-validation*, is computationally more feasible, although it still requires fitting the estimator n times, at least in principle. Note that leave-one-out CV is equivalent to n-fold CV, and the corresponding risk estimate is also known as *prediction residual error sum-of-squares* (PRESS).

Remark 21.35 Analytical expressions for leave-q-out CV are available in some situations [5], and their use can completely bypass the need for burdensome computations.

21.5 Further Topics

21.5.1 Signal and Image Denoising

Denoising is an important 'low-level' task in the context of signal and image processing. (Object recognition is an example of a 'high-level' task.) It is important in a wide array of contexts including astrophysics, satellite imagery, various forms of microscopy, as well as various types of medical imaging.

Signal or image denoising can be seen as a special case of regression analysis, the main specificity being that X is typically not random; rather, the signal or image is sampled on a regular grid. For example, in dimension 1, this could be $x_i = i/n$ for a signal supported on $[0, 1]$. In particular, in the signal and image denoising literature, kernel regression is known as *linear filtering*. (Many other parallels exist with the regression literature.)

R corner In the context of signal or image processing, the function kernel provides access to a number of well-known kernel functions. Having chosen such a kernel, the function kernapply computes the corresponding kernel regression estimate. (Numerically, this involves computing a convolution between two numerical vectors, and this can be done by an application of the discrete Fast Fourier Transform.)

21.5.2 Additive Models

Additive models are an alternative to linear models. Although nonparametric, they do *not* suffer from the curse of dimensionality.

We assume throughout that $\mathcal{X} = \mathbb{R}^p$. We say that $f: \mathbb{R}^p \to \mathbb{R}$ is an *additive function* if it is of the form

$$f(x) = \sum_{j=1}^p f_j(x_j), \quad \text{for } x = (x_1, \ldots, x_p). \tag{21.21}$$

Additive Models for Regression

Let $\mathcal{F}_0$ be any model class of univariate functions. The corresponding model class of additive functions of p variables is

$$\mathcal{F} := \{ f \text{ as in (21.21) with } f_j \in \mathcal{F}_0 \}.$$

Such models do not tend to suffer from the curse of dimensionality because, in essence, all happens on the axes.

ERM over an additive class $\mathcal{F}$ can be done via *backfitting*, described in Algorithm 21.1, which is based on being able to perform ERM over the class of univariate functions $\mathcal{F}_0$ defining $\mathcal{F}$.

Algorithm 21.1   Backfitting Algorithm

Input: data $d = \{(x_i, y_i)\}$ with $x_i = (x_{i1}, \ldots, x_{ip})$, univariate model $\mathcal{F}_0$
Output: fitted additive model

Initialize: $\hat{f}_j = 0$ for all j
Repeat until convergence:
For $j = 1, \ldots, p$
(i)   Compute the residuals $r_i \leftarrow y_i - \sum_{k \neq j} \hat{f}_k(x_{ik})$
(ii)   Compute the ERM estimator for $\mathcal{F}_0$ on $\{(x_{ij}, r_i)\}$ and update $\hat{f}_j$

Problem 21.36 Define a backfitting procedure based on kernel regression. In R, write a function taking in the data and a bandwidth, and fitting an additive model based on kernel regression with the corresponding bandwidth and the Gaussian kernel.

Additive Models for Classification

With an additive function class, we can obtain a class of classifiers as described in Section 21.3.4. In particular, when using the logistic loss, this is sometimes called *additive logistic regression*.

21.5.3 Isotonic Regression

Assume that $\mathcal{X} = \mathbb{R}$ and that $\mathcal{Y} = \mathbb{R}$. Define

$$\mathcal{F} = \big\{ f : \mathbb{R} \to \mathbb{R} \text{ non-decreasing} \big\}. \tag{21.22}$$

ERM based on this class is called *isotonic regression*.[77]

Proposition 21.37. *ERM is consistent for the class* (21.22).

The *pooled adjacent violators algorithm* (PAVA) computes the isotonic regression estimate. The paper [40] describes PAVA (and also presents an alternative convex optimization formulation).

R corner Isotonic regression is available from the package isotone.

Remark 21.38 Notice the parallel with density estimation based on an assumption of monotonicity (Section 16.10.6).

21.5.4 Prediction Intervals

In Section 16.10.4, we saw how to produce a prediction interval for a new observation based on an iid sample. The construction of this prediction interval relies on the exchangeability of the sample together with the new observation, an idea which is at the core of *conformal prediction* [194, 165].

The same idea can be applied in the context of regression in situations where the (X_i, Y_i) are iid from an underlying distribution – unlike in a typical signal or image processing setting in which the predictor variable is fixed by design (Section 21.5.1). We assume that the predictor and the response are both numerical, with values in $\mathbb{R}^p$ and $\mathbb{R}$, respectively. The exposition here follows that in [111].

---

[77] One can as easily work with the class of non-increasing functions, and ERM based on this class is called *antitonic regression*.

We have available to us data as in (21.1). Anticipating a new response at X_{n+1}, denoted Y_{n+1} (and not yet available), we want to build an interval, or more generally a set, based on the available observations, meaning based on D and X_{n+1}, denoted $I(X_{n+1}; D)$, such that

$$\mathbb{P}\big(Y_{n+1} \in I(X_{n+1}; D)\big) \geq 1 - \alpha, \tag{21.23}$$

where α is chosen beforehand, as usual. Such a construction will be based on a regression estimator. Relying on the (strong) assumption that the observations are iid, it is possible to build an accurate prediction interval without really any additional assumptions.

Let $\hat{f}$ be *any* regression estimator. For $y \in \mathbb{R}$, form the augmented dataset $d(y) := d \cup (x_{n+1}, y)$, and consider the residuals when fitting $\hat{f}$ on $d(y)$, meaning

$$r_i(y) := |y_i - \hat{f}_{d(y)}(x_i)|, \text{ for } i \leq n; \quad r_{n+1}(y) := |y - \hat{f}_{d(y)}(x_{n+1})|.$$

Then define

$$\mathsf{pv}(y) := \frac{1}{n+1} \sum_{i=1}^{n+1} \{r_i(y) \leq r_{n+1}(y)\}.$$

As already discussed in Remark 16.81, this can be seen as a permutation p-value for testing the null hypothesis that (x_{n+1}, y) comes from the same distribution as the available data d, and inverting the underlying test yields the following set

$$I(x_{n+1}; d) := \left\{ y : \mathsf{pv}(y) \leq \frac{[(1-\alpha)(n+1)]}{n+1} \right\}. \tag{21.24}$$

Proposition 21.39. *Suppose that*

$$\{(X_i, Y_i) : i = 1, \ldots, n+1\} \text{ are iid on } \mathbb{R}^p \times \mathbb{R}. \tag{21.25}$$

Then (21.23) *holds for the set defined in* (21.24).

Problem 21.40 Prove Proposition 21.39.

The procedure above is computationally intensive as it needs to be repeated for a grid of values for the response (meaning different values of y in the notation used above). However, there is another variant based on sample splitting which is computationally lighter. It goes like this:

(i)   split the sample into t (train) and v (validation), with $|v| \geq (1 - \alpha)/\alpha$;
(ii)   fit the regression estimator $\hat{f}$ on t;

(iii)   evaluate the regression estimate $\hat{f}_t$ on v by computing the residuals

$$|y - \hat{f}_t(x)|, \quad \text{for } (x, y) \in v;$$

(iv)   set δ to be the value of the $\lceil (1 - \alpha)(|v| + 1) \rceil$-th smallest residual;
(v)   define the interval

$$I(x; d) := [\hat{f}_t(x) - \delta, \hat{f}_t(x) + \delta]. \tag{21.26}$$

Proposition 21.41. *Under the same basic assumption* (21.25), *the require-ment* (21.23) *holds for the interval defined in* (21.26).

Problem 21.42 Prove Proposition 21.41.

21.5.5 Classification Based on Density Estimation

A less direct way of approximating the Bayes classifier (21.7) is via the estimation of the class proportions, $\pi_y := \mathbb{P}(Y = y)$, and the class densities, ϕ_y. Note that π defines the marginal of Y and ϕ_y is the conditional density of $X \mid Y = y$. The Bayes classifier can be expressed as

$$f_*(x) \in \arg\max_{y \in \mathcal{Y}} \pi_y \phi_y(x).$$

Problem 21.43 Show this using the Bayes formula.

Thus, if we have estimates, $\hat{\pi}_y$ and $\hat{\phi}_y$ for all $y \in \mathcal{Y}$, we can estimate the Bayes classifier by plugging them in, to obtain

$$\hat{f}(x) \in \arg\max_{y \in \mathcal{Y}} \hat{\pi}_y \hat{\phi}_y(x).$$

The class proportions are typically estimated by the sample class proportions, namely

$$\hat{\pi}_y = \frac{\#\{i : y_i = y\}}{n}.$$

The class densities can be estimated by applying any procedure for density estimation to each sample class.

Problem 21.44 Derive the classifier that results from applying a kernel density estimation procedure (Section 16.10.5) to each sample class. Compare with a local majority vote (with same bandwidth).

Discriminant Analysis

When the class distributions are modeled as normal and fitted by maximum likelihood, the resulting procedure is called *quadratic discriminant analysis* (QDA).

Problem 21.45 Show that for QDA the *classification boundaries*, meaning the sets separating the classes, are quadratic surfaces.

If in addition to being modeled as normal, the class distributions are assumed to share the same covariance matrix, then procedure is called *linear discriminant analysis* (LDA).

Problem 21.46 Show that for LDA the classification boundaries are affine surfaces.

Naive Bayes

Density estimation by local methods (e.g., kernel density estimation), also suffers from a curse of dimensionality. For this reason, some structural assumptions are sometimes made.

A *naive Bayes* approach is analogous additive modeling (Section 21.5.2). The corresponding assumption here is that each class density ϕ_y is the product of its marginals, in the sense that

$$\phi_y(x) = \prod_{j=1}^{p} \phi_{j,y}(x_j),$$

for $x = (x_1, \ldots, x_p)$, where if $X = (X_1, \ldots, X_p)$, then $\phi_{j,y}$ is the density of $X_j \mid Y = y$. This leads to estimating, for each class, the marginal densities separately and then taking the product to obtain an estimate for the class density.

21.5.6 Density Estimation as a Regression Problem

Regression and density estimation seem quite distinct tasks: most prominently, in regression there is a response, while in density estimation there is no response. And the response is what allows the use of a loss function.

Despite appearances, however, the two problems are very closely related. In fact, we already know of the strong resemblance between kernel density estimation (Section 16.10.5) and kernel regression (Section 21.2.1), and the possibility available in both situations to use CV to choose the bandwidth.

We draw a more systematic parallel below. We consider a standard density estimation problem: we have a sample $X := \{X_1, \ldots, X_n\}$ assumed iid from some density f that needs to be estimated.

Manufacturing a Response

Even though there is no response, it is in fact possible to 'manufacture' one. One way to do that is via kernel density estimation (Section 16.10.5). For some bandwidth h chosen in some way later (e.g., by cross-validation, as it is a tuning parameter of the resulting method), let $\hat{f}^h_{X_{-i}}$ denote the kernel density estimator with bandwidth h and kernel K computed on the sample $X_{-i} := X \setminus \{X_i\}$. Then compute $Y_{i,h} = \hat{f}^h_{X_{-i}}(X_i)$, which is the kernel estimator based on all observations except X_i evaluated at X_i. A regression method can then be applied to $(X_1, Y_{1,h}), \ldots, (X_n, Y_{n,h})$.

Problem 21.47 Argue that these paired variables are identically distributed but *not* independent.

Note that we have

$$\mathbb{E}[Y_{i,h} \mid X_i] = f_h(X_i), \quad f_h(x) := \mathsf{K}_h * f,$$

as in (16.23), so that the regression method in the second stage is implicitly estimating f_h and not f itself – but the two are close (Proposition 16.82).

Empirical Risk Minimization

With a response, it is possible to define a loss as we did in the present chapter for the regression problem. This is essentially what the previous approach does. It is in fact possible to define a risk directly, together with an empirical risk meant to estimate that risk. We saw how to do that in our discussion of kernel density estimation in Section 16.10.5, as this is what makes the implementation of cross-validation possible.

For a procedure $\hat{f}$, consider its *mean integrated squared error*, defined as

$$\mathcal{R}(\hat{f}) := \mathbb{E}\left[\int (\hat{f}_X(x) - f(x))^2 \mathrm{d}x\right]. \tag{21.27}$$

Developing the square in the integral, and using the linearity of the integral, and the Fubini–Tonelli Theorem, we have

$$\mathcal{R}(\hat{f}) = \int \mathbb{E}[\hat{f}_X(x)^2]\mathrm{d}x - 2\int \mathbb{E}[\hat{f}_X(x)]f(x)\mathrm{d}x + \int f(x)^2\mathrm{d}x. \tag{21.28}$$

When it comes to estimating the risk, for the first integral, we can estimate it by $\int \hat{f}_X(x)^2\mathrm{d}x$, guided by the fact that $\hat{f}_X(x)$ should have low variance if it is a 'reasonable' estimator and the sample size is suitably large. For the second integral, we can follow the same guideline, which results in approximating the integral by $\int \hat{f}_X(x)f(x)\mathrm{d}x$. In turn, this integral needs to be estimated because it still involves f, which is, of course, unknown. We use $\frac{1}{n}\sum_i \hat{f}_X(X_i)$ for that purpose.

Problem 21.48 Relate $\frac{1}{n}\sum_i \hat{f}_X(X_i)$ to $\int \hat{f}_X(x)f(x)\mathrm{d}x$.

However, it is not clear how to estimate the third integral in (21.28). Thankfully, its estimation is not needed for empirical risk minimization. Indeed, all that is needed is to be able to compare two different estimators, and this can be based on an estimate of

$$\mathcal{R}_0(\hat{f}) := \int \mathbb{E}[\hat{f}_X(x)^2]\mathrm{d}x - 2\int \mathbb{E}[\hat{f}_X(x)]f(x)\mathrm{d}x.$$

We use the following for that purpose

$$\widehat{\mathcal{R}}_0(\hat{f}) := \int \hat{f}_X(x)^2\mathrm{d}x - \frac{2}{n}\sum_{i=1}^{n} \hat{f}_X(X_i).$$

Remark 21.49 For the purpose of performing cross-validation, the following estimator is proposed in [25]

$$\widetilde{\mathcal{R}}_0(\hat{f}) := \frac{1}{n}\sum_{i=1}^{n} \int \hat{f}_{X_{-i}}(x)^2\mathrm{d}x - \frac{2}{n}\sum_{i=1}^{n} \hat{f}_{X_{-i}}(X_i),$$

where, as before, $X_{-i} = \{X_1, \ldots, X_{i-1}, X_{i+1}, \ldots, X_n\}$.

Problem 21.50 The reasoning above is quite specific to our choice of risk in (21.27), which was based on the mean integrated squared error. The same reasoning, however, applies to the relative entropy (Problem 7.101). Detail this reasoning and relate the resulting ERM procedure to maximum likelihood estimation.

21.6 Additional Problems

Problem 21.51 Show that, when X and Y are independent, under some mild assumption on the loss, a constant function minimizes the risk. Conversely, suppose that a constant function minimizes the risk. Is it true that X and Y must be independent in that case?

Problem 21.52 Let g be a bounded and continuous function on $\mathbb{R}^p$. Let $\mathcal{B}_h(x_0)$ denote the ball centered at x_0 of radius $h > 0$ in $\mathbb{R}^p$ with respect to some norm. Show that

$$\frac{1}{|\mathcal{B}_h(x_0)|}\int_{\mathcal{B}_h(x_0)} g(x)\mathrm{d}x \to g(x_0), \quad \text{as } h \to 0.$$

Problem 21.53 Assume that X has a density ϕ on $\mathbb{R}^p$ and that (X, Y) have a joint density ψ on $\mathbb{R}^p \times \mathbb{R}$. Show that

$$\mathbb{E}[Y \mid X \in \mathcal{B}_h(x_0)] = \frac{\int_{\mathcal{B}_h(x_0)} \int_{\mathbb{R}} y\psi(x, y)\mathrm{d}y\mathrm{d}x}{\int_{\mathcal{B}_h(x_0)} \phi(x)\mathrm{d}x}.$$

Using Problem 21.52 and assuming continuity as needed, show that

$$\mathbb{E}[Y \mid X \in \mathcal{B}_h(x_0)] \to \mathbb{E}[Y \mid X = x_0], \quad \text{as } h \to 0,$$

thus justifying the approximation (21.9).

Problem 21.54 Bound the mean-squared error of the local average (21.10), adapting the arguments given in Section 16.10.5. Do the same for the Nadaraya–Watson method (21.11) if you can. The analysis should provide some insights on how to choose the bandwidth h (at least in theory).

Problem 21.55 Adapt the discussion of local methods for regression under squared error loss (Section 21.2.1) to the setting where the loss is the absolute loss instead.

Problem 21.56 (Local polynomial regression) Placing ourselves in the context of Section 21.2.1, if we assume that the regression function is m times differentiable, it becomes reasonable to locally estimate its Taylor expansion of order m. This results in the so-called *local polynomial regression* estimator of order m. Define this estimator in analogy with the local linear regression estimator.

Problem 21.57 Local polynomial regression of order $m = 0$ amounts to fitting a constant locally. Compare that with kernel regression (with the same kernel function and the same bandwidth).

Problem 21.58 Provide an analysis of local polynomial regression of order m when X is uniform on some interval, say the unit interval, and the regression function is $m + 1$ times continuously differentiable on that interval.

Problem 21.59 (ERM over Lipschitz classes) Consider the case where $\mathcal{X} = [0, 1]^p$ and $\mathcal{Y} = \mathbb{R}$. Assume that X has a continuous distribution. For $f : [0, 1]^p \to \mathbb{R}$, let

$$|f|_\infty := \sup_x |f(x)|, \qquad L(f) := \sup_{x \neq x'} \frac{|f(x) - f(x')|}{\|x - x'\|}.$$

For $c_0, c_1 > 0$, define

$$\mathcal{F}_{c_0, c_1} := \left\{ f : |f|_\infty \leq c_0, L(f) \leq c_1 \right\}.$$

It is known that ERM is consistent for $\mathcal{F}_{c_0,c_1}$.

(i)   Show that, unless Y is a deterministic function of X, this is *not* the case when c_1 is unspecified, meaning that ERM is inconsistent for the class

$$\mathcal{F}_{c_0,*} := \left\{ f : |f|_\infty \le c_0, L(f) < \infty \right\}.$$

(ii)   Argue that, even when c_0, c_1 are given, ERM over the class $\mathcal{F}_{c_0,c_1}$ suffers from the curse of dimensionality.

Problem 21.60 Define an additive model where each component is monotonic (i.e., either non-decreasing or non-increasing). Propose a way to fit such a model, and implement that method in R.

22

Foundational Issues

Randomization was presented in Chapter 11 as an essential ingredient in the collection of data, both in survey sampling and in experimental design. We argue here that randomization is the essential foundation of statistical inference: It leads to *conditional inference* in an almost canonical way, and allows for *causal inference*.

22.1 Conditional Inference

We already saw in previous chapters a number of situations where inference is performed conditional on some statistics. Invariably, these statistics are not informative. This includes testing for independence as discussed in Chapters 15 and 19, as well as all other situations where permutation tests are applicable.

22.1.1 Re-Randomization Tests

Consider an experiment that was designed to compare a number of treatments, and in which randomization (Section 11.2.4) was used to assign treatments to experimental units (or said differently, to assign units to treatment groups). The design could be one of the classical designs presented in Section 11.2.5, or any other design that utilizes randomization. Suppose we are interested in testing the null hypothesis that the treatments are equally effective. As we saw in Chapter 17, this can be formalized as a goodness-of-fit testing problem: the null hypothesis is that the joint distribution of the response variables is exchangeable with respect to re-randomization of the treatment group labels.

More formally, suppose there are g treatments and n experimental units (i.e., human subjects in clinical trials), and let Π denote the possible treatment assignments under the randomization scheme employed in the experiment. (Note that Π often depends on characteristics of the experimental units, such as gender or age in clinical trials involving humans.) Let $\pi^0 = (\pi_1^0, \ldots, \pi_n^0) \in \Pi$ denote the assignment used in the experiment, where $\pi_i^0 \in \{1, \ldots, g\}$ denotes the treatment assigned to unit $i \in \{1, \ldots, n\}$. The experiment results in response y_i for unit i. We know that we can organize the data as $(y_1, \pi_1^0), \ldots, (y_n, \pi_n^0)$, written henceforth as $(\boldsymbol{y}, \boldsymbol{\pi}^0)$ and seen as a two-column array. For example, in a completely randomized design, Π is the set of all permutations of $\{1, \ldots, n\}$. In general, though, Π can be quite complicated.

Let T be a test statistic, with large values weighing against the null hypothesis. For example, T could be the treatment sum-of-squares. In the present context, the *randomization p-value* is defined as

$$\frac{\#\{\pi \in \Pi : T(\boldsymbol{y}, \pi) \geq T(\boldsymbol{y}, \pi^0)\}}{|\Pi|}. \tag{22.1}$$

This is an example of conditional inference, where the conditioning is on the responses.

In most experimental designs, if not all of them, any $\pi \in \Pi$ is a permutation of π^0. If seen as a subset of permutations Π forms a subgroup, in the sense that it is stable by composition, the quantity defined in (22.1) is a valid p-value in the sense of (12.12). This is a consequence of Proposition 22.3 below.

As usual, this p-value may be difficult to compute as the number of possible treatment assignments (which varies according to the design) tends to be large. In such situations, one typically resorts to estimating the p-value by Monte Carlo simulation, which only requires the ability to sample uniformly at random from Π – an ability the experiment must have in order to produce the initial randomization.

Problem 22.1 Verify that the re-randomization p-value corresponds to the permutation p-value in the settings previously encountered.

Remark 22.2 Re-randomization is quite natural, as the randomness is present by design and exploiting that randomness is a rather safe approach to inference. This strategy is quite old and already mentioned in the pioneering works of Ronald Fisher and Edwin Pitman in the 1930s. However, at the time, the Monte Carlo approach outlined above was impractical as there were no computers and a normal approximation was used instead. Over the years, this normal approximation became canon and is, to this day, better known than the re-randomization approach.

22.1.2 Randomization P-Value

Consider a general statistical model $(\Omega, \Sigma, \mathcal{P})$, where $\mathcal{P} = \{\mathbb{P}_\theta : \theta \in \Theta\}$. When needed, we will let $\theta_* \in \Theta$ denote the true value of the parameter. Our goal is to test a null hypothesis $\mathcal{H}_0 : \theta_* \in \Theta_0$, for some given $\Theta_0 \subset \Theta$.

Let Π denote a finite set of one-to-one[78] transformations $\pi : \Omega \to \Omega$ such that

$$\mathbb{P}_\theta(\pi^{-1}(\mathcal{A})) = \mathbb{P}_\theta(\mathcal{A}), \quad \text{for all } \mathcal{A} \in \Sigma, \quad \text{for all } \theta \in \Theta.$$

Crucially, we work under the assumption that Π forms a *group*, meaning that $\text{id} \in \Pi$; that if $\pi \in \Pi$ then also $\pi^{-1} \in \Pi$; and that if $\pi_1, \pi_2 \in \Pi$ then $\pi_1 \circ \pi_2 \in \Pi$. (id is the identity transformation $\text{id} : \omega \mapsto \omega$.)

Under these circumstances, Π can be used to obtain a p-value for a given test statistic T. Assuming that large values of T are evidence against the null, the following is the *randomization p-value* for T with respect to the action of Π:

$$\text{pv}(\omega) := \frac{\#\{\pi \in \Pi : T(\pi(\omega)) \geq T(\omega)\}}{|\Pi|}, \tag{22.2}$$

where ω represents the observed data. This is another instance of conditional inference where the conditioning is on $\{\pi(\omega) : \pi \in \Pi\}$ (called the *orbit* of ω under the action of Π).

Proposition 22.3. *In the present context, the quantity defined in (22.2) is a valid p-value in the sense of (12.12).*

This proposition implies that (22.1) is a valid p-value when Π corresponds to a subgroup of permutations. In particular, this applies to the permutation goodness-of-fit tests, as well as the permutation tests for independence and the tests for symmetry seen in previous chapters.

In order to prove Proposition 22.3, we use the following result, which implies that any finite group is isomorphic to a group of permutations.

Theorem 22.4 (Cayley). *Suppose that Π is a finite group with distinct elements denoted $\pi_1, \ldots, \pi_N$. Then, for each j, $\{\pi_1 \circ \pi_j, \ldots, \pi_N \circ \pi_j\}$ is a reordering of $\{\pi_1, \ldots, \pi_N\}$. Let σ_j denote the corresponding permutation of $\{1, \ldots, N\}$, so that $\pi_{\sigma_j(i)} = \pi_i \circ \pi_j$. Then $\mathcal{S} := \{\sigma_1, \ldots, \sigma_N\}$ is a group of permutations. Moreover, the transformation from Π to $\mathcal{S}$ that sends π_j to σ_j is an isomorphism.*

---

[78] We also require that any $\pi \in \Pi$ be bi-measurable, meaning that both π and π^{-1} are measurable functions.

Proof of Proposition 22.3   We use the notation of Theorem 22.4. Assume without loss of generality that π_1 = id. Define $T_j = T \circ \pi_j$ and $T = (T_1, \ldots, T_N)$. Note that (22.2) can be written $\#\{j : T_j \geq T_1\}/N$. We are looking at applying the conclusions of Problem 8.54.

Consider a null distribution, $\mathbb{P}_\theta$ for some $\theta \in \Theta_0$. For any i and j, $T_{\sigma_j(i)} = T \circ \pi_{\sigma_j(i)} = T \circ \pi_i \circ \pi_j = T_i \circ \pi_j$, so that $(T_{\sigma_j(1)}, \ldots, T_{\sigma_j(l)}) = T \circ \pi_j$, and this has same distribution as T since $\mathbb{P}_\theta$ is invariant with respect to π_j, by assumption. Hence, the distribution of T is invariant with respect to any permutation σ_j in the subgroup S. Moreover, for any j and k distinct, let m be such that $\pi_m = \pi_j^{-1} \circ \pi_k$. Then $\sigma_m(j) = k$. Therefore, the conditions of Problem 8.54 are satisfied and an application of the conclusions of the same problem allow us to conclude here. $\square$

Remark 22.5 (Balanced permutations) The group structure is needed besides being used in the proof above. To illustrate that, consider the case of two treatments being compared in a completely randomized design. Assume the group sizes are the same, so that the design is balanced. Suppose we want to calibrate a test statistic by permutation. Based on power considerations, it is rather tempting to consider permutations that move half of each group over to the other group. These are called *balanced permutations* in [174]. However, one should resist the temptation because, although power is indeed improved, the level is not guaranteed to be controlled, as shown in [174].

Problem 22.6 Show that the set of balanced permutations of $\{1, \ldots, 2k\}$ is *not* a group unless $k = 1$.

Monte Carlo Estimation

In many instances, the randomization p-value (22.2) cannot be computed exactly because the group of transformations Π is too large. In that case, one can resort to an estimation by Monte Carlo simulation, which as usual requires the ability to sample from the uniform distribution over Π. In detail, in the same setting as before, we sample $\pi_1, \ldots, \pi_B$ iid from the uniform distribution on Π and return

$$\widehat{\text{pv}}(\omega) := \frac{\#\{b : T(\pi_b(\omega)) \geq T(\omega)\} + 1}{B + 1}.$$

Proposition 22.7. *In the context of Proposition 22.3, this Monte Carlo p-value is a valid p-value in the sense of* (12.12).

Problem 22.8 Prove this by adapting the proof of Proposition 22.3.

22.1.3 *Goodness-of-Fit Testing*

Besides re-randomization testing, conditional inference may also be used in a situation where a sufficient statistic is available. A general approach, in that case, consists in conditioning on the sufficient statistic and then examining the remaining randomness.

In detail, consider a general statistical model $(\Omega, \Sigma, \mathcal{P})$, where $\mathcal{P}$ is left unspecified and often taken to be the class of all distributions on Σ. Suppose that we have data $\omega \in \Omega$ and want to test whether ω was generated from a distribution in a given family of distribution $\mathcal{P}_0 \subset \mathcal{P}$. For clarity, we work with a parameterization of this family, $\mathcal{P}_0 = \{\mathbb{P}_\theta : \theta \in \Theta_0\}$. Based on the particular alternatives we have in mind, we decide to reject for large values of some statistic S. But how can we obtain a p-value for S?

Suppose that T is sufficient statistic for $\mathcal{P}_0$. Let $t = T(\omega)$ denote the observed value of that statistic. If the null hypothesis is true, meaning if ω was generated from a distribution belonging to $\mathcal{P}_0$, then the conditional distribution of S given $T = t$ is independent of $\theta \in \Theta_0$, and is therefore known (at least in principle). Letting $s = S(\omega)$ denote the observed value of the test statistic, we may thus define a conditional p-value for S as follows:

$$\mathsf{pv}(s \,|\, t) := \mathbb{P}_{\theta_0}(S \geq s \mid T = t),$$

where θ_0 is a fixed but arbitrary element of Θ_0.

For example, a p-value for a test of randomness (Section 15.7) is, in the discrete setting, obtained by conditioning on the counts, and the resulting distribution is simply the permutation distribution.

Problem 22.9 Consider an experiment yielding a numerical sample of size n denoted $X_1, \dots, X_n$. We want to test whether the sample was generated iid from the uniform distribution on $[0, \theta]$ for some (unknown) $\theta > 0$. We choose to reject for large values of S, defined as the largest *spacing*

$$S = \max_i (X_{(i+1)} - X_{(i)}),$$

where $X_{(1)} \leq \cdots \leq X_{(n)}$ are the order statistics. Describe how you would obtain, via Monte Carlo simulations, a p-value for this statistic based on the approach described above. (A more direct approach, not based on computer simulations, is proposed in [202].)

We provide further examples below. The last three are quite similar, even though they are motivated by completely different applications.

22.1.4 Genetic Selection

Consider n individuals taken from a population where a particular gene is undergoing neutral selection. Let the number of alleles (variants of that gene) that appear j times be denoted M_j. Under some simplifying assumptions, and assuming a mutation rate $\theta > 0$, Ewens in [60] derived

$$\mathbb{P}_\theta(M_1 = m_1, \ldots, M_n = m_n) = \frac{n!}{\theta(\theta+1)\cdots(\theta+n-1)} \prod_{j=1}^{n} \frac{\theta^{m_j}}{j^{m_j} m_j!},$$

under the constraint that $m_1, \ldots, m_n$ are non-negative integers such that $\sum_{j=1}^{n} jm_j = n$. (For more on the Ewens formula, see [35].)

Problem 22.10 Show that, when $\theta = 0$, all alleles represented in the sample are the same with probability one. Show that, when $\theta \to \infty$, they are all distinct with probability tending to 1.

Problem 22.11 Show that the number of distinct alleles represented in the sample, namely $K := M_1 + \cdots + M_n$, is sufficient for this model.

Watterson [201] suggests to test for neutral selection based on

$$S(m_1, \ldots, m_n) := m_1^2 + \cdots + m_n^2.$$

A test based on S can be one-sided or two-sided.

Problem 22.12 Show that S is maximum when all alleles represented in the sample are distinct, and minimum when they are all the same.

If k denotes the observed number of distinct alleles in the sample, a p-value is then obtained based on the distribution of S given $K = k$.

22.1.5 Rasch Model

Consider an exam which consists of multiple questions, taken by a number of individuals. If Individual i answers Question j correctly, set $x_{ij} = 1$, otherwise set $x_{ij} = 0$. Assuming there are m individuals and n questions, the data are organized in the m-by-n data matrix $\boldsymbol{x} := (x_{ij})$. The *Rasch model* [150] presumes that, as random variables, the X_{ij} are independent with

$$\mathbb{P}(X_{ij} = 1) = \frac{\exp(a_i - b_j)}{1 + \exp(a_i - b_j)},$$

where a_i is the ability of Individual i and b_j is the difficulty of Question j. These are the parameters of the model. Let $X = (X_{ij})$ denote the (random) data matrix.

Problem 22.13 Show that this model is *not* identifiable. Then show that fixing the average subject ability $\frac{1}{m} \sum_{i=1}^{m} a_i$ or the average problem difficulty $\frac{1}{n} \sum_{j=1}^{n} b_j$ makes the model identifiable. (Identifiability is not of concern in what follows.)

Let $R_i = \sum_{j=1}^{n} X_{ij}$ denote the row sum for Individual i (which corresponds to the number of questions that individual answered correctly) and let $C_j = \sum_{i=1}^{m} X_{ij}$ denote the column sum for Question j (which corresponds to the number of individuals that answered that question correctly). Set $R = (R_1, \ldots, R_m)$ and $C = (C_1, \ldots, C_n)$.

Proposition 22.14. *The row and column sums are jointly sufficient for the individual ability and question difficulty parameters and, conditioning on these, the data matrix is uniformly distributed in the set of binary matrices with these row and column sums.*

Problem 22.15 Prove Proposition 22.14.

Suppose that we simply want to know whether the data are compatible with such a model, which we formalize as testing the null hypothesis that the data matrix was generated from an (unspecified) Rasch model. Based on the class of alternatives we have in mind, we choose to work with a test statistic, denoted T, whose large values provide evidence against the null hypothesis and in favor of the alternative hypothesis. Having observed the data matrix, x, we are left with the problem of obtaining a p-value for $T(x)$.

Inspired by Proposition 22.14, we fix the margins. Let $r = (r_1, \ldots, r_m)$ and $c = (c_1, \ldots, c_n)$ denote the vectors of observed row and column sums. If the null hypothesis is true, then conditional on $R = r$ and $C = c$, the data matrix X is uniformly distributed in the set, denoted $\mathcal{X}(r, c)$, of binary matrices with row and column sums given by r and c. The p-value conditional on the row and column sums is consequently obtained as follows:

$$\text{pv}(x) := \frac{\#\{x' \in \mathcal{X}(r, c) : T(x') \geq T(x)\}}{|\mathcal{X}(r, c)|}. \tag{22.3}$$

Problem 22.16 Show that (22.3) is a valid p-value in the sense of (12.12).

The p-value (22.3) is hard to compute in general due to the fact that the set $\mathcal{X}(r, c)$ can be very large and even difficult to enumerate. In fact, the mere computation of the cardinality of $\mathcal{X}(r, c)$ is challenging [45].

22.1.6 Species Co-Occurrence

In Ecology, co-occurrence analysis refers to the study of how different species populate some geographical sites.[79] Various species are observed, or not, in some geographical sites. These findings are stored in a so-called *presence-absence* matrix where rows represent species and columns represent sites, and the (i, j) entry is 1 or 0 according to whether Species i is found in Site j or not. For an example, see Table 22.1.

We adopt the notation of Section 22.1.5. This time $x_{ij} = 1$ if Species i is present in Site j, and $x_{ij} = 0$ otherwise, and the row sum r_i corresponds to the total number of sites where Species i was observed, while the column sum c_j corresponds to the total number of species observed at Site j.

A longstanding controversy and source of conflict in the Ecology community has surrounded the analysis and interpretation of such data. In the 1970s, Diamond [47] collected presence-absence data for various species of birds in the Bismarck Archipelago (where each island was considered a site). Based on these data, he formulated a number of 'assembly rules' having to do with competition for resources (e.g., food, shelter, breeding grounds, etc.) and implying that some pairs of species would not inhabit the same site. However, Connor and Simberloff [31] questioned the basis upon which these rules where formulated. They claimed that the presence-absence patterns that Diamond attributed to his assembly rules could, in fact, be attributed to 'chance'.[80] An important part of the resulting (ongoing?) controversy has to do with how to interpret 'chance', meaning, what probability model to use for statistical inference.

A simple version of the original null model of Connor and Simberloff [31] amounts to the uniform distribution after conditioning on the margins. This is exactly the model we discussed in Section 22.1.5, and a test statistic of interest (perhaps chosen to test the validity of some assembly rule) is calibrated based on this model. Other null models are possible and, in fact, the relevance of the model just described is controversial and has been fiercely debated.

---

[79] Related concepts of co-occurrence exist in other areas such as in Linguistics and the analysis of textures in Image Processing.

[80] Part of their criticism involved questions of how the data were handled and analyzed. In essence, they claimed that Diamond had simply selected some pairs of species to support his theory. We will not elaborate on these rules or the controversy surrounding them as these are domain specific. We refer the curious reader to [82, Ch 7] for further details.

Table 22.1 *Darwin's finch data, collected by Charles Darwin when he visited the Galápagos [182, 161]. Each row corresponds to a species of finch, while each column corresponds to an island in the archipelago.*

	Seymour	Baltra	Isabella	Fernandina	Santiago	Rábida	Pinzón	Santa Cruz	Santa Fe	San Cristóbal	Española	Floreana	Genovesa	Marchena	Pinta	Darwin	Wolf
Large Ground	0	0	1	1	1	1	1	1	1	1	0	1	1	1	1	1	1
Medium Ground	1	1	1	1	1	1	1	1	1	1	0	1	0	1	1	0	0
Small Ground	1	1	1	1	1	1	1	1	1	1	1	1	0	1	1	0	0
Sharp-Beaked Ground	0	0	1	1	1	0	0	1	0	1	0	1	1	0	1	1	1
Cactus Ground	1	1	1	0	1	1	1	1	1	1	0	1	0	1	1	0	0
Large Cactus Ground	0	0	0	0	0	0	0	0	0	0	1	0	1	0	0	0	0
Large Tree	0	0	1	1	1	1	1	1	1	0	0	1	0	1	1	0	0
Medium Tree	0	0	0	0	0	0	0	0	0	0	0	1	0	0	0	0	0
Small Tree	0	0	1	1	1	1	1	1	1	1	0	1	0	0	1	0	0
Vegetarian	0	0	1	1	1	1	1	1	1	1	0	1	0	1	1	0	0
Woodpecker	0	0	1	1	1	0	1	1	0	1	0	0	0	0	0	0	0
Mangrove	0	0	1	1	0	0	0	0	0	0	0	0	0	0	0	0	0
Warbler	1	1	1	1	1	1	1	1	1	1	1	1	1	1	1	1	1

22.1.7 Ising Model

Recall the Ising model described in Example 10.21.

Problem 22.17 Show that $\xi(X)$ and $\zeta(X)$ are jointly sufficient and, further, that conditional on $\xi(X) = s$ and $\zeta(X) = t$, X is uniformly distributed among m-by-n spin matrices satisfying these constraints.

As before, suppose that we simply want to test the null hypothesis that the data matrix was generated from an (unspecified) Ising model, and that we choose a test statistic T whose large values provide evidence against this hypothesis.

Problem 22.18 Based on Problem 22.17 and Section 22.1.5, propose a way to obtain a p-value for T. As before, implementing the method will involve serious computational challenges – describe these challenges.

22.1.8 MCMC P-Value

Besag and Clifford [14] propose a Markov chain Monte Carlo (MCMC) approach to generate samples from the null distribution in the context of testing for the Rasch model – which we saw coincides with the null model of Connor and Simberloff used in the species co-occurrence problem – and in the context of testing for the Ising model. They then build on that to propose a way to obtain a valid p-value for a given test statistic.

Rasch Model

Recall that we want to sample from the uniform distribution on the set of m-by-n binary matrices with row sums $r_1, \ldots, r_m$ and column sums $c_1, \ldots, c_n$. This model was already considered in Section 10.4.1, and we saw there how to design an Markov chain (on the set of such matrices) with stationary distribution the uniform distribution.

Remark 22.19 The Ecology community struggled for years to design a method for sampling from Connor and Simberloff's null model. The original algorithm of Connor and Simberloff [31] was quite ad hoc and inaccurate. Manly [124] used the work of Besag and Clifford [14] but mistakenly forced the chain to move at every step (Problem 10.17), a flaw that was left unnoticed for some years and upon which others built [81]. The error was apparently only discovered almost 10 years later [130]. Some other efforts to sample from this null model are reviewed in [208], which goes on to propose a weighted average approach based on the ergodic theorem (Theorem 10.18).

Ising Model

Recall that we want to sample from the uniform distribution on the set of spin matrices with given values for ξ and ζ. To do so, we also use an MCMC approach, this time based on the following Markov chain: Given such a matrix, choose a pair of distinct sites (i_1, j_1) and (i_2, j_2) uniformly at random and switch their spins if it preserves ζ; otherwise, stay put. Such a switch always preserves ξ, so that the chain remains in the set of matrices of interest.

Problem 22.20 Show that this chain has stationary distribution the uniform distribution on the set of spin matrices with given values for ξ and ζ. [See Problem 10.16.]

Obtaining a P-Value

Suppose the data are denoted X as above, which under the null hypothesis of interest is uniformly distributed on a finite set denoted $\mathcal{X}$. Having observed $X = x$, the p-value corresponding to a test statistic T is defined as

$$\operatorname{pv}(x) := \frac{\#\{x' \in \mathcal{X} : T(x') \geq T(x)\}}{|\mathcal{X}|}. \tag{22.4}$$

Assume that a Markov chain on $\mathcal{X}$ is available with stationary distribution the uniform distribution. In view of the ergodic theorem, we can run the chain from state, obtaining $x_1, x_2, \ldots, x_B$, and then estimate the p-value by

$$\frac{\#\{b = 1, \ldots, B : T(x_b) \geq T(x)\}}{B}. \tag{22.5}$$

Problem 22.21 Show that this is indeed a consistent estimator of (22.4) for any choice of $x_1 \in \mathcal{X}$, where consistency is as $B \to \infty$.

Alternatively, we can obtain a valid p-value as follows. Let K be uniformly distributed in $\{1, \ldots, B\}$ and independent of the data. Assuming $K = k$, do the following:[81]

- if $k = 1$, let $x_1 = x$ and run the chain $B - 1$ steps (forward) from x_1, obtaining $x_2, \ldots, x_B$;
- if $k = B$, let $x_B = x$ and run the chain $B - 1$ steps (backward) from x_B, obtaining $x_{B-1}, \ldots, x_1$;
- if $1 < k < B$, let $x_k = x$, run the chain $B - k$ steps (forward) from x_k, obtaining $x_{k+1}, \ldots, x_B$, and run the chain $k - 1$ steps (backward) from x_k, obtaining $x_{k-1}, \ldots, x_1$.

Having done this, estimate the p-value as in (22.5).

Proposition 22.22. *The resulting p-value is valid in the sense of (12.12).*

Problem 22.23 Prove Proposition 22.22 as follows. Show that, under the null hypothesis (where X has the uniform distribution on $\mathcal{X}$), the resulting random variables $X_1, \ldots, X_B$ are distributed as if the first state were drawn from the uniform distribution and the chain were run $B - 1$ steps from there.

Problem 22.24 The method for obtaining a valid p-value described here is the 'serial method' introduced in [14]. Read enough of this paper to understand the other method, called the 'parallel method', and prove the analogue of Proposition 22.22 for that method.

---

[81] The two chains described in this section are reversible in which case the process is the same for running the chain backward or forward.

22.2 Causal Inference

I have no wish, nor the skill, to embark upon philosophical discussion of
the meaning of 'causation'.

Sir A. Bradford Hill [95]

The concept of *causality* has been, and continues to be, a contentious area
in Philosophy. Yet, at a very practical level, establishing cause-and-effect
relationships is central to, and in many cases the ultimate goal of, most
sciences. For example, in the context of Epidemiology, according to [79],
"causal inference is implicitly and sometimes explicitly embedded in public
health practice and policy formulation".

22.2.1 Association vs Causation

It is widely accepted that properly designed experiments (that invariably use
some form of randomization) can allow for causal inference. We elaborate
on that in Section 22.2.2. In contrast, drawing some causal inference
from observational studies is not possible in general, unless one is able
to convincingly argue that there are no unmeasured confounders. We study
this situation in Section 22.2.3, where we examine how matching attempts
to mimic what randomization does automatically. (Even if there are no
unmeasured confounders, the validity of the causal inference may also rest
on the validity of the assumed model, which is often difficult to ascertain.)

In general, though, observational studies can only lead to inference about
association. Indeed, take Example 15.40 on graduate admissions at UC
Berkeley in the 1970s. Surely, there is a clear association between gender
and overall admission rate. (The statistical significance is overwhelming.)
However, inferring causation (which would imply some gender bias) would
appear to be misleading [15].

In conclusion, we warn the reader that drawing causal inferences from
observational studies is fraught with pitfalls and remains controversial [72,
7, 69, 4]. We adopt this prudent stance in our discussion, which can be
encapsulated in the following.

> Association is not causation.

22.2.2 Randomization

It may be said that the simple precaution of randomisation will suffice to guarantee the validity of the test of significance, by which the result of the experiment is to be judged.

Ronald A. Fisher [63]

We describe a simple model for causal inference called the *counterfactual model*, attributed to Neyman[82] [175] and Rubin [156]. Within this model, randomization allows for causal inference.

We describe a simple setting where two treatments are compared based on n subjects sampled uniformly at random from a large population. Each subject receives only one of the treatments. Let R_{ij} denote the response of Subject i to Treatment j. The sampling justifies our working with the assumption that the R_{i1} are iid (with distribution that of R_1) and, similarly, that the R_{i2} are iid (with distribution that of R_2). The rub is that the experiment results in observing a realization of either R_{i1} or R_{i2}, but not both since Subject i receives only one of the two treatments.

Comparing the Means

We discuss this model in the context of comparing the mean response to the two treatments, called the *average causal effect* and defined as

$$\theta := \mathbb{E}[R_2] - \mathbb{E}[R_1].$$

We interpret $\theta \neq 0$ as a causal effect: the change in treatment causes a change in average response in the entire population. Our immediate interest is to learn about θ from the experiment.

Let $X_i = j$ if Subject i receives Treatment j and let Y_i denote the response observed on Subject i, so that $Y_i = R_{ij}$ if $X_i = j$. We observe a realization of $(X_1, Y_1), \ldots, (X_n, Y_n)$. The *association* between treatment and response is defined as

$$\lambda := \mathbb{E}[Y \mid X = 2] - \mathbb{E}[Y \mid X = 1]. \tag{22.6}$$

There is a natural estimator for λ, namely the difference in sample means

$$D := \bar{Y}_2 - \bar{Y}_1,$$

where $\bar{Y}_j$ is the average of $\{Y_i : X_i = j\}$.

Problem 22.25 Assume that the X_i are iid with distribution X satisfying $\mathbb{P}(X = 1) > 0$ and $\mathbb{P}(X = 2) > 0$. Show that D is a consistent estimator for λ in that case.

---

[82] Neyman's original paper dates back to 1923 and was written in Polish.

In causal inference, however, our target is θ and not λ. But the two coincide when treatment assignment (namely X) is independent of the response to treatment (namely R_1 and R_2). This comes from writing

$$\lambda = \mathbb{E}[R_2 \mid X = 2] - \mathbb{E}[R_1 \mid X = 1],$$

and corresponds to an ideal situation where there is no confounding between assignment to treatment and response to treatment.

Problem 22.26 Prove that a completely randomized block design fulfills this condition and thus allows for causal inference. In this case, show that D is unbiased for θ and compute its variance.

Problem 22.27 Show that, in general, one cannot infer causation from association, by providing an example where D is a terrible estimate for θ.

22.2.3 Matching

Continuing with the same notation, assume now that another variable is available, denoted Z, and may be a confounder.

Problem 22.28 Argue that randomization allows us to effectively ignore Z.

Here we want to examine whether we can do away with randomization, and, in particular, if matching allows us to do that. We use matching on Z with the intent of removing any confounding it might induce. To simplify the discussion, we interpret matching as simply conditioning on Z in addition to conditioning on X. See Remark 22.32.

The punchline is that matching works as intended if the dependency of Y on (X, Z) is properly modeled and there are no other (unmeasured) confounding variables at play. (Both conditions are highly nontrivial and difficult to verify in practice.) To avoid modeling issues, we assume that Z has a finite support, denoted $\mathcal{Z}$ below.

Our access to Z allows us to consider a refinement of λ above, namely

$$\lambda(z) := \mathbb{E}[Y \mid X = 2, Z = z] - \mathbb{E}[Y \mid X = 1, Z = z].$$

Note that λ in (22.6) is the expectation of $\lambda(Z)$.

Problem 22.29 Provide a consistent estimator for $\lambda(z)$ when the (X_i, Z_i) are iid and such that $\mathbb{P}(X = j, Z = z) > 0$ for any $j \in \{1, 2\}$ and any $z \in \mathcal{Z}$.

In order to be able to estimate θ, we require that, conditional on Z, R_1 and R_2 be independent of X.

Problem 22.30 Under this assumption, show that

$$\theta = \sum_{z \in \mathcal{Z}} \lambda(z) \, \mathbb{P}(Z = z).$$

Problem 22.31 For any $z \in \mathcal{Z}$, propose a consistent estimator for $\mathbb{P}(Z = z)$.

Remark 22.32 When $\mathcal{Z}$ has small cardinality, $\lambda(z)$ may be estimated based on $\{(X_i, Z_i) : Z_i = z\}$. When $\mathcal{Z}$ is large, or even infinite, this simple approach may not be feasible. In such situations, one can simply stratify Z, which amounts to binning the Z_i into a few bins, in essence reducing the situation to the case where $\mathcal{Z}$ is of small cardinality. Another approach consists in modeling Y as a function of (X, Z). The validity of the causal inference in that case depends on whether the assumed model is accurate, which may be hard to verify in practice [70, 114].

References

[1] A. Agresti and B. Presnell. Misvotes, undervotes and overvotes: The 2000 presidential election in Florida. *Statistical Science*, 17(4):436–440, 2002. (Cited on page 141.)

[2] F. M. Aguilar, J. Nijs, Y. Gidron, N. Roussel, R. Vanderstraeten, D. Van Dyck, E. Huysmans, and M. De Kooning. Auto-targeted neurostimulation is not superior to placebo in chronic low back pain: a fourfold blind randomized clinical trial. *Pain Physician*, 19(5):E707, 2016. (Cited on page 148.)

[3] M. G. Akritas. Bootstrapping the Kaplan–Meier estimator. *Journal of the American Statistical Association*, 81(396):1032–1038, 1986. (Cited on page 258.)

[4] K. Arceneaux, A. S. Gerber, and D. P. Green. Comparing experimental and matching methods using a large-scale voter mobilization experiment. *Political Analysis*, 14(1):37–62, 2005. (Cited on page 367.)

[5] S. Arlot and A. Celisse. A survey of cross-validation procedures for model selection. *Statistics Surveys*, 4:40–79, 2010. (Cited on pages 345 and 346.)

[6] R. Arratia, L. Goldstein, and L. Gordon. Two moments suffice for Poisson approximations: The Chen–Stein method. *The Annals of Probability*, pages 9–25, 1989. (Cited on page 224.)

[7] P. C. Austin. A critical appraisal of propensity-score matching in the medical literature between 1996 and 2003. *Statistics in Medicine*, 27(12):2037–2049, 2008. (Cited on page 367.)

[8] D. Bassler, V. M. Montori, M. Briel, P. Glasziou, and G. Guyatt. Early stopping of randomized clinical trials for overt efficacy is problematic. *Journal of Clinical Epidemiology*, 61(3):241–246, 2008. (Cited on page 201.)

[9] Y. Benjamini and Y. Hochberg. Controlling the false discovery rate: A practical and powerful approach to multiple testing. *Journal of the Royal Statistical Society: Series B*, 51(1):289–300, 1995. (Cited on pages 315 and 316.)

[10] R. H. Berk and D. H. Jones. Goodness-of-fit test statistics that dominate the Kolmogorov statistics. *Zeitschrift für Wahrscheinlichkeitstheorie und verwandte Gebiete*, 47(1):47–59, 1979. (Cited on page 259.)

[11] D. Bernoulli. Specimen theoriae novae de mensura sortis. *Commentarii Academiae Scientiarum Imperialis Petropolitanae*, 5:175–192, 1738. (Cited on page 95.)

[12] D. Bernoulli. Exposition of a new theory on the measurement of risk. *Econometrica*, 22(1):23–36, 1954. (Cited on page 95.)

[13] M. Bertrand and S. Mullainathan. Are Emily and Greg more employable than Lakisha and Jamal? A field experiment on labor market discrimination. *American Economic Review*, 94(4):991–1013, 2004. (Cited on page 228.)

[14] J. Besag and P. Clifford. Generalized monte carlo significance tests. *Biometrika*, 76(4):633–642, 1989. (Cited on pages 133, 365, and 366.)

[15] P. J. Bickel, E. A. Hammel, and J. W. O'Connell. Sex bias in graduate admissions: Data from Berkeley. *Science*, 187(4175):398–404, 1975. (Cited on pages 221 and 367.)

[16] N. Black. Why we need observational studies to evaluate the effectiveness of health care. *British Medical Journal*, 312(7040):1215–1218, 1996. (Cited on page 154.)

[17] J. Bohannon. I fooled millions into thinking chocolate helps weight loss. Here's how. Gizmodo, May 27th 2015. (Cited on page 146.)

[18] R. Bonita, R. Beaglehole, and T. Kjellström. *Basic Epidemiology*. World Health Organization, 2006. (Cited on page 157.)

[19] S. Boucheron, O. Bousquet, and G. Lugosi. Theory of classification: A survey of some recent advances. *ESAIM: Probability and Statistics*, 9:323–375, 2005. (Cited on page 343.)

[20] J.-P. P. Briefcase. Insuring the uninsured. Poverty Action Lab, 2014. (Cited on page 159.)

[21] G. Bronfort, R. Evans, B. Nelson, P. D. Aker, C. H. Goldsmith, and H. Vernon. A randomized clinical trial of exercise and spinal manipulation for patients with chronic neck pain. *Spine*, 26(7):788–797, 2001. (Cited on page 152.)

[22] D. Calnitsky. "More normal than welfare": The Mincome experiment, stigma, and community experience. *Canadian Review of Sociology/Revue canadienne de sociologie*, 53(1):26–71, 2016. (Cited on page 154.)

[23] A. Canziani, A. Paszke, and E. Culurciello. An analysis of deep neural network models for practical applications. *arXiv preprint arXiv:1605.07678*, 2016. (Cited on page 330.)

[24] D. Card, C. Dobkin, and N. Maestas. Does Medicare save lives? *The Quarterly Journal of Economics*, 124(2):597–636, 2009. (Cited on page 160.)

[25] A. Celisse and S. Robin. Nonparametric density estimation by exact leave-p-out cross-validation. *Computational Statistics & Data Analysis*, 52(5):2350–2368, 2008. (Cited on page 353.)

[26] I. Chalmers and P. Glasziou. Avoidable waste in the production and reporting of research evidence. *The Lancet*, 374(9683):86–89, 2009. (Cited on page 146.)

[27] D. L. Chen, T. J. Moskowitz, and K. Shue. Decision making under the gambler's fallacy: Evidence from asylum judges, loan officers, and baseball umpires. *The Quarterly Journal of Economics*, 131(3):1181–1242, 2016. (Cited on page 23.)

[28] M. A. Cliff, M. C. King, and J. Schlosser. Anthocyanin, phenolic composition, colour measurement and sensory analysis of bc commercial red wines. *Food Research International*, 40(1):92–100, 2007. (Cited on page 289.)

[29] C. J. Clopper and E. S. Pearson. The use of confidence or fiducial limits illustrated in the case of the binomial. *Biometrika*, 26(4):404–413, 1934. (Cited on page 194.)

[30] W. G. Cochran and S. P. Chambers. The planning of observational studies of human populations. *Journal of the Royal Statistical Society: Series A*, 128(2):234–266, 1965. (Cited on page 157.)

[31] E. F. Connor and D. Simberloff. The assembly of species communities: Chance or competition? *Ecology*, 60(6):1132–1140, 1979. (Cited on pages 363 and 365.)

[32] T. M. Cover. Pick the largest number. In *Open Problems in Communication and Computation*, pages 152–152. Springer, 1987. (Cited on page 99.)

[33] P. Craig, C. Cooper, D. Gunnell, S. Haw, K. Lawson, S. Macintyre, D. Ogilvie, M. Petticrew, B. Reeves, M. Sutton, and S. Thompson. Using natural experiments to evaluate population health interventions: New medical research council guidance. *Journal of Epidemiology and Community Health*, pages 1182–1186, 2012. (Cited on page 158.)

[34] H. Cramér. On the composition of elementary errors: Statistical applications. *Scandinavian Actuarial Journal*, 1928(1):141–180, 1928. (Cited on page 287.)

[35] H. Crane. The ubiquitous Ewens sampling formula. *Statistical Science*, 31(1):1–19, 2016. (Cited on page 361.)

[36] X. Cui and G. A. Churchill. Statistical tests for differential expression in cdna microarray experiments. *Genome biology*, 4(4):210, 2003. (Cited on page 309.)

[37] D. A. Darling and P. Erdős. A limit theorem for the maximum of normalized sums of independent random variables. *Duke Mathematical Journal*, 23(1):143–155, 1956. (Cited on page 121.)

[38] A. Dasgupta. Right or wrong, our confidence intervals. *IMS Bulletin*, Dec. 2012. (Cited on page 270.)

[39] A. J. De Craen, T. J. Kaptchuk, J. G. Tijssen, and J. Kleijnen. Placebos and placebo effects in medicine: Historical overview. *Journal of the Royal Society of Medicine*, 92(10):511–515, 1999. (Cited on page 147.)

[40] J. de Leeuw, K. Hornik, and P. Mair. Isotone optimization in R: Pool-adjacent-violators algorithm (pava) and active set methods. *Journal of Statistical Software*, 32(5):1–24, 2009. (Cited on page 348.)

[41] A. Dechartres, L. Trinquart, I. Boutron, and P. Ravaud. Influence of trial sample size on treatment effect estimates: meta-epidemiological study. *British Medical Journal*, 346:f2304, 2013. (Cited on page 321.)

[42] A. Di Sabatino, U. Volta, C. Salvatore, P. Biancheri, G. Caio, R. De Giorgio, M. Di Stefano, and G. R. Corazza. Small amounts of gluten in subjects with suspected nonceliac gluten sensitivity: a randomized, double-blind, placebo-controlled, cross-over trial. *Clinical Gastroenterology and Hepatology*, 13(9):1604–1612, 2015. (Cited on page 153.)

[43] P. Diaconis. Statistical problems in ESP research. *Science*, 201(4351):131–136, 1978. (Cited on pages 138 and 202.)

[44] P. Diaconis and D. Freedman. Finite exchangeable sequences. *The Annals of Probability*, 8(4):745–764, 1980. (Cited on page 110.)

[45] P. Diaconis and A. Gangolli. Rectangular arrays with fixed margins. *IMA Volumes in Mathematics and its Applications*, 72:15–15, 1995. (Cited on pages 133 and 362.)

[46] P. Diaconis and F. Mosteller. Methods for studying coincidences. *Journal of the American Statistical Association*, 84(408):853–861, 1989. (Cited on page 45.)

[47] J. Diamond. Assembly of species communities. In M. Cody and J. Diamond, editors, *Ecology and Evolution of Communities*, pages 342–444. Harvard University Press, 1975. (Cited on page 363.)

[48] C. J. DiCiccio and J. P. Romano. Robust permutation tests for correlation and regression coefficients. *Journal of the American Statistical Association*, 112(519):1211–1220, 2017. (Cited on page 302.)

[49] K. Dickersin and Y.-I. MIN. Publication bias: The problem that won't go away. *Annals of the New York Academy of Sciences*, 703(1):135–148, 1993. (Cited on page 322.)

[50] D. Donoho and J. Jin. Higher criticism for large-scale inference, especially for rare and weak effects. *Statistical Science*, 30(1):1–25, 2015. (Cited on page 313.)

[51] R. Dorfman. The detection of defective members of large populations. *The Annals of Mathematical Statistics*, 14(4):436–440, 1943. (Cited on page 150.)

[52] H. Doucouliagos and T. D. Stanley. Publication selection bias in minimum-wage research? *British Journal of Industrial Relations*, 47(2):406–428, 2009. (Cited on page 322.)

[53] D. Du, F. K. Hwang, and F. Hwang. *Combinatorial Group Testing and its Applications*. World Scientific, 2000. (Cited on page 151.)

[54] C. Dutang and D. Wuertz. A note on random number generation, 2009. Vignette for the randtoolbox package. (Cited on page 137.)

[55] A. Dvoretzky, J. Kiefer, and J. Wolfowitz. Asymptotic minimax character of the sample distribution function and of the classical multinomial estimator. *The Annals of Mathematical Statistics*, 27(3):642–669, 1956. (Cited on page 233.)

[56] R. Eckhardt. Stan Ulam, John von Neumann, and the Monte Carlo method. *Los Alamos Science, Special Issue*, 15:131–137, 1987. (Cited on page 128.)

[57] R. J. Ellis, W. Toperoff, F. Vaida, G. Van Den Brande, J. Gonzales, B. Gouaux, H. Bentley, and J. H. Atkinson. Smoked medicinal cannabis for neuropathic pain in HIV: a randomized, crossover clinical trial. *Neuropsychopharmacology*, 34(3):672–680, 2009. (Cited on page 153.)

[58] P. Erdős and A. Rényi. On a classical problem of probability theory. *Magyar Tudományos Akadémia Matematikai Kutató Intézetének Közleményei*, 6:215–220, 1961. (Cited on page 111.)

[59] P. Erdős and A. Rényi. On a new law of large numbers. *Journal d'Analyse Mathématique*, 23(1):103–111, 1970. (Cited on page 224.)

[60] W. J. Ewens. The sampling theory of selectively neutral alleles. *Theoretical Population Biology*, 3(1):87–112, 1972. (Cited on page 361.)

[61] W. Feller. *An Introduction to Probability Theory and its Applications*, volume 1. John Wiley & Sons, 3rd edition, 1968. (Cited on page 119.)

[62] T. S. Ferguson. Who solved the secretary problem? *Statistical Science*, 4(3):282–289, 1989. (Cited on page 53.)

[63] R. A. Fisher. *The Design of Experiments*. Oliver And Boyd; Edinburgh; London, 2nd edition, 1937. (Cited on pages 216 and 368.)

[64] J. D. Forbes. XIV.—Further experiments and remarks on the measurement of heights by the boiling point of water. *Transactions of the Royal Society of Edinburgh*, 21(02):235–243, 1857. (Cited on page 301.)

[65] D. Fouque, M. Laville, J. Boissel, R. Chifflet, M. Labeeuw, and P. Zech. Controlled low protein diets in chronic renal insufficiency: meta-analysis. *British Medical Journal*, 304(6821):216–220, 1992. (Cited on page 323.)

[66] D. Freedman, R. Pisani, and R. Purves. *Statistics*. WW Norton & Company, 4th edition, 2007. (Cited on pages 139, 142, and 289.)

[67] D. Freedman and P. Stark. What is the chance of an earthquake? *NATO Science Series IV: Earth and Environmental Sciences*, 32:201–213, 2003. (Cited on page 3.)

[68] D. A. Freedman. Statistical models and shoe leather. *Sociological Methodology*, pages 291–313, 1991. (Cited on page 159.)

[69] D. A. Freedman and R. A. Berk. Weighting regressions by propensity scores. *Evaluation Review*, 32(4):392–409, 2008. (Cited on page 367.)

[70] D. A. Freedman, D. Collier, J. S. Sekhon, and P. B. Stark. *Statistical Models and Causal Inference: A Dialogue with the Social Sciences*. Cambridge University Press, 2010. (Cited on page 370.)

[71] D. A. Freedman and K. W. Wachter. Census adjustment: Statistical promise or illusion? *Society*, 39(1):26–33, 2001. (Cited on page 139.)

[72] N. Freemantle, L. Marston, K. Walters, J. Wood, M. R. Reynolds, and I. Petersen. Making inferences on treatment effects from real world data: Propensity scores, confounding by indication, and other perils for the unwary in observational research. *British Medical Journal*, 347:f6409, 2013. (Cited on page 367.)

[73] L. M. Friedman, C. D. Furberg, D. L. DeMets, D. M. Reboussin, and C. B. Granger. *Fundamentals of Clinical Trials*. Springer, 5th edition, 2015. (Cited on page 148.)

[74] M. Friedman. The use of ranks to avoid the assumption of normality implicit in the analysis of variance. *Journal of the American Statistical Association*, 32(200):675–701, 1937. (Cited on page 297.)

[75] M. Gardner. *The Second Scientific American Book of Mathematical Puzzles and Diversions*. Simon and Schuster, New York, 1961. (Cited on page 32.)

[76] A. Gelman and E. Loken. The garden of forking paths: Why multiple comparisons can be a problem, even when there is no "fishing expedition" or "p-hacking" and the research hypothesis was posited ahead of time. Technical report, Columbia University, 2013. (Cited on page 145.)

[77] C. Genovese and L. Wasserman. Operating characteristics and extensions of the false discovery rate procedure. *Journal of the Royal Statistical Society: Series B*, 64(3):499–517, 2002. (Cited on pages 315 and 317.)

[78] A. Gerber, N. Malhotra, et al. Do statistical reporting standards affect what is published? Publication bias in two leading political science journals. *Quarterly Journal of Political Science*, 3(3):313–326, 2008. (Cited on page 326.)

[79] T. A. Glass, S. N. Goodman, M. A. Hernán, and J. M. Samet. Causal inference in public health. *Annual Review of Public Health*, 34:61–75, 2013. (Cited on page 367.)

[80] L. J. Gleser and I. Olkin. Models for estimating the number of unpublished studies. *Statistics in Medicine*, 15(23):2493–2507, 1996. (Cited on page 326.)

[81] N. J. Gotelli and G. L. Entsminger. Swap and fill algorithms in null model analysis: Rethinking the knight's tour. *Oecologia*, 129(2):281–291, 2001. (Cited on page 365.)

[82] N. J. Gotelli and G. R. Graves. *Null Models in Ecology*. Smithsonian Institution Press, 1996. (Cited on page 363.)

[83] U. Grenander. On the theory of mortality measurement: Part II. *Scandinavian Actuarial Journal*, 1956(2):125–153, 1956. (Cited on page 267.)

[84] C. M. Grinstead and J. L. Snell. *Introduction to Probability*. American Mathematical Society, 2nd edition, 1997. Available online on the authors' website. (Cited on pages 13, 24, 31, 44, 46, 113, and 124.)

[85] S. P. Gutthann, L. A. G. Rodriguez, J. Castellsague, and A. D. Oliart. Hormone replacement therapy and risk of venous thromboembolism: population based case-control study. *British Medical Journal*, 314(7083):796, 1997. (Cited on page 156.)

[86] P. Halpern. Isaac Newton vs Las Vegas: How physicists used science to beat the odds at roulette. *Forbes Magazine*, May 2017. (Cited on page 21.)

[87] S. D. Halpern, J. H. Karlawish, and J. A. Berlin. The continuing unethical conduct of underpowered clinical trials. *Journal of the American Medical Association*, 288(3):358–362, 2002. (Cited on page 146.)

[88] B. B. Hansen. Full matching in an observational study of coaching for the SAT. *Journal of the American Statistical Association*, 99(467):609–618, 2004. (Cited on page 158.)

[89] C. R. Harvey and Y. Liu. Evaluating trading strategies. *The Journal of Portfolio Management*, 40(5):108–118, 2014. (Cited on page 328.)

[90] T. Hastie, R. Tibshirani, and J. Friedman. *The Elements of Statistical Learning: Data Mining, Inference, and Prediction*. Springer Science & Business Media, 2nd edition, 2009. (Cited on page 329.)

[91] R. Heffernan, F. Mostashari, D. Das, A. Karpati, M. Kulldorff, and D. Weiss. Syndromic surveillance in public health practice, New York City. *Emerging Infectious Diseases*, 10(5), 2004. (Cited on page 309.)

[92] K. J. Henning. What is syndromic surveillance? *Morbidity and Mortality Weekly Report*, 53:7–11, 2004. (Cited on page 309.)

[93] P. J. Henry. College sophomores in the laboratory redux: Influences of a narrow data base on social psychology's view of the nature of prejudice. *Psychological Inquiry*, 19(2):49–71, 2008. (Cited on page 145.)

[94] J. P. Higgins and S. Green, editors. *Cochrane Handbook for Systematic Reviews of Interventions*. John Wiley & Sons, 2011. (Cited on page 322.)

[95] A. B. Hill. The environment and disease: Association or causation? *Proceedings of the Royal Society of Medicine*, 58(5):295–300, 1965. (Cited on pages 160 and 367.)

[96] W. Hoeffding. A non-parametric test of independence. *The Annals of Mathematical Statistics*, 19(4):546–557, 1948. (Cited on page 306.)

[97] S. Holm. A simple sequentially rejective multiple test procedure. *Scandinavian Journal of Statistics*, 6(2):65–70, 1979. (Cited on page 318.)

[98] M. L. Huber, A. Laesecke, and D. G. Friend. Correlation for the vapor pressure of mercury. *Industrial & Engineering Chemistry Research*, 45(21):7351–7361, 2006. (Cited on page 303.)

[99] K. Imai and K. Nakachi. Cross sectional study of effects of drinking green tea on cardiovascular and liver diseases. *British Medical Journal*, 310(6981):693–696, 1995. (Cited on page 157.)

[100] S. Iyengar and J. B. Greenhouse. Selection models and the file drawer problem. *Statistical Science*, 3(1):109–117, 1988. (Cited on page 326.)

[101] D. Jaeschke. The asymptotic distribution of the supremum of the standardized empirical distribution function on subintervals. *The Annals of Statistics*, 7(1):108–115, 1979. (Cited on page 314.)

[102] G. James, D. Witten, T. Hastie, and R. Tibshirani. *An Introduction to Statistical Learning*, volume 112. Springer, 2013. (Cited on page 329.)

[103] R. I. Jennrich. Asymptotic properties of non-linear least squares estimators. *The Annals of Mathematical Statistics*, 40(2):633–643, 1969. (Cited on page 270.)

[104] Z. N. Kain, F. Sevarino, G. M. Alexander, S. Pincus, and L. C. Mayes. Preoperative anxiety and postoperative pain in women undergoing hysterectomy: a repeated-measures design. *Journal of Psychosomatic Research*, 49(6):417–422, 2000. (Cited on page 153.)

[105] T. J. Kaptchuk and F. G. Miller. Placebo effects in medicine. *New England Journal of Medicine*, 373(1):8–9, 2015. (Cited on page 147.)

[106] A. S. Keller, B. Rosenfeld, C. Trinh-Shevrin, C. Meserve, E. Sachs, J. A. Leviss, E. Singer, H. Smith, J. Wilkinson, and G. Kim. Mental health of detained asylum seekers. *The Lancet*, 362(9397):1721–1723, 2003. (Cited on page 299.)

[107] A. J. Kinderman and J. F. Monahan. Computer generation of random variables using the ratio of uniform deviates. *ACM Transactions on Mathematical Software*, 3(3):257–260, 1977. (Cited on page 132.)

[108] A. Kolmogorov. Sulla determinazione empirica di una legge di distribuzione. *Giornale dell'Istituto Italiano degli Attuari*, 4:89–91, 1933. (Cited on pages 251 and 268.)

[109] M. Lee. Passphrases that you can memorize – but that even the NSA can't guess. *The Intercept*, March 26th 2015. (Cited on page 98.)

[110] E. L. Lehmann and J. P. Romano. *Testing Statistical Hypotheses*. Springer, 2005. (Cited on page 290.)

[111] J. Lei, M. G'Sell, A. Rinaldo, R. J. Tibshirani, and L. Wasserman. Distribution-free predictive inference for regression. *Journal of the American Statistical Association*, 113(523):1094–1111, 2018. (Cited on page 348.)

[112] S. D. Levitt and J. A. List. Field experiments in economics: The past, the present, and the future. *European Economic Review*, 53(1):1–18, 2009. (Cited on page 158.)

[113] Z. Li, S. Nadon, and J. Cihlar. Satellite-based detection of canadian boreal forest fires: Development and application of the algorithm. *International Journal of Remote Sensing*, 21(16):3057–3069, 2000. (Cited on page 309.)

[114] S. Lieberson and F. B. Lynn. Barking up the wrong branch: Scientific alternatives to the current model of sociological science. *Annual Review of Sociology*, 28(1):1–19, 2002. (Cited on page 370.)

[115] T. Liptak. On the combination of independent tests. *Magyar Tud Akad Mat Kutato Int Kozl*, 3:171–197, 1958. (Cited on page 312.)

[116] C. R. Loader. Local likelihood density estimation. *The Annals of Statistics*, 24(4):1602–1618, 1996. (Cited on page 336.)

[117] R. H. Lock. A sequential approximation to a permutation test. *Communications in Statistics: Simulation and Computation*, 20(1):341–363, 1991. (Cited on page 229.)

[118] N. K. Logothetis. What we can do and what we cannot do with fMRI. *Nature*, 453(7197):869, 2008. (Cited on page 310.)

[119] G. L. Lu-Yao and S.-L. Yao. Population-based study of long-term survival in patients with clinically localised prostate cancer. *The Lancet*, 349(9056):906–910, 1997. (Cited on page 147.)

[120] A. M. Lucas and I. M. Mbiti. Effects of school quality on student achievement: Discontinuity evidence from kenya. *American Economic Journal: Applied Economics*, 6(3):234–63, 2014. (Cited on page 160.)

[121] R. H. Lustig, L. A. Schmidt, and C. D. Brindis. Public health: The toxic truth about sugar. *Nature*, 482(7383):27, 2012. (Cited on page 155.)

[122] Z. Ma, C. Wood, and D. Bransby. Soil management impacts on soil carbon sequestration by switchgrass. *Biomass and Bioenergy*, 18(6):469–477, 2000. (Cited on page 149.)

[123] A. Maercker, T. Zöllner, H. Menning, S. Rabe, and A. Karl. Dresden PTSD treatment study: Randomized controlled trial of motor vehicle accident survivors. *BMC Psychiatry*, 6(1):29, 2006. (Cited on page 153.)

[124] B. F. Manly. A note on the analysis of species co-occurrences. *Ecology*, 76(4):1109–1115, 1995. (Cited on page 365.)

[125] D. Marshall, O. Johnell, and H. Wedel. Meta-analysis of how well measures of bone mineral density predict occurrence of osteoporotic fractures. *British Medical Journal*, 312(7041):1254–1259, 1996. (Cited on page 321.)

[126] M. Matsumoto and T. Nishimura. Mersenne twister: A 623-dimensionally equidistributed uniform pseudo-random number generator. *ACM Transactions on Modeling and Computer Simulation*, 8(1):3–30, 1998. (Cited on page 137.)

[127] R. G. McGee and A. C. Dawson. Fake news and fake research: Why meta-research matters more than ever. *Journal of Paediatrics and Child Health*, 56(12):1868–1871, 2020. (Cited on page 146.)

[128] Q. McNemar. Note on the sampling error of the difference between correlated proportions or percentages. *Psychometrika*, 12(2):153–157, 1947. (Cited on page 214.)

[129] D. Michaels. *Doubt is their Product: How Industry's Assault on Science Threatens your Health*. Oxford University Press, 2008. (Cited on page 155.)

[130] I. Miklós and J. Podani. Randomization of presence-absence matrices: Comments and new algorithms. *Ecology*, 85(1):86–92, 2004. (Cited on page 365.)

[131] R. G. Miller. *Simultaneous Statistical Inference*. Springer, 1981. (Cited on page 281.)

[132] V. M. Montori, P. Devereaux, N. K. Adhikari, K. E. Burns, C. H. Eggert, M. Briel, C. Lacchetti, T. W. Leung, E. Darling, D. M. Bryant, et al. Randomized trials stopped early for benefit: a systematic review. *Journal of the American Medical Association*, 294(17):2203–2209, 2005. (Cited on page 201.)

[133] A. Moscovich, B. Nadler, and C. Spiegelman. On the exact Berk–Jones statistics and their p-value calculation. *Electronic Journal of Statistics*, 10(2):2329–2354, 2016. (Cited on page 259.)

[134] F. Mosteller. Note on an application of runs to quality control charts. *The Annals of Mathematical Statistics*, 12(2):228–232, 1941. (Cited on page 224.)

[135] M. G. Myers, M. Godwin, M. Dawes, A. Kiss, S. W. Tobe, F. C. Grant, and J. Kaczorowski. Conventional versus automated measurement of blood pressure in primary care patients with systolic hypertension: randomised parallel design controlled trial. *British Medical Journal*, 342:d286, 2011. (Cited on page 271.)

[136] B. Nalebuff. Puzzles: The other person's envelope is always greener. *The Journal of Economic Perspectives*, 3(1):171–181, 1989. (Cited on page 31.)

[137] R. Newcombe. Two-sided confidence intervals for the single proportion: comparison of seven methods. *Statistics in Medicine*, 17(8):857, 1998. (Cited on page 202.)

[138] D. C. Norvell and P. Cummings. Association of helmet use with death in motorcycle crashes: a matched-pair cohort study. *American Journal of Epidemiology*, 156(5):483–487, 2002. (Cited on page 156.)

[139] G. W. Oehlert. *A First Course in Design and Analysis of Experiments*. 2010. Available online on the author's website. (Cited on pages 144 and 150.)

[140] N. Oreskes and E. M. Conway. *Merchants of Doubt*. New York: Bloomsbury Press, 2010. (Cited on page 155.)

[141] K. Pearson. On the criterion that a given system of deviations from the probable in the case of a correlated system of variables is such that it can be reasonably supposed to have arisen from random sampling. *Philosophical Magazine Series 5*, 50(302):157–175, 1900. (Cited on page 226.)

[142] K. Pearson and A. Lee. On the laws of inheritance in man: I. Inheritance of physical characters. *Biometrika*, 2(4):357–462, 1903. (Cited on page 289.)

[143] C. D. Pilcher, B. Louie, S. Facente, S. Keating, J. Hackett Jr, A. Vallari, C. Hall, T. Dowling, M. P. Busch, and J. D. Klausner. Performance of rapid point-of-care and laboratory tests for acute and established HIV infection in San Francisco. *PLOS ONE*, 8(12):e80629, 2013. (Cited on page 16.)

[144] D. E. Powers and D. A. Rock. Effects of coaching on sat i: Reasoning test scores. *Journal of Educational Measurement*, 36(2):93–118, 1999. (Cited on page 158.)

[145] V. K. Prasad and A. S. Cifu. *Ending Medical Reversal: Improving Outcomes, Saving Lives*. Johns Hopkins University Press, 2015. (Cited on page 154.)

[146] M. L. Radelet and G. L. Pierce. Choosing those who will die: Race and the death penalty in Florida. *Florida Law Review*, 43:1, 1991. (Cited on page 222.)

[147] L. Rahmathullah, J. M. Tielsch, R. Thulasiraj, J. Katz, C. Coles, S. Devi, R. John, K. Prakash, A. Sadanand, and N. Edwin. Impact of supplementing newborn infants with vitamin A on early infant mortality: Community based randomised trial in southern India. *British Medical Journal*, 327(7409):254, 2003. (Cited on page 204.)

[148] J. Randi. *Flim-Flam!: Psychics, ESP, Unicorns, and other Delusions*, volume 342. Prometheus Books Buffalo, NY, 1982. (Cited on page 201.)

[149] J. Randi. *The Truth about Uri Geller*. Pyr Books, 1982. (Cited on page 138.)

[150] G. Rash. Probabilistic models for some intelligence and attainment tests. *Copenhagen: Danish Institute for Educational Research*, 1960. (Cited on page 361.)

[151] J. J. Reilly, J. Armstrong, A. R. Dorosty, P. M. Emmett, A. Ness, I. Rogers, C. Steer, and A. Sherriff. Early life risk factors for obesity in childhood: cohort study. *British Medical Journal*, 330(7504):1357, 2005. (Cited on page 155.)

[152] J. Ridley, N. Kolm, R. Freckelton, and M. Gage. An unexpected influence of widely used significance thresholds on the distribution of reported p-values. *Journal of Evolutionary Biology*, 20(3):1082–1089, 2007. (Cited on page 326.)

[153] A. Rosengren, L. Wilhelmsen, E. Eriksson, B. Risberg, and H. Wedel. Lipoprotein(a) and coronary heart disease: a prospective case-control study in a general population sample of middle aged men. *British Medical Journal*, 301(6763):1248–1251, 1990. (Cited on page 156.)

[154] J. Rosenhouse. *The Monty Hall Problem: The Remarkable Story of Math's Most Contentious Brain Teaser*. Oxford University Press, 2009. (Cited on page 13.)

[155] R. Rosenthal. The file drawer problem and tolerance for null results. *Psychological Bulletin*, 86(3):638, 1979. (Cited on page 324.)

[156] D. B. Rubin. Estimating causal effects of treatments in randomized and nonrandomized studies. *Journal of educational Psychology*, 66(5):688, 1974. (Cited on page 368.)

[157] M. Rudemo. Empirical choice of histograms and kernel density estimators. *Scandinavian Journal of Statistics*, 9(2):65–78, 1982. (Cited on pages 236 and 266.)

[158] A. Rukhin, J. Soto, J. Nechvatal, E. Barker, S. Leigh, M. Levenson, D. Banks, A. Heckert, J. Dray, and S. Vo. Statistical test suite for random and pseudorandom number generators for cryptographic applications, NIST special publication. Technical Report (SP 800-22), National Institute of Standards and Technology (NIST), 2010. Revised by L. Bassham III. (Cited on page 223.)

[159] E. Rutherford, H. Geiger, and H. Bateman. LXXVI. The probability variations in the distribution of α particles. *The London, Edinburgh, and Dublin Philosophical Magazine and Journal of Science*, 20(118):698–707, 1910. (Cited on page 50.)

[160] M. Sageman. *Misunderstanding Terrorism*. University of Pennsylvania Press, 2016. (Cited on page 16.)

[161] J. Sanderson. Testing ecological patterns a well-known algorithm from computer science aids the evaluation of species distributions. *American Scientist*, 88(4):332–339, 2000. (Cited on page 364.)

[162] E. Schlosser. Break-In at Y-12: How a handful of pacifists and nuns exposed the vulnerability of America's nuclear-weapons sites. *The New Yorker*, March 9th 2015. (Cited on page 176.)

[163] L. S. Schneider, K. S. Dagerman, and P. Insel. Risk of death with atypical antipsychotic drug treatment for dementia: meta-analysis of randomized placebo-controlled trials. *Journal of the American Medical Association*, 294(15):1934–1943, 2005. (Cited on page 325.)

[164] D. O. Sears. College sophomores in the laboratory: Influences of a narrow data base on social psychology's view of human nature. *Journal of Personality and Social Psychology*, 51(3):515, 1986. (Cited on page 145.)

[165] G. Shafer and V. Vovk. A tutorial on conformal prediction. *Journal of Machine Learning Research*, 9:371–421, 2008. (Cited on pages 263 and 348.)

[166] Z. Šidák. Rectangular confidence regions for the means of multivariate normal distributions. *Journal of the American Statistical Association*, 62(318):626–633, 1967. (Cited on page 313.)

[167] R. Sihvonen, M. Paavola, A. Malmivaara, A. Itälä, A. Joukainen, H. Nurmi, J. Kalske, and T. L. Järvinen. Arthroscopic partial meniscectomy versus sham surgery for a degenerative meniscal tear. *New England Journal of Medicine*, 369(26):2515–2524, 2013. (Cited on page 148.)

[168] R. J. Simes. An improved Bonferroni procedure for multiple tests of significance. *Biometrika*, 73(3):751–754, 1986. (Cited on page 313.)

[169] J. P. Simmons, L. D. Nelson, and U. Simonsohn. False-positive psychology: Undisclosed flexibility in data collection and analysis allows presenting anything as significant. *Psychological Science*, 22(11):1359–1366, 2011. (Cited on page 145.)

[170] N. Smirnov. Table for estimating the goodness of fit of empirical distributions. *The Annals of Mathematical Statistics*, 19(2):279–281, 1948. (Cited on page 251.)

[171] G. D. Smith and S. Ebrahim. Data dredging, bias, or confounding: They can all get you into the BMJ and the Friday papers. *British Medical Journal*, 325(7378):1437, 2002. (Cited on page 154.)

[172] R. L. Smith. A statistical assessment of Buchanan's vote in Palm Beach county. *Statistical Science*, pages 441–457, 2002. (Cited on page 141.)

[173] J. Snow. *On the Mode of Communication of Cholera.* John Churchill, 1855. (Cited on page 159.)

[174] L. K. Southworth, S. K. Kim, and A. B. Owen. Properties of balanced permutations. *Journal of Computational Biology*, 16(4):625–638, 2009. (Cited on page 359.)

[175] J. Splawa-Neyman, D. M. Dabrowska, and T. Speed. On the application of probability theory to agricultural experiments. Essay on principles. Section 9. *Statistical Science*, 5(4):465–472, 1990. (Cited on page 368.)

[176] P. Squire. Why the 1936 literary digest poll failed. *Public Opinion Quarterly*, 52(1):125–133, 1988. (Cited on page 142.)

[177] R. S. Stafford, T. H. Wagner, and P. W. Lavori. New, but not improved? Incorporating comparative-effectiveness information into FDA labeling. *New England Journal of Medicine*, 361(13):1230–1233, 2009. (Cited on page 147.)

[178] M. B. Steinberg, S. Greenhaus, A. C. Schmelzer, M. T. Bover, J. Foulds, D. R. Hoover, and J. L. Carson. Triple-combination pharmacotherapy for medically ill smokers: a randomized trial. *Annals of Internal Medicine*, 150(7):447–454, 2009. (Cited on page 256.)

[179] T. Stockwell, J. Zhao, S. Panwar, A. Roemer, T. Naimi, and T. Chikritzhs. Do "moderate" drinkers have reduced mortality risk? A systematic review and meta-analysis of alcohol consumption and all-cause mortality. *Journal of Studies on Alcohol and Drugs*, 77(2):185–198, 2016. (Cited on page 321.)

[180] J. D. Storey. A direct approach to false discovery rates. *Journal of the Royal Statistical Society: Series B*, 64(3):479–498, 2002. (Cited on page 327.)

[181] Student. On the error of counting with a haemacytometer. *Biometrika*, 5(3):351–360, 1907. (Cited on page 48.)

[182] F. J. Sulloway. Darwin and his finches: The evolution of a legend. *Journal of the History of Biology*, 15(1):1–53, 1982. (Cited on page 364.)

[183] L. P. Svetkey, V. J. Stevens, P. J. Brantley, L. J. Appel, J. F. Hollis, C. M. Loria, W. M. Vollmer, C. M. Gullion, K. Funk, and P. Smith. Comparison of strategies for sustaining weight loss: The weight loss maintenance randomized controlled

trial. *Journal of the American Medical Association*, 299(10):1139–1148, 2008. (Cited on page 271.)

[184] G. J. Székely and M. L. Rizzo. Energy statistics: A class of statistics based on distances. *Journal of Statistical Planning and Inference*, 143(8):1249–1272, 2013. (Cited on page 287.)

[185] G. J. Székely, M. L. Rizzo, and N. K. Bakirov. Measuring and testing dependence by correlation of distances. *The Annals of Statistics*, 35(6):2769–2794, 2007. (Cited on page 307.)

[186] The PLoS Medicine Editors. Ghostwriting: The dirty little secret of medical publishing that just got bigger. *PLOS Medicine*, 6(9):e1000156, 2009. (Cited on page 324.)

[187] W. C. Thompson and E. L. Schumann. Interpretation of statistical evidence in criminal trials: The prosecutor's fallacy and the defense attorney's fallacy. *Law and Human Behavior*, 11(3):167, 1987. (Cited on page 17.)

[188] S. Thornhill, G. M. Teasdale, G. D. Murray, J. McEwen, C. W. Roy, and K. I. Penny. Disability in young people and adults one year after head injury: prospective cohort study. *British Medical Journal*, 320(7250):1631–1635, 2000. (Cited on page 156.)

[189] J. Tierney. Behind monty hall's doors: Puzzle, debate and answer? *The New York Times*, 1991. (Cited on page 12.)

[190] E. H. Turner, A. M. Matthews, E. Linardatos, R. A. Tell, and R. Rosenthal. Selective publication of antidepressant trials and its influence on apparent efficacy. *New England Journal of Medicine*, 358(3):252–260, 2008. (Cited on page 322.)

[191] J. M. Utts. *Seeing through Statistics*. Brooks/Cole, 3rd edition, 2005. (Cited on pages 139 and 142.)

[192] A. Vazsonyi. Which door has the cadillac. *Decision Line*, 30(1):17–19, 1999. (Cited on page 128.)

[193] R. Von Mises. *Probability, Statistics, and Truth*. Courier Corporation, 1957. (Cited on page 3.)

[194] V. Vovk, A. Gammerman, and G. Shafer. *Conformal Prediction*. Springer, 2005. (Cited on pages 263 and 348.)

[195] V. G. Vovk and G. Shafer. Kolmogorov's contributions to the foundations of probability. *Problems of Information Transmission*, 39(1):21–31, 2003. (Cited on page 3.)

[196] A. Wald. Sequential tests of statistical hypotheses. *The Annals of Mathematical Statistics*, 16(2):117–186, 1945. (Cited on page 200.)

[197] A. Wald and J. Wolfowitz. On a test whether two samples are from the same population. *The Annals of Mathematical Statistics*, 11(2):147–162, 1940. (Cited on page 224.)

[198] B. E. Wampold, T. Minami, S. C. Tierney, T. W. Baskin, and K. S. Bhati. The placebo is powerful: Estimating placebo effects in medicine and psychotherapy from randomized clinical trials. *Journal of Clinical Psychology*, 61(7):835–854, 2005. (Cited on page 148.)

[199] J. N. Wand, K. W. Shotts, J. S. Sekhon, W. R. Mebane, M. C. Herron, and H. E. Brady. The butterfly did it: The aberrant vote for Buchanan in Palm Beach County, Florida. *American Political Science Review*, 95(4):793–810, 2001. (Cited on page 141.)

[200] W. Wang, D. Rothschild, S. Goel, and A. Gelman. Forecasting elections with non-representative polls. *International Journal of Forecasting*, 31(3):980–991, 2015. (Cited on page 143.)

[201] G. Watterson. Heterosis or neutrality? *Genetics*, 85(4):789–814, 1977. (Cited on page 361.)

[202] L. Weiss. A test of fit based on the largest sample spacing. *Journal of the Society for Industrial and Applied Mathematics*, 8(2):295–299, 1960. (Cited on page 360.)

[203] B. Welch. On the comparison of several mean values: An alternative approach. *Biometrika*, 38(3/4):330–336, 1951. (Cited on pages 281 and 287.)

[204] B. L. Welch. The generalization ofstudent's' problem when several different population variances are involved. *Biometrika*, 34(1-2):28–35, 1947. (Cited on page 273.)

[205] E. B. Wilson. Probable inference, the law of succession, and statistical inference. *Journal of the American Statistical Association*, 22(158):209–212, 1927. (Cited on page 195.)

[206] S. S. Young and A. Karr. Deming, data and observational studies. *Significance*, 8(3):116–120, 2011. (Cited on page 154.)

[207] D. Zak. *Almighty: Courage, Resistance, and Existential Peril in the Nuclear Age.* Penguin Publishing Group, 2017. (Cited on page 176.)

[208] A. Zaman and D. Simberloff. Random binary matrices in biogeographical ecology: Instituting a good neighbor policy. *Environmental and Ecological Statistics*, 9(4):405–421, 2002. (Cited on page 365.)

Index

Printed in the United States
by Baker & Taylor Publisher Services